NELSON BALANCED SCIENCE

D0768146

THE physical WORLD

Nelson

Ken Dobson

Thomas Nelson and Sons Ltd
Nelson House   Mayfield Road
Walton-on-Thames   Surrey
KT12 5PL   UK

51 York Place
Edinburgh
EH1 3JD   UK

Thomas Nelson (Hong Kong) Ltd
Toppan Building 10/F
22A Westlands Road
Quarry Bay   Hong Kong

Thomas Nelson Australia
102 Dodds Street
South Melbourne
Victoria 3205   Australia

Nelson Canada
1120 Birchmont Road
Scarborough   Ontario
MIK 5G4   Canada

First published by Thomas Nelson and Sons Ltd 1991

ISBN 0-17-438409-2
NPN 9 8 7 6 5 4 3

Printed in Hong Kong.

To the reader

The Physical World is about physics. It deals with the forces that shape the world, from the smallest particle to the galaxies of stars that the universe is made of. It also deals with the way that forces and energy are used by all of us – in living and moving, in work and play, in sending messages, in storing information and using it to control so many things in the modern world.

The book is split up into short topics. Each topic includes Activities (with a blue background) – things to do in the lab and at home – and Questions (mauve background). There are also case studies and extension exercises (with a green background) which take some of the ideas a bit further.

You'll notice that there are references to the two other books in the **Nelson Balanced Science** series. Michael Roberts' book **The Living World** is mainly about biology and John Holman's book **The Material World** is mainly about chemistry. By using our books together we hope you will enjoy finding out about science – and that you'll want to study it further.

Ken Dobson January 1991

I am grateful to many people who helped to write this book, and I would particularly like to thank John Holman and Michael Roberts. Special thanks are due to Anna Grayson for writing most of Section G: Earth. The following people helped by reading and making very useful comments on the early drafts of the book:

Mark Tweedle, Heckmondwike Grammar School
Michael Brimicombe, Cedars Upper School, Leighton Buzzard, Beds
Professor E K Walton, Department of Geography and Geology, University of St Andrews
David Fielding, Radley College, Abingdon, Oxon
RJJ Orton, ASE Lab Safeguards Committee

Contents

A1
Squashing, pulling, bending

The way materials react to forces often decides how we use them.

The pole vaulter in picture 1 uses a long glass-fibre pole to pull himself 8 metres above the ground. The pole has to be strong and light. It also has to be bendy — but not too bendy! When he comes down on the other side he relies on the softness of the foam plastic to cushion his fall.

Making and choosing materials

These materials have been carefully chosen for the job they have to do. They behave in the right way when forces are applied to them. In fact, both of these materials are **synthetic**, which means that they have been made in a factory, and don't occur naturally on Earth.

Modern materials like these are designed and made by materials scientists. They need to know what the materials will be used for. They need to know how much force they must be able to withstand, how 'heavy' (dense) they need to be, and how 'bendy' they must be. There is a lot more about materials and their uses in *The Material World*.

Useful materials

Picture 2 shows lots of synthetic materials, as well as some natural ones. Can you tell which are 'natural' and which are synthetic? In some sports grounds even the grass is artificial! The designers and makers of vaulting poles, safety mats — and even the clothes that the athletes wear — have to make a sensible choice of what materials to use.

To help them to design or choose the best materials for a purpose scientists must test them in a standard way. They must also know if changing the size of the object will make any difference. A thinner vaulting pole will be lighter, cheaper and easier to run with, but might break more easily.

Picture 1 This athlete's vaulting pole has to be strong, light and flexible

Picture 2 Natural and synthetic materials: Can you tell which is which?

Testing materials

The activities at the end of this topic will give you an idea of how materials are tested. The tests have to be 'fair' — for example the samples of different materials must be the same size, and tested under the same conditions. You could do a **flexibility** test as shown in picture 3 (see also activity A). Extra bending force is applied by simply adding weights to the end of the sample. The effect of doing this can be measured by seeing how far down the end of the beam bends.

This kind of test shows that all materials will bend to some extent — but if the sample is too thick the movement will be too small to be seen without special instruments.

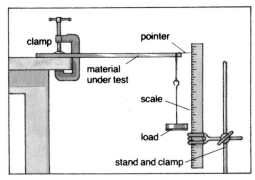

Picture 3 One way of testing materials to compare how flexible they are

Plastic and elastic

These words have a special meaning in materials science. They describe what a material does when forces are applied to it. Forces tend to change the shape of an object. If the shape stays changed even when the force is taken away the material is called **plastic**.

Mud, plasticine, clay and putty are good examples of plastic materials.

But if the object returns to its original shape when the force is removed it is called an **elastic** material. Rubber, steel, wood and many other materials behave like this. Both the foam safety mat and the vaulting pole in picture 1 are made of elastic materials.

Are plastics plastic?

Most of the materials that we call 'plastics' are synthetic materials that *were* plastic at the time they were made. Then they were heated or chemically treated to make them set hard, and they stopped being plastic.

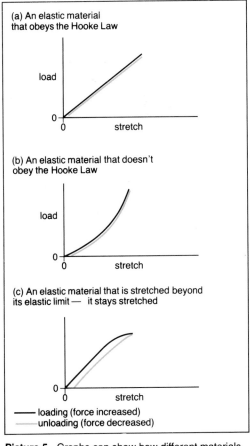

Picture 4 Steel is elastic — up to a point!

Over the limit

Many useful materials are only elastic up to a point. If too much force is applied they will not return to their original shape. Steel is quite elastic, but large forces may bend it out of shape permanently, like a car body that has been in an accident (picture 4). At some stage it stopped being elastic, and became plastic. Metals like copper and lead show this effect more easily, with less force needed.

Glass

Surprisingly, glass is also elastic. Thin glass fibres can be bent easily. But if too much force is applied glass will not just **deform plastically**, like steel or copper, but shatter. A material like this is said to be **brittle**.

Testing for elasticity

Activity B is about finding out whether a material is elastic or not, and how much force is needed to make it go beyond its elastic limit. To check if the material is elastic you need to take measurements both when forces are increased and when they are decreased.

Using graphs

Graphs are a useful way of using experimental data to compare one material with another. A graph of force applied (or **load**) plotted against the change it produces will show clearly how the materials behave.

The graphs in picture 5 show how different materials might behave in these tests. The solid line shows what happens as the load is increased. The dotted line shows what happens when they are unloaded.

(a) An elastic material that obeys the Hooke Law

(b) An elastic material that doesn't obey the Hooke Law

(c) An elastic material that is stretched beyond its elastic limit — it stays stretched

—— loading (force increased)
—— unloading (force decreased)

Picture 5 Graphs can show how different materials stretch differently

Picture 6 This force measurer relies on the Hooke Law

The Hooke Law

When objects are stretched or compressed they change their shape. For some materials length changes steadily for a steady change in the applied force. Every extra bit of force produces an equal, extra bit of length. Materials that do this are said to obey the **Hooke Law**.

As in activity B, it makes sense to measure the change in length, or the **extension**, of the sample. A material that obeys the Hooke Law will give a straight line graph. The graph produced by such a material is shown in picture 5(a). The other graphs in the picture show how other materials behave when they are stretched. Of course, the change produced could be a shortening, caused by a compression force, or an angle of bending produced by a sideways force.

Using elasticity to measure forces

Objects that obey the Hooke Law are used in making a spring balance or a newton meter. The marks on its scale show the extension or compression produced by a force. Because the spring inside obeys the Hooke Law the length changes can be marked as **force** changes. Picture 6 shows an example of a force measurer that uses the Hooke Law.

Activities

A Testing for flexibility

Picture 3 shows how to set up a test of flexibility. Make sure that the specimens of the materials you are testing are the same width, thickness and length. The specimens might be plastic, metals (steel, copper, tin) or wood.

Use a clamp to fix one end of a specimen of material firmly to the bench or table. Apply forces at the other end using either a newton meter or by adding masses. Take care to apply these forces at the same distance from the end, for each specimen.

If you load the specimen by adding masses, you need to know that the force exerted when you add a mass of 100 g is just about 1 newton. For each specimen, complete a table of results as shown:

Applied force/N	Deflection/mm

Plot graphs of deflection against applied force. Use the same pair of axes for all the materials.

Write a report on your tests, making sure that you:
■ describe the differences between the materials,
■ note any pattern of behaviour of any of the materials,
■ explain any precautions you took to ensure that the tests were fair and accurate.

B Elastic or not?

An elastic material will change its shape when a force is applied and then return to its original shape when the force is removed. In this experiment you will test one or more objects to see if they are elastic or not. In each case use a stretching force, which could be done either by using a newton meter or by hanging weights (masses) on the specimen. See picture 3.

Suitable objects to test are: an elastic band, a steel spring, a 'home-made' spring made by coiling 300 mm of copper wire around a pencil.

If you have special equipment of the kind shown in picture 8 you will be able to test a length of thin copper wire — or a wire made of other metals. This experiment needs special care — check the details with your teacher.

CARE! Wear eye protection if the specimens are wires. Put a box of waste packing material under heavy loads to catch them if they fall and prevent anyone having a foot in the way.

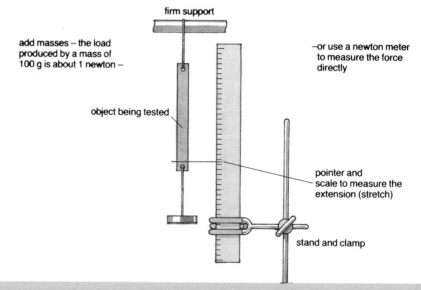

Picture 7 Testing to see if materials or objects obey the Hooke Law

firm support

add masses – the load produced by a mass of 100 g is about 1 newton –

object being tested

–or use a newton meter to measure the force directly

pointer and scale to measure the extension (stretch)

stand and clamp

Picture 8 Special apparatus for investigating how metal wire stretches

You need to measure how much each object **extends** (increases in length) when an extra load is applied. Take measurements while loading and unloading. Set out your measurements for each specimen in a table as shown:

Applied load/N	
Extension/mm when loading	
Extension/mm when unloading	

Plot graphs of extensions, both loading and unloading, for each object tested (as in picture 5 above). Write a short report on your tests:

■ describing briefly what you did and measured,
■ stating which objects behaved elastically,
■ stating which objects obeyed the Hooke Law.

C   What makes a good bouncer?

Plan an experiment to compare how good the following materials are at bouncing:

glass (e.g. a marble), steel, lead, wood and rubber.

Check with your teacher that your experiment is workable, then carry it out.

What property of the material is most important in deciding how bouncy it is? For example — density, elasticity, toughness? Use the table of values on page 9 to help you answer this question.

Questions

1   Sort the following materials into two groups:
a   elastic, b plastic:

steel, glass, wood, clay, plasticine, cotton, mud, putty, cardboard, foam plastic, polythene.

2   Use the information given in this topic, plus any ideas you already have, to answer the following questions.

a   Why is a pole used by a pole vaulter made of a material which is 'bendy, but not too bendy'?

b   Why are most springs made of metal?

c   Both steel and foam rubber obey the Hooke Law. Why are the safety mats in the gym made of foam rubber, not steel?

3   Explain as clearly as you can what the following words mean:

a   elastic,
b   plastic,
c   brittle,
d   flexible.

4   Which of the graphs in picture 9 of extension plotted against load would you expect to match the following materials:

a   rubber,
b   copper wire,
c   a piece of string,
d   plasticine.

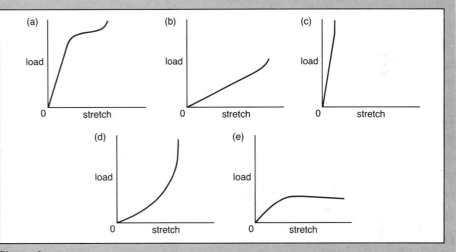

Picture 9

5   Look at picture 2. Write down the names of the synthetic materials you can identify (e.g. glass).

6   Look at a biscuit.

a   Use the words you have learned in this topic to describe its properties as a **material**.

b   Biscuits come in a variety of different packagings (see picture 10). How do the mechanical properties of the biscuit decide what kind of packaging the biscuit manufacturers need to use?

Picture 10

A2 Strong? Tough? Heavy?

Strong materials may not be tough — and needn't be heavy.

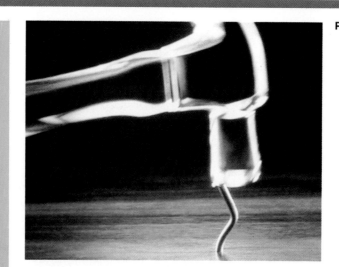

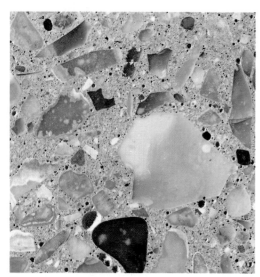

Picture 2 Concrete in close-up

Steel and glass

A strong material needs a lot of force to break it, compared with a weak material of the same size. Steel is strong — but so is glass. Think of how many milk bottles there are in the world — they get rough treatment but they last quite a long time.

The important practical difference between glass and steel is that steel is also *tough*. That is, it doesn't crack like glass does. Steel milk bottles would last much longer! When too much force is applied to steel it will bend instead of shattering.

Can glass be tough?

A special glass can be made that is very tough — but it is expensive. Picture 1 shows glass that can stand a very large force, and bends rather than breaks. This is the kind of glass that might be used in banks as a protection against thieves.

A weak material can be quite tough, like plasticine or even cardboard. A biscuit is neither strong nor tough.

Making materials stronger and tougher

Everyday objects aren't usually made of a pure material. More and more, they are made of mixtures of materials. These mixtures are called **composites**. This means that the good properties of one material can be combined with the good properties of another material.

Concrete

Concrete is a good example of a composite material. Ordinary concrete is made from cement, sand and pebbles. Sand and pebbles are very strong materials, and they are hard to break or crack. Unfortunately they aren't much use on their own — the grains or stones are separate and fall apart too easily. But thousands of years ago it was discovered that if they are stuck together by even a fairly weak material the result is a very useful building material — concrete.

A material that sticks particles together is called a **cement** or a **matrix**. The cement used in concrete is made by heating limestone with clay. When cement is mixed with water, it reacts to form a hard, strong solid.

Picture 2 shows what concrete is like inside. You can see the sand grains and parts of the pebbles, and the crystals in the cement that are holding them together.

Concrete does have a problem — it cracks too easily. The cracks start in the cement, grow bigger and move through the cement in the spaces between the grains. But if there are enough grains or pebbles of the right size they stop the cracks getting any bigger — they are too strong and tough for the cracks to get through.

Concrete cracks more easily when it is bent or stretched (in **tension**). The pulling forces make the cracks bigger. But if concrete is always used in **compression** the forces act to close the cracks up. Picture 3 shows how cracks are helped to grow by pulling forces, and may be stopped by pebbles and by compression forces.

Using fibres to make materials tougher

Concrete is a rough, heavy kind of material. It is good for making things like buildings and bridges, but is not too pleasant for making, say, indoor furniture. But the principle of using a mix of materials is used in lots of everyday objects.

In general, the idea is that one material (the matrix) gives flexibility and the other, (the reinforcing) gives strength. The combination is a composite material with both strength and toughness. Glass-reinforced plastic is another composite material. It is used to make canoes, yachts, plant containers, baths and many other things.

Glass-reinforced plastic is light, and can be flexible without losing strength. Plain plastic would bend a little and then shatter because of the cracks that develop. The glass fibres stop the cracks growing to danger level.

Carbon fibres do the same job in more expensive objects, like tennis rackets and fishing rods. They can also be used to reinforce metals.

There are more examples of composite materials in *The Material World* topic A5.

Reinforced concrete

Concrete is best used 'in compression'. This means with forces acting on it which squash it, rather than stretch it. This is because concrete is easy to crack, as explained above. This means that concrete would not be very useful as a building material except in pillars or supports, or as a road material.

But engineers have found a way of keeping concrete in compression all the time, wherever it is used. This is done by pouring the concrete mix over steel wires which are kept stretched by very large forces. When the concrete has set hard the wires are let go. As they try to return to their proper length they keep the concrete under compression (see picture 4).

This is called 'pre-stressed' concrete. The compression forces produced by the steel wires inside the concrete beam are larger than the forces the beam is designed to take in its normal use. So any small cracks that may start cannot grow large enough to split the concrete beam.

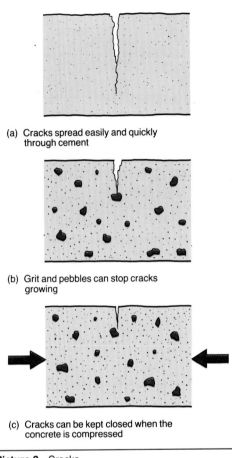

(a) Cracks spread easily and quickly through cement

(b) Grit and pebbles can stop cracks growing

(c) Cracks can be kept closed when the concrete is compressed

Picture 3 Cracks

Picture 4 (a) Making pre-stressed concrete

(b) Concrete beams

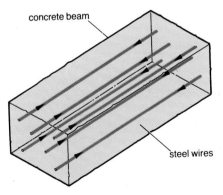

concrete beam

steel wires

(c) The steel wires are kept pulled tight until the concrete sets hard. When they are let go, the wires pull the concrete together, keeping it in tension

Picture 5 Both of these lorries are carrying the same weight

Both the lorries in picture 5 are carrying the same **load**, but one is a load of wood and the other a load of steel. The lorry with the steel looks almost empty, and it would be easy to put a lot more on it. But to do this would be illegal — and dangerous. The result would be too much force on the wheels and axles of the lorry. If both lorries are loaded to their full legal limits then they are carrying the same load — in other words the steel weighs as much as the wood.

The difference between the two materials is that steel is **denser** than wood. 20 tonnes of steel take up far less space than 20 tonnes of wood (picture 6).

Measuring density

Density means how much matter there is in a given space. In other words, density is **mass per unit volume**.

As a formula:

$$\text{density} = \frac{\text{mass}}{\text{volume}} \quad \text{or} \quad \boldsymbol{d} = \frac{\boldsymbol{m}}{\boldsymbol{v}}$$

Steel has a density of 7700 kilograms per cubic metre; wood is much less dense, even a 'heavy wood' like oak has a density of only 720 kg/m^3. We can rearrange the formula and use it to calculate how much space (volume) 20 tonnes (20 000 kg) of oak would take up:

$$\text{volume} = \frac{\text{mass}}{\text{density}}$$

$$= \frac{20\,000}{720} \quad \text{cubic metres}$$

$$= 27.8\,\text{m}^3$$

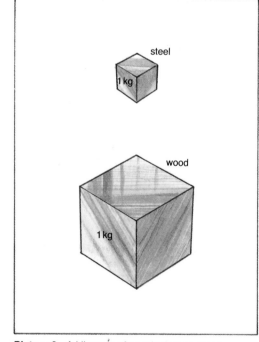

Picture 6 A kilogram of wood takes up nearly 20 times as much space as a kilogram of steel

Use the formula to check that 20 tonnes of steel would only take up 2.6 m^3.

Why is density important?

The density of a material is a very important property. Think of a structure — like a building, a bridge or the skeleton of an animal. If it was made of a dense material it would have the problem of supporting its own weight, as well as any load it had to carry. This might be difficult to do.

Bones have a hollow structure which allows them to be mostly 'empty', so combining lightness with strength. Modern buildings are designed to combine lightness with strength, using modern low-density materials.

Aircraft have to be strong, but as light as possible. They are built on an open framework, as shown in picture 7. This framework is made from special mixtures of metals (alloys). Scientists called metallurgists have had to develop new strong alloys with very low densities, such as **duralumin** (aluminium and copper) and magnesium alloys.

Why are some materials denser than others?

The basic reason for some chemical elements being denser than others is that their atoms are heavier (picture 8). An atom of lead is nearly eight times more massive than an atom of aluminium.

But lead is in fact only about four times denser than aluminium. This is because density of a material is also decided by how crowded together its atoms are. The atoms are 'packed' differently in lead than they are in aluminium. Topic C6 in *The Material World* takes a closer look at how the **structure** of a material affects its properties.

Floating and sinking

The density of a material decides whether it will sink or float. Materials denser than water will sink in it. Less dense materials will float.

The same rule applies to any fluid — helium and hydrogen balloons float in air because these gases are less dense than air. Carbon dioxide balloons fall, sinking through the air.

Of course, you can make a dense material float by giving it a hollow shape. This means that it contains air, and so the average density of the object is less than the density of water.

Picture 7 An airplane under construction showing the open framework

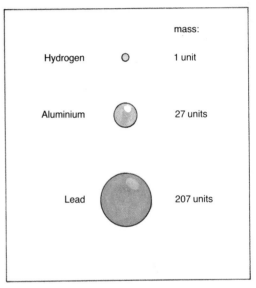

Picture 8 Atoms of different elements have very different masses

Table 1 Properties of materials

Material	density kg/m³	elasticity (arbitrary	strength units)
copper	8940	65	10
steel (high-tensile)	7800	100	100
aluminium	2700	35	1.5
aluminium alloy	2800	35-45	10
lead	11350	8	1
wood (oak)	720	5	2
wood (balsa)	200	3	2.5
glass	2500	33	3
glass-reinforced plastic	1850	37	10–70
rubber	2500	0.003	1
bone	1100	10	3
perspex	1200	1.4	12
polythene (high density)	960	0.5	6
concrete ordinary mix	2200	20	cracks
brick	1700	3.5	depends on joints
breeze block	1400	15	depends on joints

Notes:

1 Elasticity is measured in terms of how much force is needed to produce a standard percentage extension. The lower the number the easier it is to stretch.

2 Strength is measured as the force needed to break something by stretching it. The bigger the number the harder it is to break it (or deform it permanently).

3 Some materials, like brick, have variable properties. Average values are given.

Activities

A Physical properties of materials

There are many possible reasons why a designer chooses to use a particular material to make something with. This topic is about just some of the **physical** properties of materials.

1 List some other properties that you think might be important.

2 Choose any object that you can see and touch that is made of more than one material. Write down what it is made of and say why you think the designer has chosen those materials.

B Making and testing concrete

What makes the strongest concrete? It might depend on:

■ the amount of cement compared with the other ingredients (the aggregate),
■ how much water you use,
■ the size of the particles in the aggregate (sand, grit, pebbles).

In this activity you will test different concrete mixes. It is best done as a group. Decide what proportions of quantities you want to use. For example, the mix of cement to the other materials (aggregate) could be 1 to 1, 2 to 1, 1 to 2, etc. Make a clear table of the different kinds of mixes and quantities of water that you are going to make samples with.

Your teacher will tell you the size of mould to use — a reasonable size would be a small 'beam' about 20 cm long by 8 cm wide and 5 cm deep. You may have to make the moulds yourselves.

CARE: *Use goggles*. Do not touch the wet mix with your hands. It will damage sensitive skin. Use plastic gloves if possible. Put a box of waste packing material underneath to catch the loads if they fall.

Share the work out amongst the members of the group, so that everybody makes one kind of mix. Use small tins to get the volumes of the different parts of the mix in the right ratios. Put the dry ingredients in a larger tin and mix very well. Add water slowly, mixing it in as you do so. Pour the mix into the moulds and cover it with damp paper so that it doesn't dry out before it sets.

The concrete will take a few days to set properly.

Testing the samples

Picture 9 shows one way of doing this. The aim is to find out how much weight the beam can take without breaking. Make sure that all the beams are tested in the same way. Take care to wear goggles in case the concrete beam shatters with a heavy load.

Write a report on your results.

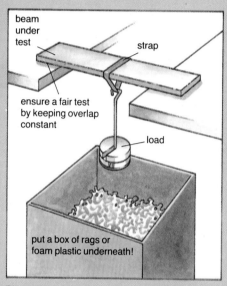

Picture 9

C Measuring density

To find the density of an object you will need to measure two things — its volume and its mass. You then calculate its density by using the formula

$$\text{density} = \frac{\text{mass}}{\text{volume}}$$

The diagrams in picture 10 show three ways for measuring volume.

You will need a suitable balance (lever arm, top-pan, spring) to measure mass.

Choose the best combination of instruments and techniques to measure the densities of some or all of the following materials:

1 a metal block,
2 a wooden block,
3 a specimen of rock,
4 oil,
5 water,
6 sand,

D Finding out about densities

Use a library or a reference book to find out:

1 What is the densest element?
2 What is the least dense element?
3 What is the densest timber?
4 Birds' bones are less dense than, say, human bones. How is this achieved?

E Floating and sinking

1 Use table 1 on page 9 and a reference book to find the densities of the following materials: oak, copper, lead, brick, balsa, candle wax.

2 Sort these materials into two groups, those that sink in water (a) and those that float (b).

3 The density of water is 1000 kg/m³. Polypropylene is a plastic with a density of 900 kg/m³. Will it float or sink in water?

4 Alcohol has a density of 800 kg/m³. Will polypropylene float in alcohol?

5 Ice floats with about 90% of its volume under water. Make a reasonable guess at the density of ice.

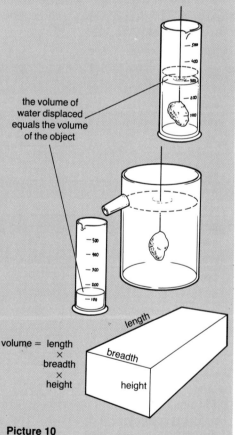

the volume of water displaced equals the volume of the object

volume = length × breadth × height

length
breadth
height

Picture 10

Questions

1 Concrete beams are best used 'in compression'.

a What does 'in compression' mean?

b What happens if the beams are used 'in tension'?

c Why are the concrete beams used in modern buildings usually 'reinforced' with strong steel wires?

2 Why are some materials made of a 'resin' or 'matrix' with fibres inside? (For example, glass fibres in plastic, carbon fibres in metals and plastics).

3 Make an estimate of how much of the material in your classroom is synthetic ('man-made') compared to the amount that is 'natural' (like wood or stone).

Choose one of the synthetic materials and give *one* reason (other than cost) why it might have been selected for its purpose.

4 Using your own experience, put the following materials in order of density, the most dense material first:

oak, expanded polystyrene, lead, glass, the human body, water.

5 Some of the key words used in this topic are: **strong, tough, brittle, weak, composite, matrix, reinforce, cracks**.

Write a sentence about three of the materials listed below, in which at least six of the above key words are used correctly. You don't need to use each word about each material!

Materials: **wood, glass, concrete, polythene**.

6 Wood is a lot less dense than steel, but most ships are made of steel. How can a steel ship float?

7 Copy out and complete the table below by calculating the missing values. You will need to use the formula

$$\text{density} = \frac{\text{mass}}{\text{volume}}$$

Material	mass	volume	density
copper	200 g	22.4 cm³	
aluminium		400 cm³	2.7 g/cm³
lead		0.5 cm	11 350 kg/m³
brick	2000 kg	1.2 m³	
steel	2000 kg		7700 kg/m³
wood	2000 kg		600 kg/m³
concrete		4.5 m³	2200 kg/m³

9 Engineers often want to use a material that is both strong and light, i.e. a material that has a good 'strength-to-weight ratio'.

a Give two examples of applications where a good strength-to-weight ratio would be useful.

b For the two examples you have given, suggest materials that might be used in practice.

c Use the table of properties on page 9 to calculate the values of strength/density for the following materials:

aluminium, steel, glass, wood, fibreglass.

Write a sentence or two commenting on the results of your calculations.

10 Describe how you might measure the density of the following:

a a metal cube,

b oil,

c a piece of rock with no definite shape,

d air.

Eureka!

The ancient Greeks were quite broadminded. But the sight of a naked man running through the streets shouting 'I've found it! I've found it!' caused a few heads to turn and probably collected a swarm of children calling out rude words.

The naked man was called Archimedes. He was already famous as a scientist, and he had just solved a physics problem! He had discovered what we now call **Archimedes' Principle**. The local king had asked him to check whether a new crown was made of pure gold or a cheaper mix of gold and silver.

The answer came to him in the bath. It was based on the fact that when you get into a bath the water level rises. Everybody knows this! But Archimedes suddenly realised that the *volume* of water the body pushes aside must equal to the volume of the body.

This breakthrough would allow him to measure the volume of the crown. By weighing it he could work out its density (see page 10). If the crown's density was the same as the density of pure gold the king would be happy. If not, a goldsmith was going to be in serious trouble.

Archimedes went on to work out that an object appears to lose weight in water. It is being supported by the water, and the weight loss is equal to the weight of water displaced. This idea is still the neatest way to measure the density of an object:

$$\text{density of object} = \frac{\text{weight in air}}{\text{loss of weight in water}}$$

Try these questions.

1 Design and carry out an experiment to check Archimedes' ideas.

2 A reel of copper wire weighs 250 g in air. When it is lowered into water it seems to weigh just 222 g. Use these results to calculate the density of copper.

3 Archimedes' rule only works if water has a density of 1 g per cubic centimetre. Why is this?

A3 Forces and structures

*Ships, bridges, buildings, animal skeletons and plants are all **structures**. They have to be strong enough to do their job, and must make economical use of materials.*

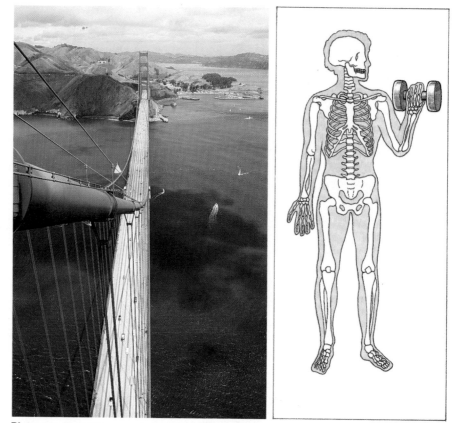

Picture 1   Most things are designed to withstand forces

Picture 1 shows both natural and synthetic objects that have to withstand forces. Of course, they also have to do other things — like carry traffic, or move at speed. They are known as **structures**.

The effects of forces on structures

Engineers describe these forces according to the effects they produce, as illustrated in picture 2. These are:

tension forces which stretch an object
compression forces which squash an object
torsion forces which twist an object
bending forces which bend an object.
shear forces which tend to tear an object

Picture 2   The forces that act on objects

Coping with forces

What the objects in picture 1 have in common is that they are made of lots of smaller parts which are joined together. The shape of the parts, and the way they are joined together, ensure that they are not damaged by the forces.

A structure may have to cope with more than one kind of deforming force. Picture 3 shows how this can be.

The building blocks of structures

Most structures are made of lots of different parts designed to cope with the various forces acting on them. The most obvious force is caused by gravity. This acts on the structure because of its own weight, or because of the load it has to carry.

Moving objects also have to cope with acceleration forces, the effects of braking — and even the effects of a collision.

Objects that might get hot have to withstand the very strong forces caused by the material trying to expand.

Picture 3 A simple plank bridge. Which part is under tension?

Structures are made from smaller structures

Most everyday structures are made up from a small range of 'building blocks'. The most used are beams, triangles, cables (or strings) and pillars (buttresses).

Beams

A beam is simply a piece of strong material used horizontally, like a plank put across a gap. They are probably the most common structure there is.

The beam may be fixed at one end or at both ends. A beam fixed at one end only is called a **cantilever** (picture 4).

A beam will always bend a little, under a load or under its own weight. When this happens one edge of the beam will be stretched (in tension), the other edge will be squashed (in compression). This is shown in picture 3. Beams tend to break by cracking at the tensioned edge, which is being pulled apart, as explained in topic A2 (page 7).

Picture 4 A cantilever structure: a stand at Watford football ground

Triangles

Many structures are made from solid rods fitted together into triangular shapes as shown in picture 5. The rods need to be strong in both tension and compression. This is because some rods are being pulled apart ('ties'), but others are being compressed ('rods'). In a complicated structure, it is sometimes quite tricky to work our which is which.

As long as the ends of the rods are fixed firmly together at the joints the structure will be very strong.

Strings, guys and cables

A piece of string is not much use in compression. It can only withstand a force when it is pulled. Then it can be very strong indeed, for its size. A **cable** is just a very thick 'string', made out of steel or some other strong material.

Cables are used in **tension**, and you will find them in some of the strongest structures in the world — suspension bridges. A suspension bridge (shown in picture 1) uses cables made from a strong material (steel) that is strong in tension. A typical cable for a bridge has to be very strong — see picture 1.

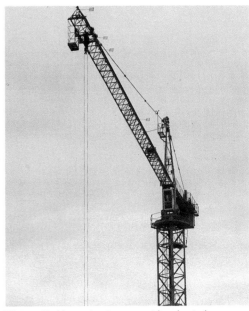

Picture 5 Many structures use triangles to keep themselves rigid

Picture 6 A Greek temple, over 2000 years old

Picture 7 Hollow structures

Cables are also used to support tall structures like TV transmitter masts. They do this by pulling down on the mast, rather than holding it up as with a suspension bridge.

Thinner cables made of rope are used to hold down tents and are called **guy ropes**. Old sailing ships used thousands of ropes to keep masts and sails in position.

Pillars

The towers of the bridge in picture 1 may be made of concrete rather than steel. This is because they are designed to withstand mainly compression forces. Concrete is strong in compression but weak in tension.

The roadway of the bridge is a **beam**. It must be heavy and has to carry heavy traffic. It would not be strong enough on its own unless the beam was made very thick. This would be expensive and would look rather ugly.

The bridge designers have used the different materials in the structure to their best advantage. They have shared the forces neatly between the beam, the cable and the pillars.

Of course, the pillars have to carry the total load. They are the strongest and thickest parts of the structure.

A problem with pillars

The pillar has to be very strong, and one way of making it strong is to make it big. This means giving it a large cross section. But this makes it even heavier. In the end most of its bulk is there to support itself rather than the rest of the bridge. Old Greek temples are very beautiful, but there do seem to be a lot of pillars for the load of the roof! (See picture 6)

Hollow pillars

A hollow pillar can be just as strong as a 'filled-in' one. This fact was a great surprise to the ancient Romans who discovered it, and is still a surprise to most people today. The Romans learned to build pillars for bridges and buildings in which the strength was in a fairly thin shell on the outside of the pillar. The inside was filled in with rubble to stop this outer shell bending.

A modern bridge may have its pillars made of hollow steel tubes, which are just as strong as concrete and weigh far less.

Steel is much denser than concrete, and a lot more expensive, but because the pillar is hollow it uses very little material.

Everyday structures

Many everyday objects make use of this idea to reduce their weight and cost, as shown in picture 7. As you can see, some of these structures are biological structures. Nature got there before the Romans!

Living structures

The human body makes use of strings (tendons) and hollow tubes (bones). The bones are strong and light, making best use of the material they are made from. The tendons carry very large forces produced by the muscles. They apply these forces to the bones, so allowing us to move and to lift things (see *The Living World*, topic D6).

Plants make use of hollow stems to support themselves, even if the hollow is often filled with a liquid of some kind. Their roots act as cables, anchoring them firmly in the ground (see picture 8).

Picture 8 Carrots are easier to pull up than grass is!

Bones

The skeleton of an animal uses bones as beams and pillars. Look at picture 1 and see if you can work out which job different bones do.

The internal structure of a bone gives strength by using combinations of triangles in a six-sided 'honeycomb' effect (picture 9). The bones of birds have to be especially light for their strength.

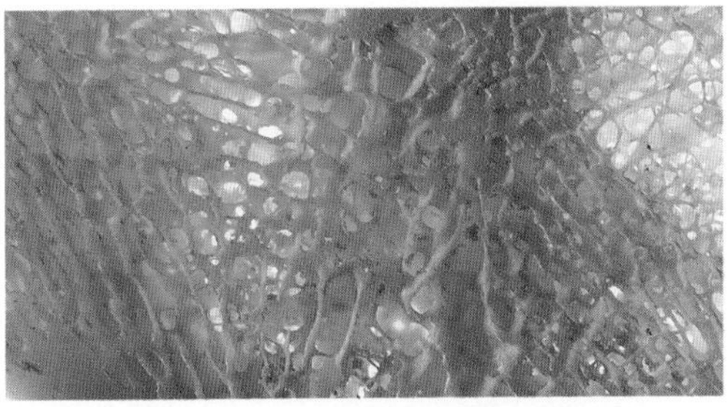

Picture 9 The honeycombed structure of a bone

Activities

A Building a strong structure

Your task is to design and make a structure, then see which structure can hold up the greatest weight. You will be given:

■ some rods (e.g. straws, lengths of dry spaghetti or macaroni, thin card),

■ a way of fixing the rods together (glue, sticky tape, bits of wire,

■ whatever tools that might be useful.

You will be told the smallest height that your support will have to be above bench or ground level.

Test your structure to see how much of a load it can hold up.

Write a brief, illustrated report explaining how successful your design was, with reasons for choosing that design.

B Studying structures

Choose one of the following:

■ a building you know well,

■ any kind of transport vehicle,

■ any household device,

■ any piece of furniture,

1 Look at it very carefully.

2 Write down what materials it is made of.

3 Write down what materials are there to give it strength.

4 Write or draw how it is designed/shaped so that it is strong.

5 The materials in the object may not all have been chosen just to give strength. Can you give one *other* reason, apart from giving it strength, for using any of the materials in the object you are looking at?

C Looking at natural structures

Use a hand lens and a microscope to look at the internal structures of the following. You may be given prepared slides, or you may have to collect the material yourself. Look at them both in cross-section and lengthways, if possible.

1 A bone,

2 fresh wood,

3 a holly leaf,

4 an insect wing,

5 a hollow plant stem,

6 the skeleton of a small animal (e.g. a mouse or frog),

7 a seed that is spread by wind (e.g. sycamore, dandelion).

There are many other things you could look at as well. For two or three of them draw and describe how the materials and the structure are suited to the job they do.

Questions

1 Explain what is meant by the following:

a a tension force,

b a shear force,

c a compression force.

2 What are the kind of forces (from the list on page 12) are the following structures designed to withstand? (Picture 10).

3 What kind of force is present:

a In your legs when you are standing up?

b In your arm when you hold a chair with your arm out level?

c In your wrist when you are using a screwdriver?

4 Give two examples of a beam being used as a cantilever in your home or in school.

5 Give an example of the use of triangles in a structure you can see in your home or in school.

6 Explain the following:

a Lightweight tents have poles which are hollow and made of an aluminium alloy.

b Plants with 'fibrous roots' get good support even if their roots don't go very deep into the soil.

c Water plants have stems which are much weaker, in compression, than land plants.

d Electricity pylons (and TV masts) are thinner at the top than at the bottom.

Picture 10

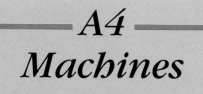

A4
Machines

Human muscles are weak. We can increase their effect by using machines...

Making work easier

Every kitchen and workshop contains tools or implements that make it easier for us to do a job of work. Most of them are used to increase the size of a force. Opening a tin is easy — with a tin opener.

Bottle openers, screwdrivers, scissors, — even door handles — are all examples of 'force multipliers': devices which 'multiply forces' (picture 1).

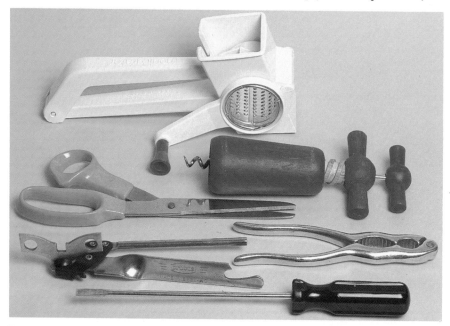

Picture 1 Household machines

Some machines are very complicated. Just look at a car engine. Even a bicycle has more to it than you might think.

But all these machines are made up of a very few quite simple ones. Some of these are levers, pulleys and sloping planes.

Levers

Picture 2 shows a small child lifting a larger one — by using a seesaw. The large weight is being moved by a smaller one, using the **lever principle**.

When you pull on one end of a pivoted bar it will turn around the pivot. Both children on the seesaw are producing a turning effect. When they are exactly in balance each of them is applying the same size turning effect, but obviously in opposite directions. The turning effect (or **moment**) of their weights is given by multiplying the force by the distance they are from the pivot. Picture 3 shows what we mean by this.

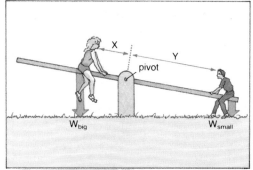

Picture 3 The lever principle: $W_{big} \times X = w_{small} \times Y$

Picture 2 See-saws are fun

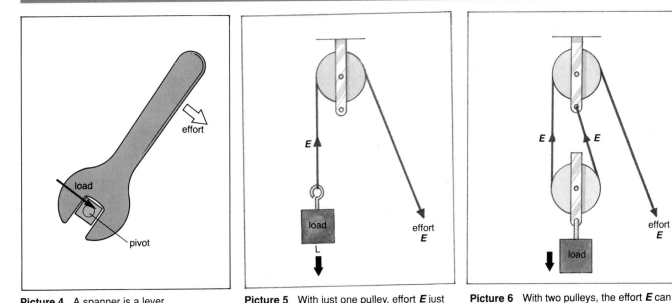

Picture 4 A spanner is a lever

Picture 5 With just one pulley, effort *E* just balances load *L*

Picture 6 With two pulleys, the effort *E* can be about twice as effective

A spanner is a simple machine based on this lever principle. The pivot is now the centre of the nut being turned (picture 4). The **effort** force is applied near the end of the handle. The force that is being 'overcome' is called the **load**. In the case of the spanner it is the friction force in the nut being turned.

When the spanner is just turning the nut the lever principle says that:

effort x effort distance = load x load distance

There are very many examples of levers in everyday life. Many of them are 'hidden' as part of more complicated looking machines. Try counting how many levers there are on a bicycle.

Pulleys

A single pulley is really a kind of moving, circular lever, as shown in picture 5. The wheel is pivoted and so keeps on turning as the load is pulled up. But a single pulley doesn't increase the applied force, because the pivot is exactly in the centre. Pulleys become useful as force multipliers when more than one of them is used.

Picture 6 is a simple two-pulley system. One string goes from the load, around both pulleys to where the effort is applied. The load can be nearly twice as large as the effort. This can be explained by the fact that the 'force in the string' is the same all through it. It is *E*, the effort. The load is supported by a double string, so the upward force is $2 \times E$. This is only exactly true if there is no friction in the system. If there is, some of the effort force has to overcome friction, so less is left to lift the load.

Sloping planes

These are the largest machines you will ever see. The principle was used by the Ancient Britons to build Stonehenge, and by the Egyptians to build the Pyramids.

Both had the task of lifting very large, heavy stones. They did it by building sloping roads or mounds of earth and sliding them up on wooden rollers (picture 7). The force needed to roll something up a slope is much less than the gravity force acting straight down on it.

The less steep the slope the smaller the force you need. Mountain roads are built on a zig-zag to make their steepness less. Other examples of slopes used as force multipliers are: wedges, knife blades, screws and bolts (picture 8).

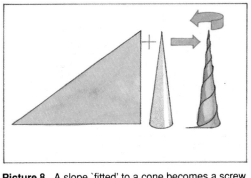

Picture 7 A slope is also a machine

Picture 8 A slope 'fitted' to a cone becomes a screw

Something for nothing?

The use of machines to multiply forces seems to be against the basic law of energy — see topic D4. Are we really getting something for nothing?

The answer, unfortunately, is no. The energy we put into any machine is always at least equal to, and usually greater than the energy we get out of it. The simple reason for this is that in every single one of the machines dealt with in this topic, *the effort moves further than the load*.

The energy supplied is equal to the work done by the effort. This is measured by 'applied force x distance moved in the direction of the applied force' (see topic D2). The work done by the effort is likely to be greater than the useful energy obtained by moving the load. At best it will be the same. There is no such thing as a machine which does more work that we put into it!

Activities

A   Investigating machines

You will be given a simple machine, or a working model of one. It could be a pulley system or a car jack, for example.

1   Measure how much effort force is needed to lift up to five different loads.

Record your results in a table, as above. You can measure the forces involved using a newton meter — or you can assume that the gravity force on 1 kilogram is approximately 10 newtons.

	1	2	3	4	5
Effort/N					
Load/N					

Choose a range of sensible load values.
Is there a pattern in your results?

2   For one of your load-and-effort combinations measure how the effort moves ('effort distance') when the load moves a certain distance (the 'load distance').

Calculate the work done by the effort, and on the load, by using the formula:
work done (in joules) = force × distance moved in the direction of action of the force

B   Looking at machines

Look at a selection of tools and kitchen utensils. Which of them are machines? Draw two that are and explain how they work, based on the principles of the four basic machines described in this topic.

Questions

1   Name ten simple machines that you might find in your home or in the laboratory.

2   Why are machines useful?

Describe one machine and explain how it works.

Why can't you 'get more work out of it' than you put in?

3   Put the following useful things into two headed columns 'machines' and 'non-machines':

a bucket, a knife, an egg whisk, a screwdriver, a door handle, a bookshelf, a spanner, a light switch, a spade, a pen-top.

For one example in each column, give your reasons for putting it there.

4   Name or describe five different levers you could find as part of a bicycle.

5   Picture 9 shows two wheels fixed on the same axle. The large wheel **A** has a radius of 50 cm, the smaller one **B** a radius of 10 cm.

a   What happens to wheel B when A turns around once?

b   What happens to the bucket on the rope wrapped around B?

c   Would this device make it easier to lift the bucket of water?

d   This is an old-fashioned kind of machine for lifting water out of a well. Can you think of any modern device that uses the same idea?

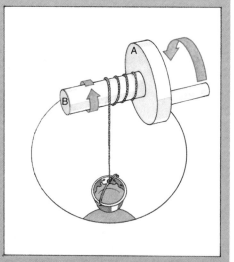

Picture 9   A wheel-and -axle is still useful for getting water from a well

Analysing machines

Machine	1 effort (N)	2 load (N)	3 4 distance moved (in metres)	
			by *effort*	by *load*
1 3-pulley block	20	56	6	2
2 6-pulley block	10	52	12	2
3 Sail-lifter on yacht (wheel-and-axle)	25	120	10	2
4 Bolt cutter	30	270	0.1	0.01
5 Rotary cheese grater	14	2	2.4	0.6

Here is some data about different types of machine. It tells you how much effort is needed to lift or move a given load. It also tells you how far the effort force moves for a typical movement of the load. Use this information to answer the questions below. You will need graph paper to answer some of them.

1   In each machine, the effort force has to move quite a lot further than the load. Considering the design of any one machine, explain why this is. (Two of the machines are illustrated in the picture).

2   Each machine 'multiplies' a small effort force into a larger load force. Use the values in the first two columns to calculate how good each machine is at doing this, by dividing the **load** by the **effort**. The result is called the *mechanical advantage* of the machine. Put the results in a list headed 'mechanical advantage'.

3   Use the data in columns 3 and 4 to calculate, for each machine, how far the effort has to move compared with the load. Do this by dividing ' distance moved by effort' by ' distance moved by load'. The result for each machine is a number that engineers call its *velocity ratio*. Put the results you obtain into a list headed 'velocity ratio'.

4   Look at the values of mechanical advantage (MA) and velocity ratio (VR) you have worked out. It is suggested that there might be a connection between how effective a machine is at moving a load (its MA) and how far the effort has to move compared with the load (its VR)

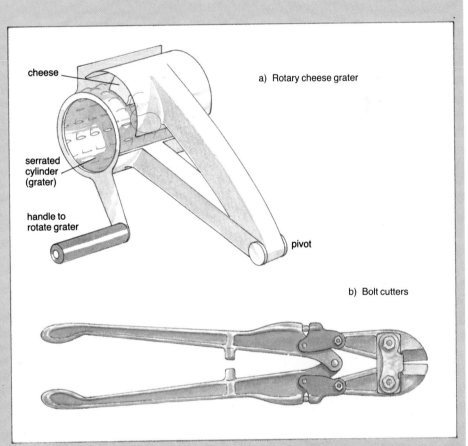

a) Rotary cheese grater

cheese

serrated cylinder (grater)

handle to rotate grater

pivot

b) Bolt cutters

How are these machines adapted to their different jobs?

a   Is there any relationship (or pattern) between mechanical advantage and velocity ratio for these five machines? Make a rough guess at what it might be and write it down.

b   Now plot the values against each other on a set of graph axes — mechanical advantage up, velocity ratio along.

c   Draw the 'best straight line' through these plotted points. Does this confirm your prediction in (a)?

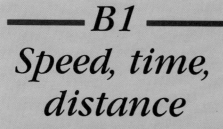

—B1—
Speed, time, distance

We live in a world of movement — the flight of birds, high-speed trains, and the high-speed molecules that keep up your bicycle tyres are all examples . . .

Picture 1 Racing against the clock

Speed

The **speed** of an object is how far it goes in a unit of time. It is measured in units such as metres per second (m/s), or kilometres per hour (km/h). In Britain you will also see the unit 'miles per hour' (mph), but this is not used in scientific measuring.

Table 1 shows the range of speeds of a number of moving things.

Table 1 Comparison of speeds

Speeds	m/s	km/h
light speed	300 000 000	1 080 000 000
Earth in orbit	29 790	107 244
typical Earth satellite	7 500	27 000
fast jet aircraft	833	3 000
Concorde (supersonic jet)	648	2 333
average speed of air molecule	500	1 800
sound in air	340	1 224
Boeing 747 Jumbo Jet	270	970
fastest bird (falcon)	97	350
high-speed train (French TGV)	60	216
motorway speed limit (UK 70 mph)	31	112
town speed limit (UK 30 mph)	13.4	48
Olympic sprinter	10.3	37
average walking speed	1.7	6
average speed of a snail	0.006	0.02

Measuring speed

Very many different instruments are used to measure the wide range of speeds shown in the chart. But all of them need to measure just two things: **time** and **distance travelled**.

If you wanted to measure the speed of a sprinter at an athletics track you would measure how long it took the runner to cover the distance of the race. For example, you could use a stop watch, and find that it took the runner 13 seconds to cover a distance of 100 metres. You could calculate the speed (distance covered per second) like this:

$$\text{speed in metres per second} = \frac{\text{distance in metres}}{\text{time in seconds}}$$

$$= \frac{100 \text{ metres}}{13 \text{ seconds}}$$

$$= 7.7 \text{ m/s}$$

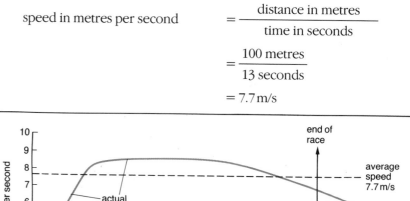

Picture 2 How a runner's speed changes during a race

Average speed

What we have worked out for the sprinter was the *average* speed over the distance. The runner wouldn't have travelled at this speed steadily for all of the 13 seconds of the race.

A runner starts slowly then accelerates to her fastest speed. She might slow down near the end when her legs get tired. A graph would show this (see picture 2).

In the same way, a car might take four hours to travel a distance of 200 km. Unless it was on a very clear motorway it isn't likely that it would have travelled at the same speed of 50 km/h all the time. It's useful to know the *average* speed of a car on different kinds of roads. When you are planning a journey you can use it to work out how long the journey is likely to take.

We can work out average speeds using a formula:

$$\text{average speed} = \frac{\text{total distance covered}}{\text{time taken for journey}}$$

Written as a formula: $v = \dfrac{s}{t}$

Instantaneous speed

Sometimes we need to know more about what happens when moving things are changing speed. To do this we need to be able to measure the speed at any given instant. This is not so easy to do. It means that we have to measure the distance an object travels in a very small interval of time, usually much less than a second.

This can be done by using **radar**, as the police are doing in picture 3. The radar 'speed gun' measures the distance the car moves in a time of less than a millionth of a second! A built-in computer works out the result. In school laboratories a useful device called a ticker-timer does the same job (picture 4). It uses a time interval of a fiftieth of a second. See activity C for more details of how to use a ticker timer.

Your school lab may also have computer-assisted speed measuring devices using interrupted light beams. (See picture 5.) All these things might look quite complicated, but all they do is what you do with a stop watch and measuring tape. They measure *time* and *distance covered*.

Picture 3 Police can measure the speed of a car using radar

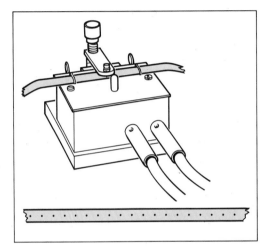

Picture 4 A ticker-timer and a sample of tape: the object was moving with a constant speed

Each length shows how far the object travelled in a fifth of a second. A 'ten-tick' length has 10 spaces

Movement formulae

$$\text{average speed} = \frac{\text{distance covered}}{\text{time taken}} \qquad v = \frac{s}{t}$$

$$\text{distance covered} = \text{average speed} \times \text{time taken} \qquad s = vt$$

$$\text{time taken} = \frac{\text{distance covered}}{\text{average speed}} \qquad t = \frac{s}{v}$$

Working things out

Suppose you wanted to estimate how long it would take to cycle from your home to a holiday area. You might be staying at a camp site or a youth hostel when you get there.

You work out on the map (picture 6) that it is 56 miles away. You now have to decide what your average cycling speed is. A reasonably fit person could cycle at 10 miles an hour on fairly flat roads.

Picture 5 These children are using a computer-assisted speed measuring device

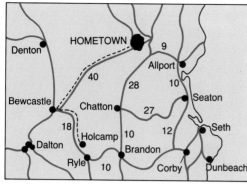

Picture 6 Which is the quickest route?

You would use the formula:

$$\text{time taken} = \frac{\text{distance covered}}{\text{average speed}}$$

$$t = \frac{s}{v}$$

put the values in:

$$t = \frac{55}{10 \text{ miles per hour}}$$

$$\text{time needed} = 5.5 \text{ hours}$$

This is just an estimate, of course. You might decide to stop for a coffee. The route might be hilly. So you'd allow an extra hour or so to make sure.

Car and lorry drivers have to make the same kind of estimates every time they take a new route. You soon learn that the hardest thing to get right is an estimate of your average speed.

How fast do you walk?

People who do a lot of walking need to have a good estimate of their average walking speed. If you go hiking in the countryside, especially in mountains, you have to make calculations very carefully (picture 7). You may be far from roads and shelter, and mistakes could be dangerous. You wouldn't want to find yourself far from shelter as night falls or the weather turns bad.

Experienced hikers use *Naismith's Rule* to help them calculate timings.

This says: *allow 1 hour for every 3 miles (5 km) you measure on the map then add 1 hour for every 2000 ft (600 m) you have to climb.*

This rule works for a fit walker, not carrying a lot of equipment. Question 4 asks you to use this rule to work out some walking times.

Picture 7 How fast do you walk?

Activities

A Timing pendulums

The first accurate clocks were based on the principle of the pendulum. A pendulum is a weight on a cord, made to swing from side to side. Does it swing from side to side in a constant time? What decides how long the weight takes to swing from side to side?

Do an investigation to answer these questions. First list the things that *might* affect the time of swing. Then think of how you can measure the time of swing accurately. Is it good enough to measure just one swing?

This investigation could take a long time if you tried everything yourself. Organise your group into small teams that can investigate different things. Make sure that they all share their results at the end.

B Measuring average speed

For this investigation you will need a stop watch. You will also need some way of measuring the distances involved. Measure the time taken to cover a measured distance and use the formula:

$$\text{average speed} = \frac{\text{distance covered}}{\text{time taken}}$$

Try some or all of the following measurements:

1 the average speed of a runner,
2 the average speed of a cyclist travelling to school,
3 the average speed of a car passing the school or your home,
4 the average speed of a bird in flight,
5 the average speed of an object falling: (i) from a height of 2 m, (ii) from a height of 5 m,
6 the average speed of flow of a river or stream.

C Using ticker-timers

Picture 4 shows a ticker-timer. A length of paper ticker tape is pulled through the timer. It prints a dot on the tape once every fiftieth of a second. It works using electromagnetism and is controlled by the mains ac frequency of 50 hertz. Thus the distance between the dots shows the distance the tape moved in just 1/50 of a second.

Use the timer to investigate one or more of the following:

1 Do you walk at a constant speed?
2 Do objects fall at a constant (steady) speed?

3 How does the speed of the weight at the end of a pendulum change as it swings from side to side?

Hints:
i) Use a fairly short piece of tape (about 1 metre long).
ii) Work with 'ten-space' lengths, not one-space lengths. (Ten spaces represent a fifth of a second, which is usually accurate enough)

iii) You can make your length of tape into a kind of graph showing the motion. Start at the beginning of the tape and count off *ten spaces*. Call this piece of tape (a 'ten-tick') number 1. Count off the next ten spaces and carry on doing this to the end of the tape, numbering each short piece. Cut the tape into your numbered ten-ticks and set them out on a base line as shown in picture 8.

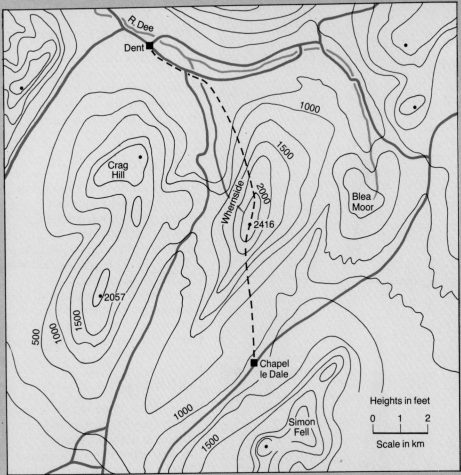

Picture 8

Questions

1 Estimating speed

Look at table 1 (page 20) showing the speeds of a number of well-known objects. Use it, and any other information you might have, to make a sensible guess at the likely (average) speeds of the following:

a a cyclist in a long-distance road race,

b a tennis ball during a serve,

c the signal from a TV transmitter,

d a marathon runner,

e the growth of grass.

2 Using formulae

Use the formulae given on page 21 to calculate the following.

a The speed of a train that covers 300 km in two hours.

b The speed of a walker who covers 24 km in five hours.

c How far a cyclist would travel in five hours at an average speed of 12 km/h.

d How far a car would travel in eight hours at an average speed of 70 km/h.

e How long it would take a ship to travel a distance of 400 km at an average speed of 25 km/h.

f How long it takes light to travel the distance of km from the Sun to the Earth, at a speed of 300 000 kilometres per second. Give the answer in minutes and seconds.

3a Name *four* different ways of measuring time.

b All timing devices make use of some kind of regular repetitive movement (e.g. a pendulum). What, do you think,

are the the things that do this job in the examples you have given in a.

4 Jane, Sally and Salman need to travel from the bridge in the village of Dent (see the map in picture 9) to meet their friends who are staying in Chapel-le-Dale. They plan to travel by the route shown.

a Use the scale of the map (and a piece of cotton, say) to work out how far they have to walk.

b How many feet do they have to climb?

c They are fit and not carrying a load. Use Naismith's Rule (page 22) to work out how long it should take them.

Picture 9

B2
Crashes and bangs

Stopping and starting can be gentle — but what about when they are not? The key idea is momentum.

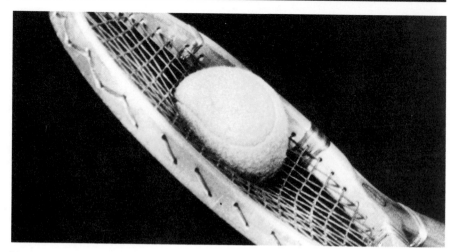

Picture 1   Hitting a tennis ball

Picture 2   Forces occur in pairs: the upward push on the rocket equals the downward push on the exhaust gases

Hitting things

The photograph in picture 1 shows a tennis ball just as it is being hit by the racket. You can see why they change the balls so often, and why rackets have to be made so very strong. The forces on both are very large.

Both the ball and the racket are distorted by the impact. The ball is squashed almost flat, and the racket strings are stretched. This is a good example of one of the basic laws to do with forces and movement.

The law of action and reaction

This was first discovered by the great scientist Isaac Newton over 300 years ago. He said that action and reaction are equal and opposite. This means simply that the force exerted on the ball by the racket — the 'action force'is the same in size as the force exerted on the racket by the ball — the 'reaction'. The forces are acting in opposite directions.

Twin forces

This law of Newton's is one of those laws of nature that always applies. Another way of looking at it is to assume that forces always occur in pairs.

Blow up a balloon and let it go. The balloon rushes forwards and the air in it rushes out backwards. In a rocket or jet the force produced by the engine acts just as much on the vehicle as on the hot gases rushing out at the back (picture 2). This means that the forces involved are equal in size and opposite in direction.

Even when you walk, your foot pushing back on the ground causes a reaction that pushes *you* forward. When a magnet attracts a piece of iron, there is an equal and opposite force pulling the magnet towards the iron. If you try this you can feel both forces.

Collisions

In a collision, forces act for a very short time. The forces are often quite large and the objects that collide run a risk of being damaged. This is what happens in traffic accidents.

In ball games the equipment is designed to avoid damage, although cricketers might not agree! (Picture 3.)

But apart from any damage that might happen in a collision, there is also a change in movement. Objects speed up or slow down. Large objects seem to be harder to speed up and slow down than smaller ones. Think of the difference between tennis and table tennis, or between pushing a pram and pushing a car.

It was again Isaac Newton again who realised that there are just two things that are important when collisions happen.

According to Newton, these are:

■ the **speed** and its **direction**,
■ the **masses** of the colliding objects.

In any collision, the more massive the object and the faster it is going the more effect it will have. The combination is a quantity called **momentum**. **Momentum is mass (*m*) multiplied by speed in a given direction.** Speed in a given direction is called **velocity** (*v*). So **momentum = *mv*.**

Mass

Mass was a very new and strange idea in 1680, and it's not that easy to grasp even now. We can think of it as a measure of how hard it is to make an object move, even when it is perfectly free to do so, like an object in space. The more matter there is in an object, the more mass it has. This idea helps to explain why table tennis bats aren't much use on a tennis court.

Force and momentum

The effect of a force is to change the momentum of an object. The change is bigger when the force is *larger*, or when the force *acts for a longer time*.

So what exactly happens in a collision?

Let's start by thinking about an easy example.

A simple collision

The simplest collision is when a moving ball collides centre to centre with another, identical, ball which is perfectly still ('at rest'). As they make an impact, the forces between them are the same all the time. The moving ball stops dead and the other one moves off with the same speed and in the same direction. In terms of momentum, all the momentum of the first ball has been given to the second one.

The law of momentum

Collision experiments produce the result that whenever a collision happens the total momentum of the colliding objects stay the same. The key formula is:

total momentum before collision = total momentum after collision

Picture 4 shows this. Try it with two coins.

Picture 3 This sportsman is prepared to withstand unwanted forces

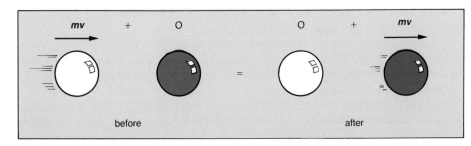

Picture 4 In collisions, momentum stays the same

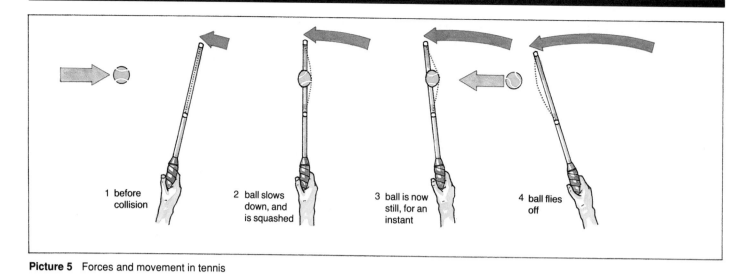

Picture 5 Forces and movement in tennis

Sticky collisions

In some collisions the two objects stick together. Guess what might happen. Then try activity A.

Tricky collisions

Let's consider a slow motion action replay of what happens when a tennis ball is hit by a tennis racket. As the ball touches the strings of the racket the action-reaction forces begin to act. The ball is crushed, and the strings are pushed back. The forces increase and at some stage the ball is stopped, probably when the forces are greatest.

The racket is only slowed down a little. This is because it is so much more massive than the ball, while the force on it is the same as the force on the ball.

But this is only half the story. The forces are still acting, even if they are getting less. The ball regains its normal shape and the strings go back to being straight. The forces are now pushing the ball forward very quickly, faster than the racket in its 'follow through'. So the ball flies away from the racket. The drawings in picture 5 summarise this.

But what has all this to do with real life? Think about safety on the road and other places where collision accidents can happen.

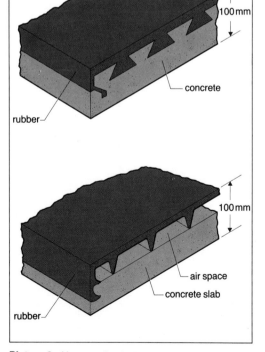

Picture 6 How a safe playing surface can be made

Stopping safely

To stop a moving object you have to take away its momentum. The only way to do this is to provide a force. Think of stopping a car. It can be done using a large force for a short time — as in a collision. Or it can be done with a small force over a longer time — by using brakes.

A safe design allows fast-moving objects to give up their momentum slowly. Cars are designed with 'crumple zones'. The front of the car is deliberately made 'softer' than it could be, so that it crumples slowly in a collision. In the same way, playground surfaces should be made of a non-rigid material. Picture 6 shows some ways of doing this. The surface 'gives', so that the falling child takes a little bit longer to stop. Picture 7 shows how far a child can fall on to a number of surfaces without danger to life.

Force and change of momentum

The key idea is that:

force x time = change of momentum.

For a given momentum change we can have a small force for a long time to get the same effect as a large force for a small time.

Example

A car with a mass of 1000 kg travelling at 30 metres per second (110 km/h) has a momentum of 30 000 kg m/s. To stop it in 5 seconds, using the brakes, needs a force F_{brakes}. We can calculate the force using this rule:

$$F_{brakes} \times 5 = 1000 \times 30$$

giving $F_{brakes} = 6000$ newtons

In a collision, the car might stop in a tenth of a second. The force needed is F_{bang}, given by:

$$F_{bang} \times 0.1 = 1000 \times 30$$

so that $\qquad F_{bang} = 300\,000$ newtons

This is 50 times bigger than F_{brakes} — the force involved in stopping the same car with the brakes. It will certainly cause damage to the car — and maybe the driver and passengers as well.

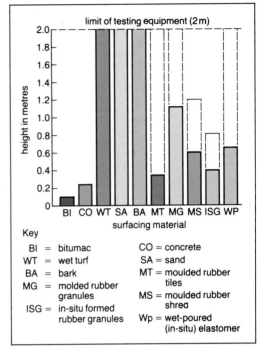

Picture 7   How safe are playgrounds? The chart shows how far a child can fall on various surfaces reasonably safely

Key

BI	= bitumac	CO	= concrete
WT	= wet turf	SA	= sand
BA	= bark	MT	= moulded rubber tiles
MG	= molded rubber granules	MS	= moulded rubber shred
ISG	= in-situ formed rubber granules	Wp	= wet-poured (in-situ) elastomer

Travel slowly — *think quickly*

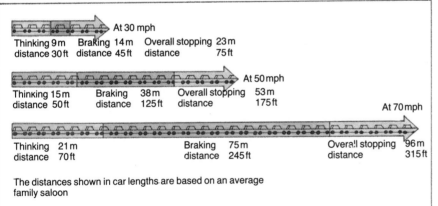

At 30 mph
Thinking distance 9 m / 30 ft   Braking distance 14 m / 45 ft   Overall stopping distance 23 m / 75 ft

At 50 mph
Thinking distance 15 m / 50 ft   Braking distance 38 m / 125 ft   Overall stopping distance 53 m / 175 ft

At 70 mph
Thinking distance 21 m / 70 ft   Braking distance 75 m / 245 ft   Overall stopping distance 96 m / 315 ft

The distances shown in car lengths are based on an average family saloon

Table 1   Official stopping distances

The chart in table 1 shows the official stopping distances for cars with good tyres and good brakes. It has been worked out by the government's Department of Transport. The faster a car is going, the more momentum it has, so it takes longer to stop. The 'braking distance' is how far the car travels in this time.

The chart also shows the 'thinking distance'. It takes time for a driver to react to an emergency — *The Living World* topic D1 explains about reaction times. The thinking distance is how far the car travels before the driver reacts by stepping on the brake pedal.

Speed limits

The physics of this topic should help you understand the reasons for speed limits on the road (picture 8). At high speeds both thinking distance and braking distance are increased. In fact the braking distance is quadrupled for a doubling in speed.

DON'T

GET

DEAD CLOSE.

ALWAYS KEEP AT LEAST TWO SECONDS BEHIND THE DRIVER IN FRONT.

Picture 8   Do posters help to save lives?

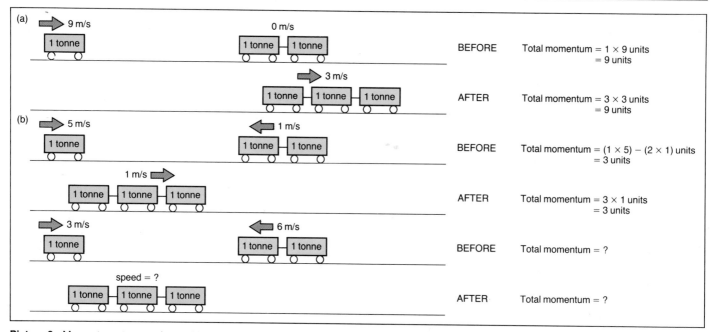

Picture 9 Momentum stays unchanged in a collision

Also, the force of a collision is greater at a greater speed, so that more damage is done. In fact, the damage also more than doubles when the speed doubles. This is because a bigger force acts for a longer time (picture 9).

Seat belts

For safety, all people travelling in a car should wear a safety belt. When they are strapped in they are part of the car. This means that they are stopped at the same rate as the car. In a crash they stay with the car, instead of flying through the windscreen.

If the car stops slowly, so do the people in it. The friction force between them and the seat is enough to slow them down. In a crash, or even very sharp braking, they will *carry on moving* unless they are strapped in. In a crash at 100 km/h they will hit the dashboard or windscreen of the car at this very high speed. Very serious injuries are caused in this way (picture 10). This is because the stopping time is so short. Seat belts are designed to increase the stopping time. The belts are made so that they 'give' a little, and the driver and passengers takes a little longer to stop than the car does.

The principle is the same when you jump from a height on to soft foam rubber, compared with rigid concrete. You take longer to stop and so the stopping force is less.

Picture 10 Why safety belts are helping to save lives

Activities

A Hitting things

This is an exercise to give you a feeling for the main ideas dealt with in this section. Do the tests carefully out of doors, or in the gym, so that you don't damage anything, and make a short note of what you experience in terms of forces.

1 Use a tennis racket to hit the following objects. The idea is to get them to move away with approximately the same speeds:

a table tennis ball, a squash ball, a tennis ball, a cricket or hockey ball.

How do the forces you experience compare? Is there a pattern relating the forces you are exerting to the 'weights' of the balls?

2 Put your hand, palm open and upward, on a bench or table. Drop each of the balls in turn from the same height (no more than half a metre) on to the palm of your hand. What do you learn from this experience?

B Investigating collisions

Design a simple form of apparatus that would allow you to investigate the speed changes that occur when things collide. The colliding objects must be able sometimes to stick together, sometimes to bounce apart. Think of ways of ensuring this, and also how you might measure or compare speeds before and after collisions.

There is a rule that says: momentum before a collision equals momentum after a collision. Do your investigation results bear this out?

C Speed and stopping distance

This activity needs care and some skill in riding a bicycle. **Don't do it without supervision.** It is to find out the stopping distance of a bicycle travelling at various speeds. The variables (things that change and need to be controlled or measured) are:

- the speed of the bicycle,
- the braking force,
- the distance the bike takes to come to a stop,
- the weight (mass) of the rider.

We have to vary the speed, measure the stopping distance and keep the other two factors the same all through.

Measuring speed: time how long it takes to cover the distance between the two lines ten metres apart (see below) and use the formula

$v = s/t$ to calculate speed,

   or use a cycle speedometer.

Keeping the brake force constant: use a block of wood of the right size between the brake handles and handlebars, to limit the amount of brake pull.

Experiment to find a reasonable working arrangement. The brakes must be in good condition, and **both brakes must be used**!

1   Make a line on a good flat surface, such as a school playground (grass or tarmac), and another one ten metres away. This is the distance used for timing to measure the speed. (You won't need to do this if you are using a cycle speedometer.)

2   Get the cyclist to cross the space between the two lines at different speeds. The brakes must be applied when the front wheel crosses the second line.

3   For each speed, measure the stopping distance.

4   Make a chart or graph of the results. Is there any pattern?

5   Would it make any difference if the riding surface was different (e.g. grass versus tarmac, wet versus dry)?

D Thinking about road safety

You will need a collection of road safety pamphlets and slogans, and a copy of the Highway Code. List some key road safety factors, such as seat belts, drunken driving, the clothing that motor cyclists wear and so on. How many of the factors are needed because of the law of physics that says:

force × time = change of momentum?

Produce a poster linking as many road safety ideas as you can with this law.

Questions

1   Predict and try to explain what happens when:

a   a car travelling at 30 mph on an icy road comes to a sharp bend,

b   a cyclist travelling downhill at 40 mph puts on the **front** brake.

2   Why do heavy container lorries need stronger, more powerful brakes than an ordinary passenger car does?

3   a   How do they stop ships (like tankers or liners)?

b   How do they slow aircraft down while they are in the air?

c   How do *you* slow down, when you are running fast?

d   Why do jumbo jet aircraft need longer airport runways than ordinary jets?

4   How does a 'crumple zone' help to protect the passengers in a car?

5   Design an experiment you could do to test a playground surface to see if it was a 'safety surface' or not.

6   Calculate the momentum of the following:

a   a man of mass 70 kg running at 10 m/s,

b   a container lorry travelling at maximum speed on a motorway (mass = 30 000 kg; speed 30 m/s),

c   an ocean liner travelling at its cruising speed (mass 50 thousand tonnes (50 000 000 kg), speed 20 m/s).

7   Use your answers to question 6 to answer the following.

a   What would be the force exerted on the runner if he bumped into a tree and stopped in a time of 1/20th second (0.05 s)?

b   What would be the braking force needed to stop the container lorry if it stopped in a time of 5 seconds? (This is the braking time assumed in the Highway Code.)

c   What might happen if the lorry had brakes which were, say, ten times as powerful?

d   How long would it take to stop the ocean liner if the stopping force was 1 MN (1 000 000 newtons)?

8   A car designer thought of an idea which he thought would cut down the injuries produced in road accidents. He suggested that cars should be made to a new and very strong design. They would use steel which was very elastic so that the cars would neither break nor crush, but just bounce off each other.

a   What would happen if two cars of this type collided with each other head on?

b   Would this kind of design be a good idea? Explain.

9   Describe what you would expect to happen to the speeds of the moving objects as a result of the following collisions.

a   A small boy running along a corridor bumps into a very large man.

b   A large oil tanker moving at speed collides with a small yacht.

c   Two equal sized cars travelling at 80 km/h in opposite directions collide with each other head on.

10   Give simple explanations of the following, using the ideas of momentum and/or of 'action and reaction'.

a   A moving snooker ball moves another one when it hits it, but doesn't seem to move the table when it hits the side.

b   One way to get off a perfectly smooth surface, such as a sheet of very slippery ice, would be to take off a boot and throw it along the ice.

c   Sprinters use spiked shoes.

d   When a gun fires a shell, the gun 'recoils'.

e   Kicking a football is quite easy, but kicking a stone cannonball of the same size could seriously damage your foot.

11   When a gun is fired the bullet leaves it at a very high speed. It is driven out by the force of expanding gases. This force acts on both the gun and the bullet (by Newton's Law 3). Why doesn't the gun move backwards at the same speed as the bullet goes forward?

12   A railway truck of mass 200 tonnes is moving at a speed of 8 m/s in a shunting yard. It bumps into a group of three similar trucks (total mass 600 tonnes) and sticks to them. At what speed will the four trucks begin to move off together?

B3
The mathematics of movement

Physics and mathematics go hand in hand. The great value of physics is that it can use mathematics to make very accurate predictions.

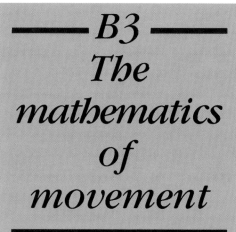

Picture 2 Adams and Leverrier — the discovers of the planet Neptune

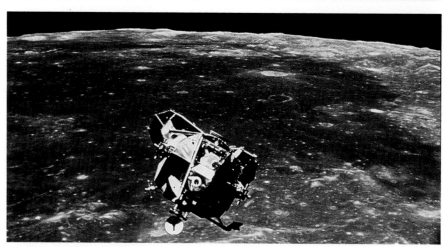

Picture 1 Accurate scientific calculations have enabled us to send capsules and satellites into space — and keep them there

Moving in space

1969. A space craft is orbiting the Moon. At exactly the right moment, to the nearest hundredth of a second, the rockets are fired. They produce a very precise force, in exactly the right direction. They are on for a precisely calculated time. The capsule, and the precious human cargo it contains, land safely on the lunar surface.

1840. The planet Uranus, orbiting the Sun moves slowly against the background of the fixed stars. Astronomers plot its path, and calculate exactly where it should be in a week's, a month's, a year's time. Then they notice that it is not quite where it should be. What has gone wrong? Perhaps the laws of physics don't work so far away from Earth? Did they do the sums correctly?

A new planet is found

Two astronomers, unknown to each other and working in different countries have the same idea. Could there be another planet out there in space that no one had seen? The gravity force from the unknown planet could be pulling the planet Uranus out of its plotted path.

One astronomer, Jean Leverrier, is French; the other, John Adams, is English. Unknown to each other, they both work out where the new planet should be. Leverrier sends his predictions to the German astronomers in Berlin, and they spot it in the sky the same night.

Physics and mathematics working together

John Adams needed four years to make his calculations, using pen and paper. The planet was named Neptune, and was discovered in 1846. The astronauts had their orbits, speeds, forces and times controlled by an on-board computer. The calculations were done in thousandths of a second. Picture 3 shows the planet as it was photographed by the Voyager Space Probe as it flew past in 1989. Although separated in time by over a hundred years, astronomers and astronauts both relied on very accurate mathematical calculations.

Their calculations were based on the same simple rules of physics. These rules were set out by Isaac Newton in the 17th century. They are the laws of *gravity* (see topic B4), and the laws obeyed by objects moving under the action of forces (*dynamics*).

The basic mathematics dealing with moving objects is summarised below. You can follow the proofs to help you understand what the formulae mean.

In this section we use the following symbols for the quantities we measure:

time	*t*	force	*F*
distance	*s*	acceleration and deceleration	*a*
speed at start	*u*	mass	*m*
speed at end	*v*		

Picture 3 The planet Neptune

The equations of movement: speed, distance, acceleration and time

To help you keep track, each formula is numbered.

Average speed

We have already seen, in topic B1 that

$$\text{average speed} = \frac{\text{total distance covered}}{\text{total time taken}}$$

Or $v = s/t$ (1)

Acceleration

Something that speeds up is **accelerating**. If it is increasing its speed steadily we say that it has a steady **acceleration**, *a*. For straight line movement, this is defined as **change in speed per second**. To measure acceleration we need to measure not only the change in speed of an object but also the time it takes to change its speed. (See picture 4).

That is:

$$\text{acceleration} = \frac{\text{change in speed}}{\text{time taken for the change}}$$

Or, acceleration $= \dfrac{\text{speed at end of timing} - \text{speed at start}}{\text{time interval}}$

As a formula : $a = \dfrac{v - u}{t}$ (2)

The unit is $\dfrac{\text{metres per second}}{\text{seconds}}$ or m/s^2

The formula (2) can be rearranged as $v = u + at$

Picture 4 How acceleration is calculated

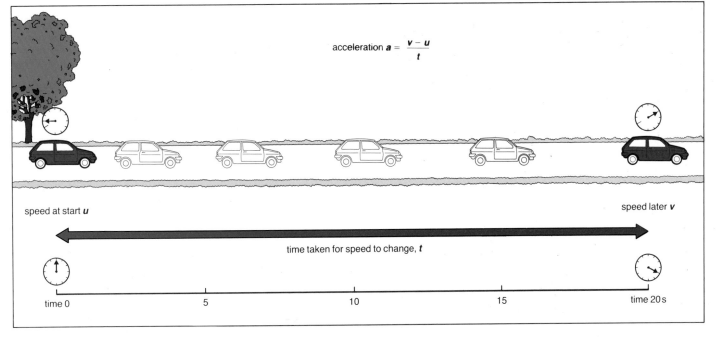

acceleration $a = \dfrac{v - u}{t}$

speed at start *u*

speed later *v*

time taken for speed to change, *t*

time 0   5   10   15   time 20 s

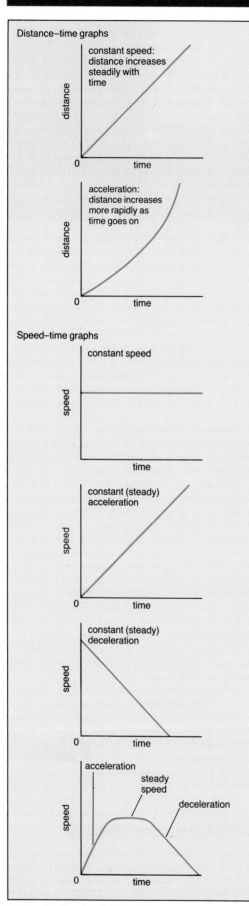

Distance–time graphs

constant speed:
distance increases
steadily with
time

distance

0   time

acceleration:
distance increases
more rapidly as
time goes on

distance

0   time

Speed–time graphs

constant speed

speed

time

constant (steady)
acceleration

speed

0   time

constant (steady)
deceleration

speed

0   time

acceleration

steady
speed

deceleration

speed

0   time

Picture 5 Graphs showing different kinds of
movement

Dynamics: momentum, force and acceleration

We have already come across momentum (mv) and the idea that it is a combination of force (F) and time (t) that causes a change in momentum:

$$Ft = \text{change in } (mv) \qquad (3)$$

Momentum changes when speed changes, so it makes sense to rewrite (3) like this:

$$Ft = \text{momentum change}$$
$$Ft = (\text{momentum with ending speed}) - (\text{momentum with starting speed})$$
$$Ft = mv - mu$$
$$Ft = m(v-u) \qquad (4)$$

Ft is called **impulse**, and the formula put into words says

$$\text{impulse} = \text{change of momentum}$$

We can rearrange formula (4) like this:

$$F = m\,\frac{(v - u)}{t} \qquad (5)$$

Now $\dfrac{(v-u)}{t}$ looks familiar. It is what we defined above (2)

as **acceleration**, a. The formula becomes simpler:

$$F = ma \qquad (5)$$

In words: force = mass × acceleration.

This is one of the most important formulae in physics. It is the basis of space flight, astronomy, car performance and indeed any kind of animal and human movement.

It will be a good idea to do some of the activities at the end of the topic so that you get a feel for what all this mathematics means in real life situations.

The newton

It is the above formula (5) that is used to define the unit of force used in science and engineering. A **newton** is the size of the force that can give a mass of 1 kilogram an acceleration of $1 \, \text{m/s}^2$.

$$1 \, \text{N} = 1 \, \text{kg} \times 1 \, \text{m/s}^2$$

Graphs

One way to help your understanding of what happens when objects are moving is to use graphs. Picture 5 shows graphs of distance plotted against time, and speed plotted against time. If you have already used ticker-time charts you will know that these are speed-time graphs. Look at the graphs and relate them to the formulae you have met in this section. The questions at the end of the topic will help you.

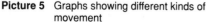

Newton's Laws of Motion

Newton wrote down three simple laws to describe how and why objects move. They summarise the ideas we have already met.

Newton 1: An object will keep still or carry on moving at a steady speed in a straight line unless a **force** acts on it.

Newton 2: The effect of a force on an object free to move is to change its **momentum** such that force × time equals the momentum change produced.

Newton 3: Forces are always found in **pairs**, equal in size but acting in opposite directions.

Newton's Laws are very reliable in everyday life and in astronomy. But in this century we have found that they break down at very high speeds and we have to use Einstein's ideas instead.

Activities

A Use a library or encyclopaedia to find out what you can about:

1 the Apollo space flights and Moon landings

2 the discovery of Neptune.

B Think up a way for producing a steady (constant) force on an object that is free to move. Discuss your idea with your friends and your teacher and agree on a good, workable method.

Plan an experiment which uses your constant force system to find out if larger masses are harder to accelerate than smaller ones. Be clear about what you need to measure.

C Extend the investigations in activity B to make measurements as accurately as you can to check the following.

(*Note*: you will need now to have a good method for measuring or comparing accelerations, such as a ticker timer).

1 For a constant mass, the acceleration is proportional to the force applied.

2 If you double the mass and halve the force, you get the same acceleration.

3 If you increase the mass, you have to increase the force in proportion if you want to get the same acceleration.

In each case explain what these statements have to do with the key formula **force = mass × acceleration.**

D You can buy a toy called 'Newton's Cradle'. When one steel ball is moved to one side and let go it hits the row of balls and one ball flies off at the other end to reach almost the same height as the first one fell from.

1 Use the idea of momentum to explain this.

2 Predict what *could* happen when you pull two balls to one side and then let them go. Try this and explain what actually happens. If you can't get hold of a proper Newton's Cradle you can use instead:

■ a row of ball bearings on a flat table,

■ a row of 2p coins,

■ a set of marbles or steel balls in a bent curtain rail track.

Questions

To answer these questions you may need to use the formulae given in this section or to draw graphs.

1a A cyclist travelled 25 km in two hours. What was the average speed of the cyclist?

b How far would a car travel in 5 hours at an average speed of 80 km per hour?

c How long would it take an athlete running at an average speed of 5 m/s to cover a distance of 2000 metres?

2 Calculate the following:

a the change in speed produced when a car accelerates at 2 m/s^2 for a time of 15 seconds,

b the acceleration required to speed up an aircraft from a speed of 100 m/s to 500 m/s in 40 seconds,

c the speed change produced in a ship when it is decelerated at a rate of 0.2 m/s^2 for 100 seconds,

d how long it would take a car travelling at 30 m/s (about 70 mph) to stop, at a deceleration of 4 m/s^2.

3 Look at Newton's Law 1 (above). It says that any moving object will carry on moving unless a force acts on it. When you hit a snooker ball it moves off at high speed, but it doesn't keep on moving for ever, even on an empty table.

a Why is this? Does this mean that Newton was wrong?

b Is there any object anywhere in the universe that moves at the same speed in a straight line, for ever?

4 Why is it harder to push start a bus than a mini-car?

5 The following measurements were taken on a specially calibrated car speedometer.

a Draw a speed/time graph of these results.

b Describe the motion of the car during this 20 second period.

c By how much did the speed change in the first 10 seconds?

d What was the acceleration of the car during the first ten seconds?

e The car had a total mass of 800 kg. What force was needed to produce this acceleration?

f How was this force actually *made to act on the car*?

time after start in s	0	2	4	6	8	10	12	14	16	18	20
speed (in m/s)	0	3	6	9	12	15	18	20	21	21	21

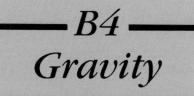

B4
Gravity

Gravity is one of the great forces of nature. It makes things fall, and keeps the planets in their orbits.

Gravity is everywhere

Everything we do is affected by gravity. Running, jumping, swimming or just standing still, our bodies are affected by the force that pulls us towards the centre of the Earth. We are so used to it that our very bones grow weaker without it. This happens to astronauts who spend a long time in 'free-fall' (see below).

We can't switch gravity off, like we can switch off an electromagnet. We can't neutralise it, like we can the forces of static electricity. We have to live with the fact that everything on the surface of the Earth is in a strong gravity field.

The design of roads, railways, buildings, aircraft and even the bodies of living things has to take gravity into account. The study of movement under gravity is vital in physics and engineering, as well as in ball games like tennis and football.

The gravity force field

Every object attracts every other object with a gravity force. The gravity force between you and the person sitting next to you is very small. You attract each other with a force of about a millionth of a newton. You will not notice this.

Gravity forces become important when at least one of the objects is very massive. The earth has a mass of about six million million million million kilograms, so its gravity field is quite strong.

The Earth's gravity field

The strength of a gravity field is measured in terms of how much force it exerts on a 1 kilogram mass.

On Earth, the force of gravity on an object of mass of 1 kilogram is about 10 newtons — or 9.8 newtons to be more exact. So the strength of the Earth's gravity field, **g**, is 9.8 N per kg.

If we put a more massive object in the gravity field it will have a bigger gravity force acting on it. A piece of iron with mass 2 kg has twice as much iron in it as a piece of iron with a mass of only 1 kilogram, so it will be pulled towards the Earth with twice the force.

A 2 kg mass feels heavier than a 1 kg mass because it is being pulled down by a force of about 20 N, compared with only about 10 N for the 1 kg mass. The force caused by gravity on a mass is called its **weight**.

We can get a rough measure of weight by just holding the object up. To measure it more accurately we need a newton meter. This is usually a spring balance marked off in newtons.

The weight of an object of mass **m** — the force **F** due to gravity on it — is given by the formula $F = mg$.

gravity force (weight) = mass × field strength

Picture 1 Playing with gravity

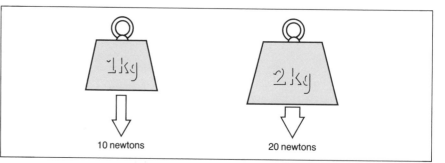

Picture 2 Gravity fields pull harder on a greater mass

What decides the strength of a gravity field?

Gravity is a force between two or more objects. The size of this force depends on two things:

■ how massive the objects are,
■ how far apart they are.

Like all forces, gravity forces occur in pairs: both objects pull equally on each other.

This means that you pull on the Earth with just as much gravity force as the Earth pulls on you. Of course, when you jump off a diving board into a swimming pool *you* do the moving, not the Earth. This is because the Earth has so much more mass than you have, and so doesn't move so easily.

When we look at objects on Earth, the Earth is so massive that it alone decides the strength of the field. By comparison, a tennis ball, an aeroplane or even a continent is too small to make much difference to the main field.

The Sun

The Sun is very much larger than the Earth and its field is so strong that it affects the Earth. The Sun is a long way away, but it is the Sun's gravity that makes the Earth travel in an orbit around the Sun. It also helps to make the tides of the sea.

The Moon

On the Moon, the gravity field is smaller than on the Earth, because the Moon is so much less massive (picture 3). Its field strength is only 1.6 N per kg — about a sixth of that on Earth. So on the Moon you weigh only one-sixth of your weight on Earth.

Gravity and distance

But distance also comes into it. The further away you get from a massive object the smaller is the strength of its gravity field (picture 4).

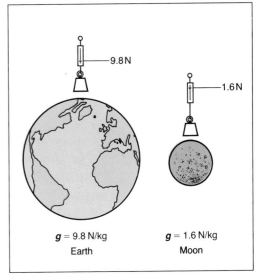

g = 9.8 N/kg
Earth

g = 1.6 N/kg
Moon

Picture 3   The Earth's gravity field is 6 times stronger than the Moon's

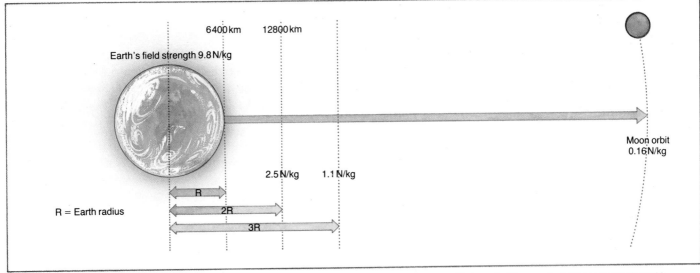

6400 km   12800 km

Earth's field strength 9.8 N/kg

2.5 N/kg   1.1 N/kg

R

R = Earth radius

2R

3R

Moon orbit
0.16 N/kg

Picture 4   The gravity field grows weaker with distance — doubling the distance cuts the field to a quarter

If the Earth was compressed to the size of the Moon the gravity field on its surface would be a lot bigger — about 14 times bigger. This is because you would be much closer to the centre of the Earth. You would weigh 14 times as much and need much stronger bones to be able to stand up and move about. Life on Earth would be very different.

Black holes

If you could get a very large mass squashed into a very small space indeed it would produce an immensely strong gravity field. This is what happens in a **black hole**. The Earth would become a black hole if it was squashed to the size of a table tennis ball. Its gravity field would be so strong that nothing could get away from it — not even light. This is illustrated in picture 5

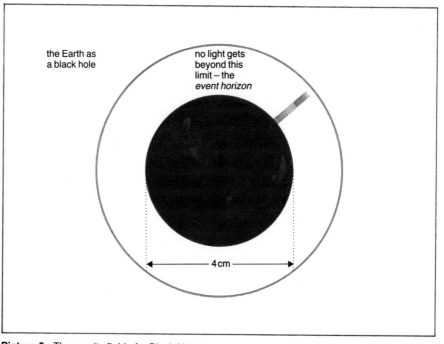

the Earth as a black hole

no light gets beyond this limit – the *event horizon*

4 cm

Picture 5   The gravity field of a Black Hole is so strong that not even light can escape from it. As a Black Hole, the earth would collapse to be about 4 cm across

Measuring the strength of the Earth's gravity field

The strength of the Earth's gravity field is defined as the force it exerts on a kilogram of matter. The easy way to measure it would be to use a newton meter as a force measurer and hang a mass of 1 kilogram on it (picture 6) But this would be cheating! The newton meter has been made at a factory and tested for accuracy by seeing if it gave the right reading when a mass was hung on it. It has been marked off on the assumption that the Earth's field at the surface is 9.8 newtons per kilogram.

A better way is to measure the *acceleration of free fall*, **g**. As explained in the next topic (topic B6), the acceleration of free fall is numerically the same as the field strength, i.e. $9.8 \, \text{m/s}^2$. Activity C in topic B6 gives ideas for doing this.

Activity C at the end of this topic gives another way of doing it, using a pendulum.

Because its gravity field is so weak the Moon doesn't have an atmosphere. Air has weight, and it is the force of gravity that holds it on the Earth. But there is more to this than just the weight of the air.

Air molecules move very fast — on Earth they move at an average speed of about 500 metres a second. Some molecules move a lot faster than the average, of course. Now to escape from the Earth, they would need to move at the speed of an Earth satellite — over 11 000 metres per second. This is called the escape speed for the Earth. Hardly any molecules move this fast, so air stays on Earth.

On the Moon the escape speed is much less, because of its weaker gravity field. It is only 2400 metres per second. The fastest molecules travel faster than this, so any atmosphere on the Moon would gradually leak away.

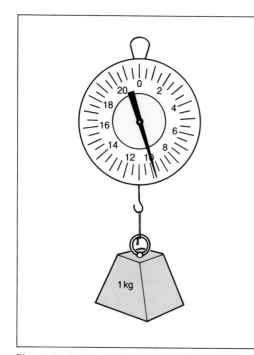

Picture 6   Measuring the strength of the gravity field

Activities

A   Agreeing about gravity

In this activity the idea is to get the whole class to agree about some statements about gravity. You will be given some paper. Write on it your own *personal* opinion about the statements below. You have to say whether you agree with them or not. You are allowed to say 'Don't know'!

Next, team up with one other person. Discuss the statements and decide for each one whether you agree or not. You have to argue until you agree with each other!

Then, find another pair of students and do the same, until all four of you agree. Then team up with another four and come to an agreement about them.

Finally, get a fresh sheet of paper and pin it up on the wall with your large group decision on it. Walk around and see what the other groups have decided.

If you all disagree your teacher will probably do something about it! If you all agree and you are *wrong*, your teacher has some work to do!

Task

Say whether or not you agree with the following statements about gravity.

1   The Earth is not affected by the gravity of the sun because it is too far away.
2   An astronaut in a space craft going around the Earth is not acted on by the earth's gravity.
3   There is no gravity above the Earth's atmosphere.

4   A heavy stone will fall faster than a light stone.
5   The air is too light to be affected by the Earth's gravity.

B   Black holes

Read up anything you can find about 'black holes'. Draw a poster explaining what they are.

C   Measuring the Earth's gravity field

Some of the most accurate measurements of strength of the Earth's gravity field made last century used a **simple pendulum**. This was a heavy ball suspended, very firmly, on a long wire string.

1   Set up a pendulum as shown in the diagram (picture 7); measure the length **L** of the string from the point of support to the middle of the ball.
2   Set the ball swinging through an arc of about 5 degrees.
3   Time 50 swings. Call the time **t** seconds. A swing is counted each time the ball goes through the middle of its swing going from left to right.
4   Divide the time **t** above by 50 to calculate the time, **T** for just one swing.
5   Use the formula to calculate the value of **g**, the strength of the Earth's gravity field.

$$g = \frac{4\pi^2 L}{t_0}$$

D   Gravity and animals

The pull of gravity on animals depends on their mass. An ant has a mass of, say, 0.1 g, while an adult African elephant has a mass of about 6 tonnes. The largest land

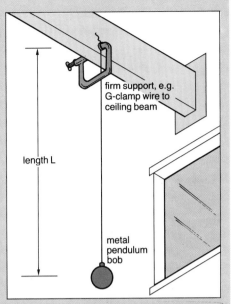

Picture 7   A simple pendulum

(labels in picture: firm support, e.g. G-clamp wire to ceiling beam; length L; metal pendulum bob)

animal known is an extinct dinosaur. This was *Brachiosaurus*, and it had a mass of about 100 tonnes. The largest animal on Earth today is the blue whale, which has a mass of about 130 tonnes. Think about the problems of being very light and very heavy. Work as a group to give answers to the following questions.

1   What problems does an animal have if it is very light?
2   What problems would a very heavy animal have? (Think about *leg size, moving fast, falling down . . . .*)
3   Why are sea animals able to be so much more massive than land animals?
4   What problems would a human being have on (a) a planet with a very strong gravity field (b) a planet with a very weak gravity field?

Questions

1   Why is the gravity field near the Earth bigger than the field near the Moon?
2a What is the difference between the *mass* of an object, and its *weight*?
b   How far would an object have to be from the Earth (or any other planet, or a star) so that it didn't have any weight?
3   An astronaut on the Moon drops a hammer. It falls to the ground more slowly than it would on Earth. This is because:
a   there is no air pressure on the Moon to push it down,

b   the hammer has less mass on the Moon than it has on Earth,
c   the gravity field of the Moon is weaker than the gravity field on Earth,
d   the force of gravity is always weaker in a vacuum.

4   Use the formula:

gravity force (weight) = mass × field strength

and the data given in the table to calculate the weight of the following:

a   a car of mass 800 kg on Earth,
b   an astronaut of mass 70 kg on the Moon,
c   an astronaut of mass 70 kg on Earth,
d   an apple of mass 0.2 kg on the planet Jupiter,
e   an astronaut of mass 70 kg on the surface of a neutron star.

Place	Earth	Moon	Jupiter	neutron star	
Gravity field strength, **g**	10	1.6	26.3	2×10^{12}	N/kg

(NB 10^{12} is a million million)

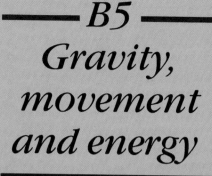

B5 Gravity, movement and energy

All movement on Earth is affected by gravity. This topic deals with the way objects move in a gravity field.

Picture 1(a) shows a ball falling freely towards the Earth. It has been taken using a flashing light that lit the ball every tenth of a second. As you can see, the ball travels a greater and greater distance in each tenth of a second. It is *accelerating*.

This is because it is being pulled down with a steady force, and this produces a constant acceleration (see topic B4). It is called the **acceleration of free fall.**

Picture 1(b) shows the ball again. This time it has been thrown sideways. But gravity still acts, and the ball is pulled downwards exactly as before. *This will happen however fast the ball is thrown sideways.*

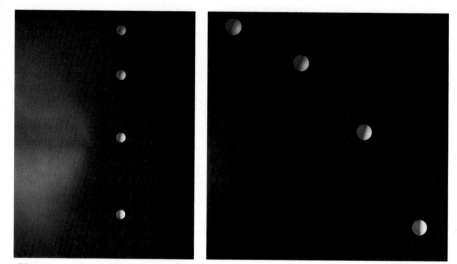

Picture 1 A freely falling object accelerates

Do heavy objects fall faster than light ones?

No — but keep on reading.

An object is **heavy** , and feels heavy, because gravity is pulling on it with a large force. This is because a heavy object has more mass than a light object.

But because it has more mass it is harder to accelerate! The extra gravity force on the heavy object exactly compensates for the extra mass. So the acceleration of free fall is exactly the same, whatever the mass of the object — see picture 2. This is shown below mathematically.

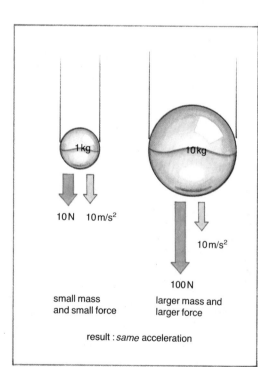

Picture 2 Freely falling objects accelerate at the same rate

Topic B4 has shown that a force causes an acceleration according to the rule

$$\text{force} = \text{mass} \times \text{acceleration}$$

or $F = ma$ (1)

The gravity force on an object in a field of strength g is

$$\text{force} = \text{mass} \times \text{field strength}$$

or $F = mg$ (2)

Putting these two ideas together (1) and (2) tell us that if the accelerating force is gravity, then

$$F = ma = mg$$

and this can only be true if $a = g$.

The acceleration of free fall in a gravity field is equal to the field strength.

How free is free fall?

There is another force that acts on a falling object, on Earth at least. This is the force of friction caused by the object moving through the air. This force depends on the size and shape of the falling object. People falling from an aeroplane accelerate quite rapidly to a high speed — unless they are using a parachute. The shape of the parachute increases the air friction, so they slow down once the parachute is opened.

But whether they wear a parachute or not the falling people eventually reach a steady speed, when they are not accelerating any more. This happens because the force of air resistance acts on them in the opposite direction to the gravity force — see picture 3.

The force caused by air resistance gets bigger the faster you go. (You can feel this when you travel fast on a bicycle.) When the air resistance becomes equal to the gravity force a falling object stops accelerating (picture 4). It has reached what is called its **terminal** velocity.

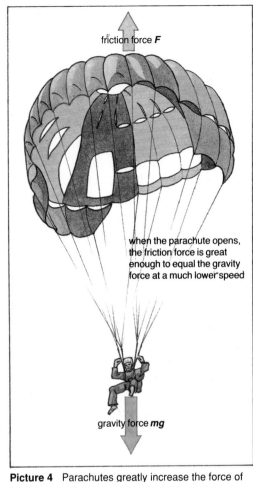

when the parachute opens, the friction force is great enough to equal the gravity force at a much lower speed

friction force *F*

gravity force *mg*

Picture 4 Parachutes greatly increase the force of air resistance

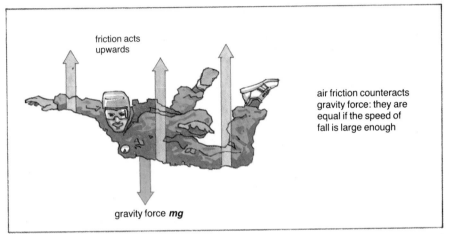

friction acts upwards

air friction counteracts gravity force: they are equal if the speed of fall is large enough

gravity force *mg*

Picture 3 A free-fall parachutist — but they don't fall freely!

Working against gravity

Climbing up a hill is usually hard work, whether we walk or ride a bicycle. We are doing work by using muscular force to lift ourselves against the force of gravity. Mountain roads and paths are built in zig-zags to make this job easier (see picture 5).

This makes the way less steep, so that we gain height more slowly, using less force at each step.

We can do this work by using part of the 'energy store' in our bodies, obtained from the food we eat. But not all of the energy is used to lift us uphill against the gravity force. We still have to keep our bodies alive, and doing such hard work makes us feel much hotter. Our bodies are quite good at getting rid of this waste heating energy. But quite a lot of the work we did hasn't been wasted. It has been transferred into a mysterious kind of energy called **gravitational potential energy.**

Potential energy

This potential energy can be a rather dangerous kind of energy for someone high up on a mountain. You can't see it and you can't feel it. But mountain climbers have more than enough of it to kill themselves, if they don't control it properly.

Picture 5 An Alpine road

The reason for this is that this potential energy can be changed to movement energy, which is usually called **kinetic energy**. If you fall off a mountain the gravity force pulls you down and you move faster and faster. The energy that you put into climbing the mountain is being given back to you as you fall!

Energy can't just disappear or appear from nowhere (see topic D4). It makes sense to think of work you do as go uphill as being stored as hidden or 'potential' energy. This turns into kinetic energy as you fall.

Of course, potential energy does you no harm at all. Neither does the kinetic energy. It is what happens when you hit the ground that causes the damage.

Picture 6 illustrates the kind of energy transfers that happen when we climb up a hill and come down again.

Don't let energy kill you!

People coming down from the mountains have to get rid of their potential energy safely, a little bit at a time. Climbers and walkers come down carefully — never running! Cyclists need to keep braking, moving the potential energy safely into the surroundings by heating the wheels and brake blocks. This is much better than storing it as more and more kinetic energy!

Measuring potential energy

Whenever we do work against the force of gravity there will be an increase in potential energy. Lifting a can of beans on to a high shelf gives them extra potential energy. You can't eat this extra energy, but it will give you a nasty bump if the can falls on your head.

Topic D2 explains how energy is measured and how the formula for calculating potential energy is obtained.

The formula is:

$$\text{potential energy} = \boldsymbol{mgh}$$

$\boldsymbol{m}$ is the mass of an object, $\boldsymbol{g}$ the strength of the gravity field and $\boldsymbol{h}$ the height through which the object has been lifted, or can fall.

When you lift a can of baked beans on to a shelf a metre above the ground you have given it some extra potential energy. If the can weighs 0.5 kg you increase its potential energy by

$$\boldsymbol{mgh} = (0.5\,\text{kg} \times 10\,\text{N/kg} \times 1\,\text{metre})$$

$$= 5\,\text{joules}$$

This is not very much, considering that the beans in the can have over 1 **million** joules of 'food energy'!

Picture 6 (a) Climbing a mountain is hot work. The climber warms up the surroundings — but also gains potential energy.
(b) If the climber fell, the change of potential energy to kinetic energy would happen much too quickly

Potential energy to kinetic energy

The energy in a moving object depends on how massive it is and how fast it is going. It is calculated using the formula:

$$\text{kinetic energy, } \boldsymbol{E} = \tfrac{1}{2}\boldsymbol{mv}^2$$

($\boldsymbol{m}$ is the mass of the object, $\boldsymbol{v}$ is its speed).

When an object falls feely in a gravity field its potential energy is getting less and its kinetic energy is increasing.

Its loss in potential energy equals its gain in kinetic energy.

So if our can of beans fell off the shelf, its kinetic energy would be 5 joules just before it hit the ground (picture 7). Its potential energy would now be zero.

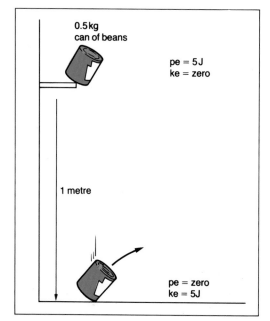

0.5 kg can of beans	pe = 5J ke = zero
1 metre	
	pe = zero ke = 5J

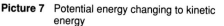

Picture 7 Potential energy changing to kinetic energy

How fast would it be going just before it hit the ground? We can use these two formulae to work this out:

$$\text{kinetic energy gained} = \text{potential energy lost}$$
$$\tfrac{1}{2}\boldsymbol{mv}^2 = \boldsymbol{mgh}$$
$$\text{or} \quad \boldsymbol{v}^2 = 2\boldsymbol{gh}$$
$$= 2 \times 10 \times 1$$
$$\text{giving speed } \boldsymbol{v} = \sqrt{20} = 4.5 \, \text{m/s}$$

Kinetic energy and road safety — speed kills

The formula for kinetic energy tells us that the energy increases as the square of the speed. This means that doubling the speed increases the energy by 2^2. It becomes four times as much.

In a road collision, or in a fall, doubling speed means that four times as much energy is available for causing damage. If the speed is tripled, e.g. from 30 mph to 90 mph, the energy becomes **nine times as much**.

Topic B2 and topic D2 also deal with aspects of kinetic energy.

Activities

A   The monkey and the hunter

A hunter sees a monkey hanging by its arms from the branch of a tree (picture 8). He points his gun straight at the monkey's head, and fires off a bullet.

The monkey sees the hunter and lets go of the branch at exactly the same time as the hunter fires the gun. The monkey hopes that when the bullet gets there he will have fallen far enough to escape the bullet.

Does the monkey escape?

Argue about it, then persuade your teacher to set up a 'model experiment' (no need to use a real monkey, or a gun!) to test your prediction.

B   Studying movement using diluted gravity

You can 'dilute' gravity by letting something roll down a sloping board instead of letting it fall straight down. Use a sloping board and a stopwatch to find out if heavier things roll down the board faster, slower or take the same time as lighter ones.

Use model cars or lorries, or special 'dynamics trolleys' of differing masses to roll down the slope. Make a careful record of your measurements and explain how they back up your conclusion.

C   Measuring the acceleration of free fall

There are lots of ways of doing this, and the details will depend upon what kind of equipment your school can give you. There are two main methods.

Picture 8

Method 1

In this method you measure the speed near the start of the fall and then near the end. The difference between them is the **change** in speed.

You also measure the **time taken** for the speed to change. See picture 9.

You calculate the acceleration

as $\dfrac{\text{change in speed}}{\text{time taken}}$

You can make these measurements using a ticker-timer and tape, or electronically.

Method 2

In this method you have to measure the **distance** an object (e.g. a metal ball) falls and the **time** it takes to fall that distance. The object has to fall from rest. You then calculate the acceleration by a formula:

acceleration of free fall,

$$g = \frac{2 \times \text{distance fallen}}{(\text{time})^2}$$

For a rough measurement you can use a fairly tall building and a stop watch. You can make more accurate measurements using electronic timing, triggered off when the ball starts and stops.

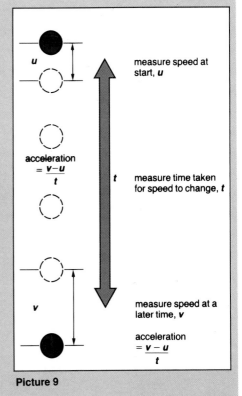

measure speed at start, u

acceleration $= \dfrac{\boldsymbol{v} - \boldsymbol{u}}{t}$

measure time taken for speed to change, t

measure speed at a later time, v

acceleration $= \dfrac{\boldsymbol{v} - \boldsymbol{u}}{t}$

Picture 9

D Giving things gravitational energy

For this activity you will need: spring balances, bathroom scales, a metre rule and/or a tape measure, a calculator.

Use the formula:

gravitational potential energy = mass x gravity field strength (**g**) x height moved

Take **g** = 10 N/kg

1 Take measurements to allow you to work out how much extra potential energy:

a you get when you climb up a flight of stairs in your school,

b you give this book when you lift it from the floor to the bench or desk top,

c you give to a stool or chair when you lift it on to the bench or desk.

2 *Some harder tasks:*

How much kinetic energy do you give a ball when you throw it straight upwards as high as you can?

If you live in a hilly area you can work out how much gravity energy you might gain (and lose) as you travel to school. You will need a map with contours or heights marked on it in metres.

What happens to all the gravitational energy you gain as you climb the hills?

Questions

1 Which of the drawings in picture 10 gives the best idea of how a ball moves when it is thrown?

2a Why does a parachute slow down the rate of someone falling from an aircraft?

b Why does a person falling without a parachute eventually reach a steady speed? (If they fall far enough!)

c Why do you think airline passengers are never issued with parachutes?

3 In theory all objects fall at the same rate, due to gravity. So why does a coin fall to the ground faster than a feather?

4 When you throw a ball across a field it eventually comes down to the ground again, however fast you throw it.

a Draw a sketch showing the path taken by the ball as it goes from your hand to the other side of the field.

b Draw the forces acting on the ball: (i) just as it leaves your hand, (ii) half way across and (iii) just before it hits the ground. Show the forces with labelled arrows and label them with their correct names.

5 The formula

$$s = \tfrac{1}{2}gt^2.$$

can be used to work out how far an object goes when it is let go and then falls freely. In this formula **s** is the distance fallen and **t** is the time it takes to fall that distance.

a What is **g**?

b A girl dropped a coin down a well and timed how long it took before she heard the splash as the coin hit the water. The coin took 2 seconds to reach the water. How far down was the water?

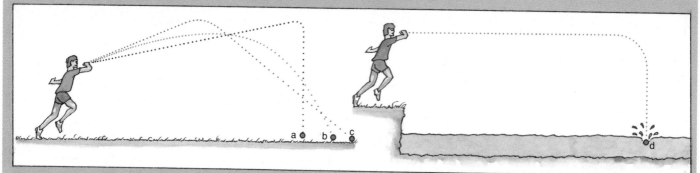

Picture 10

Take **g** = 10 N/kg for these questions

6 Name one situation in each case where gravitational potential energy might be:

a dangerous,

b useful.

7 Hydroelectric power stations need lots of water stored in dams. Why are these dams usually high up in the mountains?

8 Explain the difference between **potential energy** and **kinetic energy**.

9 Copy the diagram in picture 11. It shows the path of a car as it goes from A to E. It stops at E. What kind(s) of energy does the car have at each of the points A to E? Write your answers on your diagram.

10 Calculate the change in gravitational potential energy when:

a a car of mass 500 kg climbs a hill 400 m high,

b a climber of mass 65 kg climbs a rock face 200 m high;

c a bird of mass 0.5 kg flies from the ground to a height of 500 m.

11 Do you do any work when you stand still with a 10 kg mass in each hand?

12 A stone of mass 5 kg falls from a cliff 25 metres high to the beach.

a How much potential energy did it have before it fell? Use the formula: potential energy = **mgh**.

b What was happening to this potential energy as the stone fell?

c How much kinetic energy had the stone gained, just as it was about to hit the beach?

d Use the formula $E = \tfrac{1}{2}mv^2$ to calculate the speed with which the stone hit the beach.

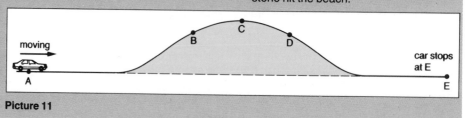

Picture 11

Weighing the Earth

How do scientists know how much the Earth weighs? Strictly speaking, the Earth doesn't weigh anything, of course . . . .

When we 'weigh' something we are really trying to find out what its **mass** is. The mass of an object is a measure of how much matter it contains. **Weight** is a measure of the gravity force on a piece of matter in a gravitational field. As the Earth is in its own gravity field it doesn't make much sense to talk of its 'weight'.

Also, as the Earth is in free fall orbit around the Sun it is 'weightless' even as far as the Sun's gravitational field is concerned.

The laws of gravity were discovered by Sir Isaac Newton (see page 54). He realised that gravity is caused by mass. The bigger the mass, the bigger is the gravity force it can exert. Two masses (of size M and m, say) exert an equal gravity force (F) on each other, given by a formula:

$$F = G\frac{Mm}{r^2}$$

r is their distance apart, in metres. G is a constant, called the **universal constant of gravitation** (see picture 1). This formula could be used to measure the mass of the Earth — M, say — if the other values are known.

Newton could measure the force on a mass m in the Earth's gravity field. In modern units it is well known to be 9.8 newtons for a 1 kilogram mass. He also had a good idea of the radius of the Earth. This had been calculated quite accurately in 1684 by the French astronomer Jean Picard. Its modern value is 64 000 000 metres (6.4×10^6 m)

But at that time no one knew the value of the constant G. Indeed, Newton died before it was measured. So he never knew the mass of the Earth that his theories had made it possible to measure. It is hard to

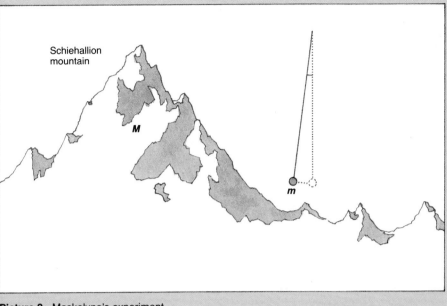

Picture 2 Maskelyne's experiment

measure G because gravity is such a weak force. The gravity force between you and the person sitting next to you, for example, is about one ten-millionth of your weight.

The first fairly accurate measurement of G was done in 1774, nearly 70 years after Newton's death. It was done by a Scot, Nevil Maskelyne. He used a simple pendulum which he set up near a cone-shaped mountain in Scotland (see picture 2). The mass of the mountain caused a small gravity force which pulled the pendulum sidways. It was a very tiny effect, but he was able to measure it.

He then worked out the volume of the mountain and measured the density of the rocks it was made of. He then calculated the mass of the mountain using the formula: **mass = density × volume**.

He now knew enough to put into Newton's formula to get a value for G. A modern experiment which measures the force between two gold spheres a small distance apart, gives a value of:

$$G = 6.7 \times 10^{-1}21 \, \text{Nm}^2/\text{kg}^2$$

This is the force in newtons between two 1 kg masses a metre apart. It is just 670 billionths of a newton!

The result of such experiments allows us to calculate the mass of the Earth, the Sun and the other planets. We also use it to calculate the masses of distant stars and galaxies.

1 'Gravity is a weak force'. Yet it seems to be quite a large force when you lift up a loaded suitcase, or try to cycle up a steep hill. Why is this ?

2 Why is it so hard to measure the value of G?

3 A kilogram mass on the Earth's surface has a gravity force of 9.8 newtons on it. The mass of the Earth acts as if it was all at the centre, as far as gravity is concerned. The centre of the Earth is 6400 000 metres from the surface.

Use a scientific calculator to check that the mass of the Earth is about 6×10^{24} kg.

Use Newton's formula, with the values:
$F = 9.8 \, \text{N}$ $m = 1 \, \text{kg}$ $r = 6\,400\,000 \, \text{m}$
$G = 6.7 \times 10^{-11} \, \text{Nm}^2/\text{kg}^2$

4 The volume of the Earth is $1.1 \times 10^{21} \, \text{m}^3$.

a Use the formula

$$\text{density} = \frac{\text{mass}}{\text{volume}}$$

to calculate the density of the Earth.

b You should have got an answer of about 5500 kg/m³ for the density of the Earth. But the density of nearly all the rocks we find on the Earth's surface is about 2500 kg/m³. How can this be explained? (See topic G1.)

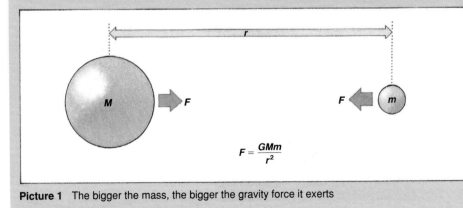

Picture 1 The bigger the mass, the bigger the gravity force it exerts

$$F = \frac{GMm}{r^2}$$

B6 The Leaning Tower of Pisa

This is about a very famous experiment that never actually happened . . .

Picture 1   The Leaning Tower of Pisa

Picture 1 shows the famous Leaning Tower of Pisa, which has been defying gravity for 600 years. But it is also famous in the history of science — for an experiment about gravity that may never have actually been done!

This strange non-event was supposed to be the work of a man called Galileo. Galileo Galilei was an Italian astronomer and physicist who lived from 1564 to 1642. Along with Newton and Einstein he ranks as one of the most famous scientists who ever lived.

Don't rock the boat!

In his day, science was something to be learned out of books, written by the old Greek philosophers two thousand years earlier. Experiments and investigations were rarely done, because educated people thought that all the answers were already known. It was considered impolite to challenge this ancient wisdom. (Picture 2.)

But Galileo was quite a rude man. He didn't believe anything unless he could see it with his own eyes or prove it by experiment or by clear mathematics. He didn't have much patience with important people who weren't very clever. He was also very witty, and this mixture of rudeness and wit made him many enemies.

In his day, important books were written in Latin, which most people couldn't read (picture 3). But Galileo wrote in Italian, the language of ordinary people. So everybody could read his attacks on other scholars.

The dangers of making a new science

Galileo's ability to challenge old ideas and replace them with exciting new ones made him a great scientist. But it also meant that he spent the last ten years of his life under arrest, and he nearly got himself sentenced to death.

The old science

The Greek philosopher Aristotle (384 BC to 322 BC) had laid down the rules of physics. For 4th century BC they were pretty good, but modern scientists like Galileo had started to question them. For example, Aristotle had said that when you throw an object it will carry on, in a straight line, until it runs out of 'force'. Then it does what comes naturally, and falls down (picture 4).

He also said that when they do fall, heavy objects fall faster than light ones.

Picture 2   Galileo, who challenged Aristotle

Picture 3   Science books used to be written in Latin

Tom and Jerry physics

But the ideas are completely wrong, even though many people still actually believe them. Watch a Tom and Jerry or Road Runner cartoon! They only fall when they 'run out of speed'! There were no cartoons like this in 16th century Italy, life was more serious. Galileo, indeed, wasn't just a 'theoretical' scientist. He was employed by the state of Venice as a military consultant. He had in fact invented a device which he called a telescope, which the Venetian navy found very useful. It allowed them to see and recognise distant ships before the other ships saw them.

Galileo gets the right answers

But Aristotle's theory of movement was not giving the right answers. This was also important in the battles of that time. Cannon balls weren't going where they were supposed to! This was really serious at a time when many cities in Italy were at war with other cities. Galileo's experiments gave him a new theory, and his skill at mathematics allowed him to make much better predictions of where cannon balls were likely to end up — see picture 5.
  Professor Galileo became more famous, and his salary was increased.

Picture 4 This is what should happen when a cannon is fired — according to Aristotle

A famous non-experiment

Take two cannonballs, a 1 kg ball and a 10 kg ball. Carry them to the top of the Leaning Tower of Pisa. Go to the overhanging side and let them both go at the same time. (Warn the people underneath first).
  If Aristotle is right, the 10 kg ball will hit the ground first. If Galileo is right both balls will hit the ground at the same time.
  Galileo was right. It's a great pity that we have no evidence that he actually did this experiment. But millions of people, watching on television, saw an astronaut carry out a version of the experiment on the Moon, in 1972. He didn't drop two cannonballs, but a hammer and a feather. Both hit the ground at the same time. Galileo would have been clever enough not to use a *feather* in his experiment, of course!

Don't rock the earth!

Galileo was getting too big for his boots. It was one thing to challenge Aristotle about cannon balls, especially if you had the military on your side. But Aristotle, and the other ancient Greeks, had also said what the Earth and the heavens were like. What they said fitted in with 16th century Christian religious ideas. The Earth was the centre of the universe, and heaven was up there in the sky. The Earth was still, and the Sun, Moon and stars moved around it. (They were right about the Moon).
  Galileo said no. The Sun was still, and the Earth moved around that. So did the planets. His naval telescope could be pointed at the heavens, and there you could see a little model of his ideas. You could see the moons of Jupiter

Picture 5 But this is what does happen, as worked out by Galileo

going around the planet, just like the Earth and other planets went around the Sun.

This was heresy! It contradicted what people thought the Bible said. Galileo was put on trial and condemned as a heretic. The punishment was to be burned at the stake. Galileo was no hero. He 'changed his mind' and denounced his ideas.

But he was still thought to be a dangerous thinker, and had to spend the rest of his life in 'internal exile', in his house in a small town in Italy.

But Galileo was right about the Sun and the planets, and his ideas were proved beyond any shadow of doubt by a man who was born in the very year that Galileo, now old and blind, died. This man was Isaac Newton (see topic B9).

Activities

A Trying Galileo's ideas

1 Plan and carry out your own 'Leaning Tower of Pisa' experiment.

   CARE! Make sure no-one can be hit by falling objects.

2 Drop a feather and a coin at the same time. Do they reach the ground at the same time? Why not? Was Galileo wrong after all? Ask your teacher to show you how to devise an experiment in which both a coin and a feather do fall at the same rate.

B Finding out about Galileo and Italy

Use a library to find out more about what Italy was like at the time of Galileo. Why did people think that his scientific ideas were so important? Why did other people disagree with him so strongly that they would have put him to death?

Questions

1 People said to Galileo: 'The Earth can't be moving! If it did, we would all be left behind! There'd be chaos!' How would *you* answer these critics, using modern science? Galileo didn't know about gravity. How could you answer these critics without using the idea of gravity pulling down on everything on Earth? (Hint: think of dropping something on a train.)

2 The diagram in picture 6 shows a cannon about to be fired at a target hidden behind a city wall. On a simple copy of this diagram draw:

a the angle of the cannon and the path a cannonball would take on Aristotle's theories,

b the angle and path assuming Galileo's (Newton's) theories.

c Why do you think that the military supported Galileo's ideas?

3 What was the key piece of evidence that Galileo used to prove that it was at least *possible* that the earth and planets went round the Sun?

4 'Of course the Sun goes round the earth! You can see it moving, every day!' . How can *you* explain the apparent movement of the Sun through the sky, from dawn to sunset?

5 Galileo's ideas about the sun and the earth were thought at the time to contradict the Bible. The Church in Italy thought that this would confuse ordinary people and make them lose their faith in God. Other scientific ideas, such as Darwin's Theory of Evolution, might have the same effect.

a Can you think of any modern examples of such 'dangerous' ideas?

b Should such ideas only be learned by people who are intelligent enough not to be confused by them?

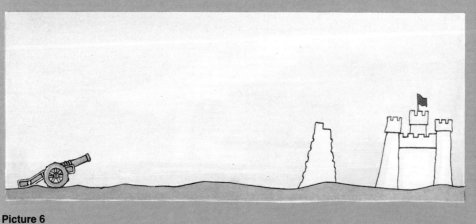

Picture 6

The biggest catapult in Britain

Picture 1   Testing the catapult

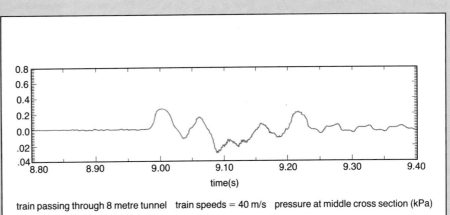

train passing through 8 metre tunnel   train speeds = 40 m/s   pressure at middle cross section (kPa)

Picture 2   Pressure changes in the tunnel

The biggest catapult in Britain — if not in the world — is owned by British Rail. It is housed in what looks like a large greenhouse at their Research Laboratories in Derby.

They use it to fire model trains at speeds of up to 200 km per hour (55.6 m/s) along a track 132 metres long. Picture 1 shows a model high-speed train coming out of a tunnel. The train is built as a 1/25th scale model.

The pressure problem

The trend in rail transport is towards faster and faster trains. One problem with trains travelling at high speed is what happens when they go into a tunnel. In the open air a well designed train has a front end that can slide easily through the air. It is specially shaped to push the air in front of it aside — it is *streamlined.*

But in a tunnel the air has nowhere to go! It is pushed forward by the train and is squashed up in front of it. The rise in pressure can be big enough to slow the train down, and the pressure change inside the train can be annoying to passengers and crew. For environmental reasons, new railway systems have to be designed to carry high speed trains through tunnels many kilometres long. The British Rail research is aimed at finding out exactly what happens when trains enter and leave tunnels at high speeds. Picture 2 shows the results from one experiment which measured the pressure changes in the air as the model train went through a tunnel.

Human bodies can cope with quite large pressure changes. Normal air pressure is about 1 **kilopascal**. When you dive 3 metres deep in water the pressure on your body increases by about 0.3 kPa, which is about the same as the maximum change shown on the graph in picture 2. But what annoys people is how *quickly* the pressure changes. A quick change can cause your ears to 'pop'. A very quick change may damage the eardrum.

The catapult

The catapult is very simple. It is made from ordinary rubber cords 26 mm in diameter, just like the ones used as luggage straps ('bungees') for car roof racks. They are connected to an undercarriage which holds the train, then pulled back to nearly double their length (see picture 3). The rubber cords are let go, using a magnetic release, and the carriage with the train on it accelerates. At the right time the undercarriage is stopped and the train carries on along the part of the track used as the test bed. This is 46 metres long, leaving about 80 metres in which to stop the train.

Picture 3   The catapult

The train is braked by simply squashing a plastic tube, which it hooks on to at the end of the test bed. If it misses the waiting hook it crashes into a pile of foam rubber at the end of the track. Answer the following questions.

1   A full-size train might be 100 metres long. How long would a 1/25th model train be?

2   Use the graph and the other information given in picture 2 to answer the following questions.

a   What was the maximum **increase** in pressure in the tunnel?

b   How long did it take for the air pressure to go from normal to this maximum value?

c   The graph shows how the pressure changed at one point — the middle of the tunnel.

   i)   Use the value for the length of the model train (you worked it out in question 1 above) to show that it would take the model train 0.1 seconds to pass this mid-point.

   ii)   Suggest why the air pressure kept changing even after the train had gone past. (Hint — think of blowing across a glass tube, or pushing a weight on the end of a string . . . . .)

d   Many new railway systems are being built in Europe and the UK, and parts of them are put in tunnels for *environmental reasons.* What do the words in italics mean? Give an example of one of these reasons, and name any tunnel that you have heard about which has been built for an 'environmental' reason.

e   Describe the energy transfers that take place when the model train is accelerated, travels down the track and misses the correct braking system.

B7
Satellites

No engines. No wings. What keeps satellites up there?

Picture 2 Modern communication satellite

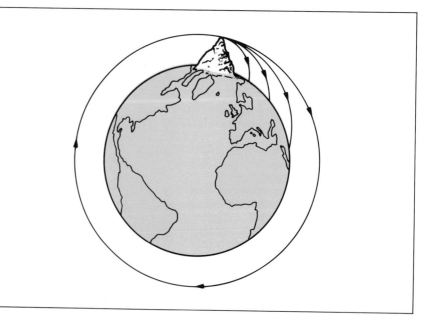

Picture 1 Sputnik — the first artificial satellite

'I just don't believe it!'

The first artificial satellite to orbit the Earth was launched in 1957. It was a Russian satellite, named **Sputnik**, shown in picture 1. It had a mass of 84 kg, and it moved in an orbit between 217 km and 944 km above the Earth's surface.

The Astronomer Royal at the time didn't believe it! He didn't think that a rocket engine could provide enough energy to lift any object that far above the Earth. Since then many thousands of satellites have been launched, some as big and as heavy as a bus. Picture 2 shows a typical modern Earth satellite.

Newton again

Earth satellites had been predicted by Isaac Newton in 1666. The way he thought of getting them into orbit was not very practical, as he well knew. Picture 3 shows his idea. But we can learn from this how actual satellites do in fact stay in orbit.

Picture 3 Newton imagined firing a cannon ball from the top of a high mountain. The faster it went, the further the cannon ball would travel before it hit the ground. At a high enough speed, it wouldn't hit the ground at all!

Any object fired sideways (horizontally) from a tall mountain will not only move sideways but also fall towards the centre of the Earth. It is pulled there by the force of gravity (see topic B4). The faster it is fired the further it will get before reaching the ground. Newton realised that at a certain speed it *will never reach the ground*, simply because the Earth itself is curved. At this speed the curved path of the falling satellite would exactly match the curve of the Earth.

How are satellites launched?

No mountain on Earth is high enough for Newton's method to work. There is no gun powerful enough to fire a satellite at the speed needed to stay in orbit. And if there were, the force needed to accelerate the satellite inside the gun barrel would squash it flat!

Instead a rocket system is used to lift the satellite to the top of an 'invisible mountain'. This is called the **injection point**, and it is at least 200 km high (see picture 5).

Picture 4 shows the rocket system used to launch the American Space Shuttle. Most of what you see of the rocket is simply a hollow tank filled with fuel. Most of this fuel is used to lift *itself* through the atmosphere. There is just enough spare fuel to accelerate the shuttle sideways when it gets high enough to be put into orbit.

The rocket system has three *stages*, each with its own engine and fuel supply. The first stage contains the most fuel and has the biggest engine. It lifts itself and the next two stages as high as it can. Then it falls off. Stages two and three take over, in turn, and in their turn are thrown away. The Space Shuttle is left travelling in orbit at the speed required to stop it falling closer to earth (see picture 5)

Why do some satellites fall down?

Newton explained why satellites stay up — but they don't stay up for ever. The main reason for this is air friction. The Earth's atmosphere gets thinner and thinner the higher you go, but it never thins away to nothing. Even at a height of 1000 km there is enough air left to cause a drag on a satellite which slows it down. Eventually it is travelling too slowly to stay in orbit. It re-enters the thicker part of the atmosphere where friction becomes so great that the satellite 'burns up'.

The energy transferred from movement energy by this air friction heats up the satellite until it melts and burns away.

Picture 4 The rocket system used to launch the Space Shuttle

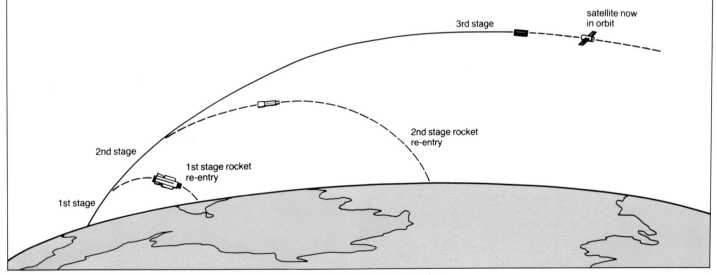

Picture 5 Geostationary orbit: the earth turns at the same rate as the satellite. This means that the satellite is always positioned over the same point on Earth

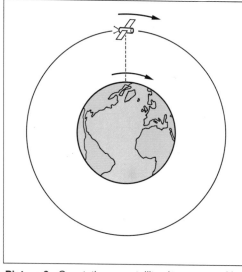

Picture 6 Geostationary satellites keep pace with the spinning Earth

Geostationary orbits

Sputnik went around the earth once every 90 minutes. This would have made it useless as a TV satellite. It would be in view at any one place on Earth for only a few minutes in each orbit. The TV aerial would need to be motor-driven to follow it. A neat solution would be to have a satellite which always seems to be in the same place in the sky. This in fact is what is done.

The satellite is put into a very high orbit. It still moves, but its speed is just enough to follow the movement of the spinning Earth (see picture 6). This means that it is always over the same place on earth.

What are satellites used for?

Most earth satellites are used for military purposes. The big 'satellite powers' (Russia and the USA) keep an eye on each other's movements using spy satellites which can pick up radio and other signals. They have special cameras that can take very clear pictures. They can photograph things as small as a human being from hundreds of kilometres up in space. They can even read newspaper headlines!

But other satellites can photograph clouds — and see through clouds. They can show us what the great air streams of the Earth are doing, and so help forecast the weather. Picture 7 shows a typical weather picture that you can see every day on TV. Satellites can carry sensors that detect infra-red radiation. This makes it possible to measure the temperature of the soil and of vegetation. They can find places where crops may be unhealthy or short of water and nutrients (see picture 8).

Other satellites can sense small changes in gravity forces that might show where the deep-lying rocks are different. This might help us to discover oil or other minerals.

Picture 7 A satellite weather picture

Picture 8 A satellite picture of the Nile Delta

Communication satellites

Communication satellites are the ones most people know about. They are put into geostationary orbits and are used to collect and retransmit TV and radio signals to give us instant world-wide communication.

There are also scientific satellites, which are used to pick up radiations from space — light, X-rays, UV and infra red — which are teaching us more about the planets, stars and galaxies.

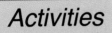

Activities

A Computing your way into orbit

You might be able to get a computer simulation of satellite launching. If you can, try using different speeds and 'angles of launch' to see what happens.

B Have you ever seen a satellite in orbit?

Some newspapers list the times when important satellites cross overhead at night. Use this information to try and spot a satellite. Explain why you can see it as a shiny spot, although the satellite does not carry any lights of its own.

C Some things to find out

Use a library or any other source of information to answer the following questions.

1 What satellite has travelled furthest from Earth? What task did it have to do?

2 How are messages sent to and from satellites?

3 When did humans first land on the moon? Do you think it was worth sending them there?

D Space travel

Copy out the following passage about travelling in space, filling in the missing words.

Satellites have not escaped the pull of Earth's _____. This is in fact the force that keeps them in _____ around the Earth. But the astronauts do not feel this force. This is because they are in a state of _____ – __ _____.

This makes life difficult. The astronauts, and every loose object, will _____ about in the satellite. If astronauts have to be in space for a long time, it would be a good idea to have some artificial _____ . This could be obtained by _____ the satellite around its own axis. The faster it _____, the bigger the _____ force they would feel.

The main problem about space travel is that space is so huge that it years to travel to the nearest _____, even if we could travel at the speed of light. At best, spacecraft could travel at about a tenth light speed, so it would take at least _____ years to get there. A return trip would be impossible. Also, think of the supplies they would need. They would have to recycle _____ and _____, and try to grow their own _____. They would also need a good supply of _____ so that they could manouver the spacecraft when they arrived. A starship, with a crew of 16, would need 300 tonnes of consumables, and another 3000 tonnes for the ship itself.

You can read more about the problems of space travel in a science fiction paperback by Robert L Forward, *Dragonfly* (New English Library, 1985). It's a good read!

Questions

1 Why wouldn't Newton's 'big gun' idea for getting a satellite into orbit work?

2a Why do many satellites eventually fall back to Earth?

b Why do we not need to worry about being hit on the head by a satellite when it does fall?

c Give five uses of Earth satellites.

3 Copy the diagrams which show the Earth and the Moon in space (picture 9). On your copy, draw your idea of:

a the likely orbit of a 'spy satellite',

b the likely orbit of a TV broadcast satellite,

c the path that a rocket might take to get into orbit around the Moon.

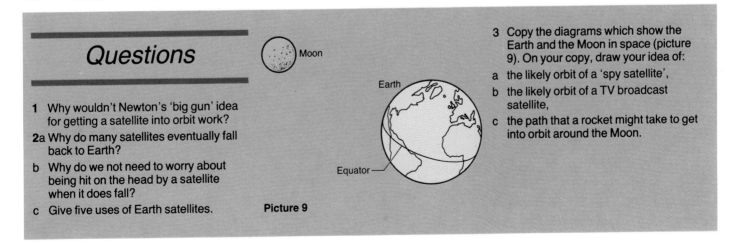

Picture 9

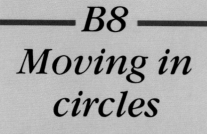

B8
Moving in circles

Satellites aren't the only things that move in circles . . . .

Picture 1 Gravity provides the centripetal force **F** that keeps a satellite in orbit

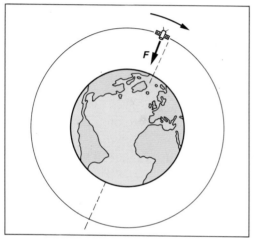

Picture 2 On an icy road, there is no friction to provide the centripetal force. So there will be no turn

Forces and orbits

Satellites stay in orbit because they move under the action of a **central force** (picture 1). This is the force of gravity, which acts towards the centre of the earth. But there are other things that move in circles.

The same basic rules apply whether it is the movement of wheels, of planets orbiting the Sun, or cars going around corners. In topic B3 we looked at what happens when forces are only involved in straight-line movements. The result is an acceleration along the line of action of the force, obeying the rule $F = ma$, or if the force is gravity, $W = mg$ (see page 38).

These rules still apply when the force acts in a different direction to the line of movement of the object. This time the force also causes a change in the **direction** of movement. When a car has to go around a bend in the road it has to change direction. A force is needed to make it do this. The force is produced by changing the angle of the wheels and relying on the **friction** between the tyres and the road to act on the car. No friction — no turn! (See picture 2.)

Conkers and fairgrounds

The natural path of an object, free to move, with no force acting on it, is a straight line. To move it off a straight line a force of some kind is needed. When you whirl something around your head on a piece of string it is the inward pull from the string that makes the object move out of its straight path. If you let go of the string it will carry on in the direction in which it was already moving (see picture 3). Not many people believe this — so try it, but out in the open where it is safe!

Fairgrounds are great places for scaring people using Newton's Laws applied to circular motion. Picture 4 shows a ride which spins people around at a speed of 80 km/h, then rises up to a vertical position.

The rods holding the seats to the axis of spin have to be very strong! This is because they have to carry the central force that keeps turning the seats, and the people in them, in a circle.

What you *feel* is a force in your back, as picture 5 explains. This is the inward force that keeps you 'in orbit'. The force needed to do this is quite large — about four times your own weight!

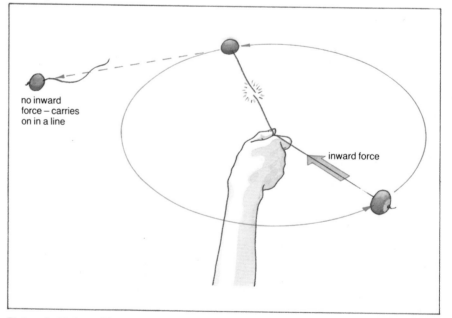

no inward force – carries on in a line

inward force

Picture 3 No inward force, so no circular motion

Picture 4 Having fun with Newton's Laws

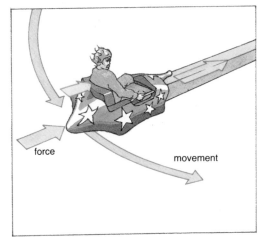

Picture 5 As the ride spins you feel a strong force in your back — the centripetal force that keeps the ride going in a circle

Centripetal force

The name given to the force that keeps an object moving in a circle is **centripetal force**. It acts inwards towards the centre of the circle. Some complicated mathematics shows that the force is given by the formula:

$$F = mv^2/r$$

where m is the mass of the object, v is its orbital speed and r is the radius of the orbit (picture 6).

This formula tells us that the faster the object goes the bigger the centripetal force needed to keep it in orbit. In fact, doubling the speed means increasing the force by a factor of 4. This helps to make fairground rides even more terrifying!

Spin dryers

Spin dryers use this effect to get water out of clothes. As the clothes are spun at high speed, a large force is needed to keep them 'in orbit'. This is provided by the outer wall of the dryer. But the wall has holes in it. Any drop of water next to a hole has nothing to keep it in orbit, so it carries on in a straight line and escapes. Drops of water in the middle of the clothes escape through the holes in the cloth. Eventually they get to the outer, holed wall and escape completely.

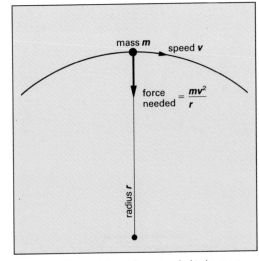

Picture 6 The centripetal force needed to keep an object moving in a circle is $F = \dfrac{mv}{r}$

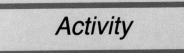

Questions

1 What provides the central (centripetal) force on each of the following objects going in a circle:

a The Moon orbiting the Earth,

b the Earth orbiting the Sun,

c a car going around a circular race track,

d the rim of a bicycle wheel,

e you, just standing on the Earth's surface?

2 Why is a car going around a corner on an icy road in danger of skidding?

3 If you are confident enough, and suitably dressed, you can whirl a bucket full of water in a vertical circle without losing any water. Explain why the water doesn't fall out when the bucket is whirling upside down.

Activity

Investigating circular motion

You can investigate circular movement by using a spinning turntable. This could be turned by hand, or could be an old record player.

1 Cover the turntable with a disc of smooth paper or card. Place a coin near the edge of the turntable and slowly increase the speed of turning. What happens to the coin? (Observe very carefully!)

2 How could you stop the coin from coming off the turntable so easily? You should be able to think of at least two ways of doing this.

B9 Newton and Einstein

Two men who changed our view of the universe

Picture 1   Time and the Stars

Humans are very curious

As far as we can tell, human beings have always wondered about the Sun, the Moon and the stars. They have built temples, pyramids and great circles of standing stones. They have lined them up with the main stars, or the rising and setting of the Sun at important times like spring and harvest. They have worshipped the Sun and the Moon, and put their heroes and heroines into the star patterns. They have used them to make clocks and calendars, and to find their way across seas and deserts.

Making theories

And from the time of ancient Greeks, over two thousand years ago, they have made theories about what these heavenly objects really are. The Sun is hot, the giver of life, a great ball of fire. The Moon is cold and pale, and just a little dangerous. Stars are tiny sparks, of different colours, in sky patterns that are always the same.

But there are also wandering stars, the planets, which travel through the star patterns, sometimes bright, sometimes pale. Sometimes they disappear altogether, only to return, months later. How can this be explained?

What makes a good theory good?

It isn't too hard to think up a theory for something. The hard bit is to prove it to other people. One way of doing this is to use the theory to predict what is going to happen.

Astrology is a theory that makes predictions. You can look up 'what the stars foretell' in magazines and newspapers. Do they get it right? Do astrologers always win the football pools?

The job of **astronomy** was also to make some predictions: 'If you follow this star you will get to Damascus'. Or 'If you plant your crops when a certain star is just setting at dawn the crops will grow well'. Astronomy always gave better results than astrology, even if it was less fun. But the next step was to find out *why* the stars moved so regularly, and why the wandering stars (called **planets**) were different. In other words, could you *predict* what the stars and planets were going to do?

To do this people needed a theory about what stars and planets actually were.

Why are the stars useful?

Over the many thousands of years that people watched the skies, the patterns of movements became well known (picture 2). Astronomers taught sailors how to use the changing positions of sun, stars and planets to find their way across seas out of sight of land.

By the 15th century, ships and the skills of navigation became good enough for traders to travel further than they had ever done before. The Arabs and the Chinese sailed across the Indian Ocean and the West Pacific, bringing rich trade to and from the islands of the 'Indies'.

Picture 2   Columbus discovered America, even though he thought that the Earth was as the centre of the Universe

Picture 3 An old map of the world

The age of discoveries

Then Europeans ventured around the great barrier of Africa to join in this trade. The Portuguese were the first great navigators and the first Europeans to sail around Africa into the Indian Ocean.

Some of these Portuguese sailors were swept off course by wind and current and 'discovered a new land', which we now call Brazil. But they didn't know where they had been, and lost it again. (Picture 3.)

Christopher Columbus

It wasn't until the end of the 15th century, in the year 1492, that a good navigator called Christopher Columbus found South America again, or at least the islands where the tribe called the Caribs lived. However, he thought that he had in fact reached India. He called these islands of the Caribs the 'West Indies'. But India was a good 4000 miles further on, across the huge Pacific Ocean.

Money in the stars

By the middle of the 16th century great fleets of Spanish galleons were carrying tonnes of gold and silver to the King of Spain. The Spanish investment in Christopher Columbus had paid off handsomely.

All this hope for trade and wealth made the study of astronomy even more valuable. It became more than a hobby or a way of 'seeing your future in the stars'. Kings and emperors employed astronomers to work out more carefully the positions and movements of stars and planets useful for navigation (picture 4).

Isaac Newton (1642–1727)

Isaac Newton was born in 1642, exactly a hundred and fifty years after Columbus bumped into the West Indies. Knowledge of stars and planets had improved in that time, but no one knew what they were, or how and why the moved or stayed still.

Navigation was better, but was still more of an art than a science. Many ships were still lost, their sailors and cargoes never to be seen again.

Picture 4 It was here in Tycho Brahe's observatory in Prague, Czechoslovakia, that the astronomer Kepler first proved that the Earth and planets went around the Sun

Picture 5   Isaac Newton

Picture 6   Newton had to invent new mathematics, the calculus, to go with his new physics

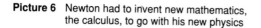
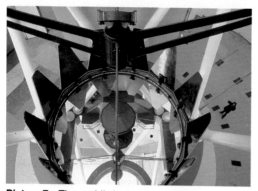

Picture 7   The world's largest telescope at Zelenchuk, Caucasus, is based on Newton's design

The young Isaac Newton was very bright. He was taught at home until he was twelve, when he was sent away from his family farm in Woolsthorpe, Lincolnshire, to the local grammar school. He had to learn Latin, Greek and mathematics, but he became well known for the toys and working models he made. He was as good with his hands as with his brain.

At the age of 18 he went to the University of Cambridge, and was made professor of mathematics when he was only 27. (Picture 5.)

Newton's year of discoveries — 1665

As a young man, before he even got a degree or became a professor, Newton made so many discoveries that we still marvel at him. In 1666 the University was closed down because of the Great Plague and he spent a year at home. It was then, the story goes, that he first thought about the force that made an apple fall off a tree. He thought it might also reach as far as the moon, and keep it in orbit. In a few months he had worked out his first ideas about the Law of Gravity and his Laws of Motion.

During the next years he had to invent completely new mathematics (the **calculus**) which he needed to prove and check his results (picture 6).

All mysteries solved?

The mystery of the solar system was solved. Scientists now knew why and where the planets moved in their orbits (see topic I1). They were able to work out where the planets would be in the future, and when eclipses of planets, sun and moon would occur. Sailors could use these calculations to navigate across the widest seas with great accuracy.

More discoveries

Newton also invented a new kind of telescope, which is still the one most used by astronomers today (picture 7). He discovered the spectrum and so explained how rainbows are formed. He produced new theories of light, and heat. He became very famous, and his ideas changed the way people thought about the world.

Scientists began to believe that the world — even the whole universe — must be very simple. Everything must obey simple, clear laws of nature — although they hadn't all been discovered yet. Everything, they thought, could then be predicted.

But at the height of his fame, Newton lost interest in science. He left the University and was given the job of looking after the Royal Mint, where the coinage of Britain is made.

It wasn't a very difficult job, and he had plenty of spare time for his new interest in life, which was working out the dates of when things happened in the Old Testament of the Bible. He was made *Sir* Isaac.

Newton had always been very shy and lonely. He had quarrelled with most of the other scientists he knew, and his friends found it very hard to get him to publish the books he had written about his great discoveries. He worried that he might have made mistakes and he did not like to be proved wrong!

Einstein's new universe

Albert Einstein (picture 8) was one of the few scientists to become as famous as Newton. At the height of his fame thousands of people would crowd into theatres to hear him explain his theories. He knew that most of them didn't understand a word of what he was saying, so he used to play them a couple of tunes on his violin to make up for it.

Try harder, Einstein!

Like Newton, Einstein was very bright. But he was easily bored.

He was born in Germany (in 1879), and went to school there. He didn't do too well at school at subjects he didn't like, and was unpopular with his teachers. He left school at 15 without proper qualifications and taught himself, while he spent a year hiking and climbing in the mountains of Italy.

Then at 16 he failed the entrance exam to get into the university in Switzerland where he was now living. But he did so well at maths that the professor invited him to join the class anyway. A year later he had swotted up enough of the boring subjects to pass the exam.

But even university was boring — especially the physics lectures! He nearly failed his exams again, and couldn't get a job. After two years he succeeded and became a civil servant, an examiner of inventions for the Swiss Government.

It was a nice easy job, and he was good at it. In his spare time he was a genius. Like Newton, he taught himself because there was no one else who had thought about things as hard as he had.

Picture 8 Albert Einstein, aged 25

Einstein's year of discoveries — 1905

At the age of 25, still a civil servant, he wrote three scientific papers which changed the world of physics. The ideas that had been worked out by Newton, and by generations of physicists since, had to be looked at in a new light.

Like Newton's great works, they dealt with space and time, light and heat. In one, he showed that light was not only a wave (see topic C8) but also a particle. In another, he produced his first 'Theory of Relativity'. In the third, he worked out from something as simple as the way smoke spreads out, or sugar dissolves in tea, that atoms must really exist.

Relativity

Einstein is most famous for his two theories of relativity, which have changed the way we look at the Universe. For Newton, matter was 'mass', the unchangeable cause of gravity. For Einstein, matter can be changed into energy — and back again — in accordance with the formula $E = mc^2$. For Newton, time rolled on at the same rate everywhere. For Einstein, how long something takes to happen depends on how fast you are travelling. At the speed of light, time stands still.

For Newton, gravity was a **force**: for Einstein, it is a **curve** of 4-dimensional space-time.

The new universe of modern physics

The universe described by physics has always been hard to understand, and probably always will be. After Einstein, we can't even *imagine* it, even when we understand it. Newton's ideas made possible the improved navigation of the 18th and 19th centuries. They gave us an understanding of the engineering needed to make and use aircraft, rockets and space satellites. Einstein's ideas have led to our modern world. His theories have given us the engineering underlying nuclear energy — and the nuclear bomb. They have led to the ideas needed for lasers, the understanding of the genetic code, the strange world of sub-atomic particles, the reason why the Sun is hot, why black holes exist, and why the Universe is expanding.

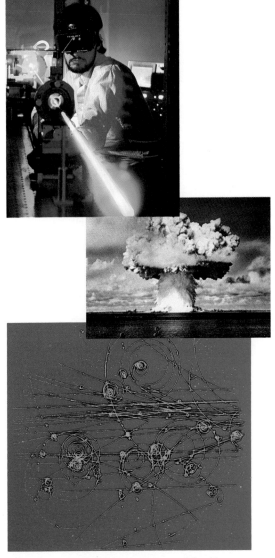

Picture 9 All of these can only be explained by Einstein's work

C1
Signals and codes

Human beings need information. We also send out information. This topic is about how information is coded, carried and controlled.

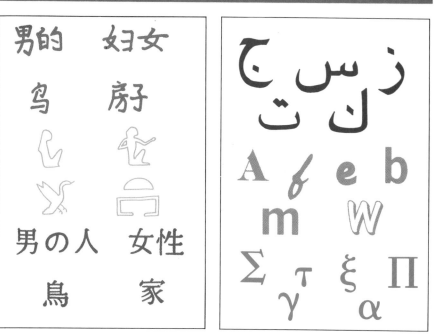

Picture 1 Codes for ideas

électron	French
electrón	Spanish
Elektron	German
elettrone	Italian
ηλετρόνιο	Greek
elétron	Portugese
אלקטרון	Hebrew
电[电]子	Chinese
elektron	Dutch
электрон	Russian
elektrono	Esperanto
ėlěctrŏn	phonetic
ইলেক্ট্রন:	Bengali

Picture 2 Different codes for the same word

Codes

The language you speak is a code, and not everybody in the world understands it!. Writing is a code. Picture 1 shows how different languages have tackled the problem of putting sounds into 'pictures'. The very oldest, like Ancient Egyptian, used drawings of what the sounds meant. The word for 'house' was drawn to look like a house. But this means having a different symbol for each word. It is hard to learn, slow to write and to read, and needs thousands of different code symbols.

It is easier to break the words up into their different sound parts, and have a symbol for each of these. In English we can just about manage with 26 of these symbols — the letters of the alphabet. Of course, we do use more than 26 sounds, but we can double up letters (ff, gh, etc) to help us cope with the extra sounds. Picture 2 shows some of the codes used in the world today.

When we learn to read we are learning which sounds go with which symbols. In Western languages we 'scan' the letters from left to right. In Arabic, Hebrew and some other languages the symbols are scanned from right to left.

The earliest written books from Ancient Greece show that they were read from right to left on one line and then left to right on the next, and so on. To save time, computer printers print in this way, every other line being printed backwards.

Carrying messages

Before writing was invented messengers needed very good memories.

Even after the invention of writing they also needed strong legs, like the messenger who carried the news of the battle of Marathon to Athens in 490 BC. He ran so hard that he died after delivering the news, and so never knew that he had just invented marathon running. But sending a messenger was a slow way of carrying information.

Light travels a lot more quickly — at 300 million metres a second. The ancient Romans used light to send messages very long distances. The Roman army built a network of signal stations criss-crossing Europe. Each station had large wooden 'flags' to send messages many kilometres across country (see picture 3).

A Roman signal station

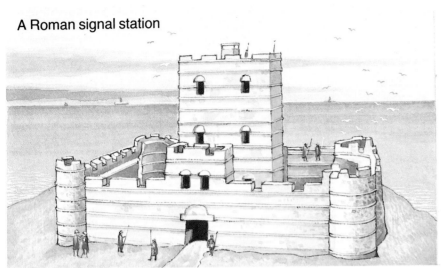

Picture 3   The Romans sent messages using light, over 2000 years ago

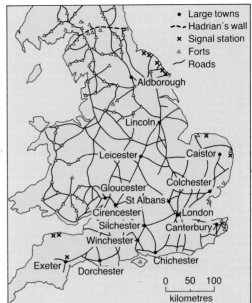

To make use of 'light messages' new codes had to be invented. To have a different flag movement for each letter of the alphabet would have been a very slow way of doing it. Standard messages, like HELP!, would be given one flag movement. But even so, messages had to be kept very simple. Long, chatty letters were still sent by messengers on ship or horseback.

Electric messages

Electric current was discovered at the beginning of the 19th century. It was soon used as a message carrier. In fact that was its first main use, in the form of the **electric telegraph**. A new code — the **Morse Code** — was invented for this kind of message carrier (see picture 4).

Then in 1876 a Scotsman, Alexander Graham Bell, invented the first artificial **transducers**, which allowed speech to transmitted over long distances. He had invented the **telephone**.

A transducer is a device that transfers signals from one energy system to another. For example, our ears change sound waves into electric signals that the brain can understand. Bell had found another way of changing sound signals into electric signals — the **microphone**. He also had to invent the transducer at the other end to change the electric signals back into sound — a **receiver** or **loudspeaker**. These devices are described in topic C5.

Even after the telephone was invented, light was still one of the main means of sending messages long distances. Armies and navies used flags, lamps and flashing mirrors (heliographs). They were cheap, quiet and didn't need a network of wires to carry the message.

Then a young Italian called Guglielmo Marconi (picture 5) took up an idea that university physicists had already been experimenting with, and made it practical. He invented a message carrying system that didn't need wires — a **'wireless' telegraph**. To carry messages, this system used what we now call radio waves. These are an invisible part of the **electromagnetic spectrum**, the family of waves (see topic C8), which also includes light.

Picture 5   Guglielmo Marconi

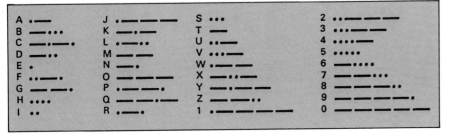

Picture 4   The earliest electric message carrier and the code it used

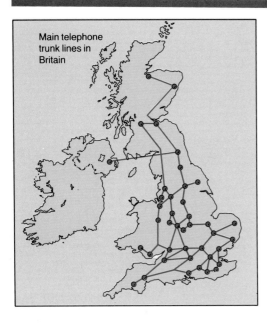

Main telephone trunk lines in Britain

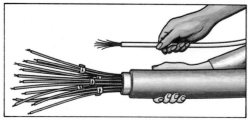

Picture 6 Modern communications are going back to using light — for very good reasons

This was one of the most important inventions of the 20th century. It led to the development of radio, radar and TV, and the discovery of new facts about the universe as scientists began picking up radiations from outer space.

Back to light

So much information is now being sent from place to place that it is hard to find room for it. The air is getting more and more crowded with radio and TV signals, criss-crossing each other and getting in each others' way. Telephone lines can only carry so many conversations at once, even when they are specially coded.

This has forced scientists to think of other ways of sending messages. Their solution to the problem is to use light, but light sent down 'wires'. The 'wires' are made of glass drawn out into very thin fibres — **optical fibres** (see topic C6). Very soon, the main 'trunk' telephone lines in the UK, which carry messages between the main cities, will carry the signals coded into pulses of light instead of electric currents (see picture 6(a)).

A standard copper cable can carry up to a thousand coded conversations at the same time. The optical fibre that replaces it can carry 11 000 conversations, using present coding systems. These can be improved to allow the cable to carry five times as many, if necessary. Using optical systems also improves the quality of the signals, so that they can carry complicated 'computer data' without losing its accuracy. The optical fibre system is also lighter and smaller (see picture 6(b)).

Digital codes

Most modern communication systems use the same code. Information, like speech, music, data from experiments, pictures and computer files can all be put into numbers, using a simple binary (two-number) **digital** code. These

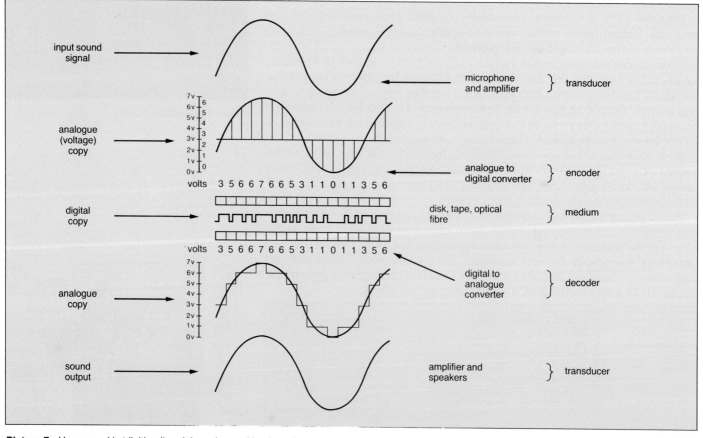

Picture 7 How sound is 'digitised' and then changed back again

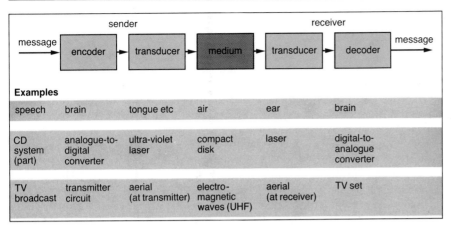

speech	brain	tongue etc	air	ear	brain
CD system (part)	analogue-to-digital converter	ultra-violet laser	compact disk	laser	digital-to-analogue converter
TV broadcast	transmitter circuit	aerial (at transmitter)	electro-magnetic waves (UHF)	aerial (at receiver)	TV set

Picture 8 The main parts of an information transmission system

numbers are sent through a medium (wire, optical cable, the air, empty space) using some form of carrier. The carrier could be an electric current, or light, radio or TV waves.

The signals are picked up by some kind of receiver. How digital coding works is shown in picture 7.

A block diagram of a typical information transmission system is shown in Picture 8.

All information transmission systems need the same basic parts:

Table 1 Parts of an information transmission system

Part	Example
an *encoder* to translate data, ideas, words etc. into a suitable form of message (information)	*brain and tongue* change words into spoken sounds
a *sending transducer* to change the message into something the medium can carry current	*microphone* to change sounds into a changing electric current
a *medium* to carry the message from one place to another	*copper wire* to carry the electric current
a *receiving transducer* to change the form of the carried message into a form in which it can be decoded	*telephone earpiece* to change the current back into sounds
a *decoder* to translate the message into a form that can be understood or used	*ear and brain* to change sounds into words you can make sense of

Physics and information

This topic has given you an overview of how information is transmitted. How the various parts of the different kinds of systems actually work will be explained in the rest of the topics in this section and in section F.

Humans and other living things have their own ways of coding, sending, receiving and using information, and this is dealt with in more detail *The Living World*.

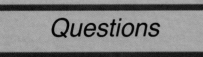

Questions

1 Name *two* devices used in your home that contain information transducers.

2 A TV set is part of an information transmission system. Look at the parts of such a system in picture 8. What jobs do:
a the screen,
b the loudspeaker,
c the aerial of a TV set, perform?

3 Picture 8 shows the main parts of an information transmission system. Copy it out and write underneath each box what you would put in it to illustrate:
a sending a letter,
b how a six-month old baby communicates with its mother.

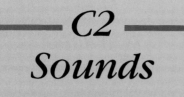

C2
Sounds

This topic is about how sounds are made and how they travel as waves.

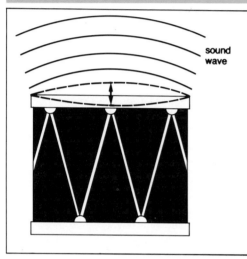

Picture 1 As the drum skin moves up and down, the air above it is squashed, then expanded

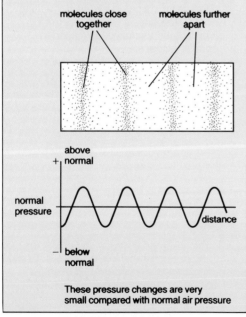

molecules close together

molecules further apart

above + normal

normal pressure

below normal

distance

These pressure changes are very small compared with normal air pressure

Picture 2 (a) A sound wave in air
(b) A graph of the pressure of the air looks like this

Making sounds

Sounds are made when objects vibrate. As the skin of the drum in picture 1 moves up and down it also moves the air next to it. When the drum skin moves up it squashes the air in front of it. When it moves back it drags air after it, so that the air has to expand. The result is a **sound wave** that travels through the air away from the drum.

The sound wave produced by the vibrating drum skin is a series of compressions (squashed air) and expansions. These move away from the drum, as each layer of air squashes the one next to it. Picture 2 shows what we imagine a sound wave in air to be like.

The speed at which these alternate layers of compressed and expanded air move away from the drum is the **speed of sound**. This is 340 metres per second for normal air.

Sound can also travel through other materials. It moves best when the material is very stiff, for example in metals. The speed of sound in different materials is given in table 1.

Table 1 The speed of sound in different materials

Material (medium)	Speed in m/s (at room temperature and pressure)
air	331
hydrogen	1286
carbon dioxide	260
wood	4200 (variable)
copper	3813
iron	5000
rubber	1600 (variable)
cork	500
water	1480

How sounds are different

We soon learn to recognise different people's voices, the sounds of moving water, the wind in the trees. Most people like music, and can easily tell the difference between one note and another, and one instrument from another. All these sounds are very different, but the differences are based on just three things: What we hear is decided by:

- pitch (how high or low?): the number of vibrations the sound makes per second (its frequency)
- loudness: the energy carried by the sound wave
- quality (e.g. from different instruments): the shape of the sound wave

Of course, what the wave is like is decided at the start by the type of object that is vibrating – the musical instrument, the engine of a car, etc. See *The Living World*, topic D5, to find out about the ear and hearing.

Frequency and pitch

Musicians use the word **pitch** to describe how 'high' or 'low' a note is. A high-pitched note is made by something vibrating very quickly. Something vibrating slowly produces a low-pitched note. We use the word **frequency** to describe the rate at which something vibrates. It measures the number of vibrations per second, in a unit called the **hertz (Hz)**.

The human ear can hear sounds from sources which vibrate from as low as about 16 vibrations per second (16 Hz) to as high as over 20 000 Hz. Activity A is about measuring your range of hearing.

Wave shapes — amplitude and quality

But there are other differences between sounds. They can be loud or quiet. They can also be different in **quality** – which means we can tell the difference between a note played on a guitar and the same note played on a flute.

Loudness is decided by the energy in the wave. In turn, this is decided by how strongly the source of the sound compresses the air. Sound waves are pressure waves in the air. One way to show a sound wave is to draw a graph of the *change* in the air pressure caused by the wave.

Picture 3 shows some of these graphs. They show how the pressure changes with time at one point in the air for different sounds. The first two (a) and (b) show the same note, but (b) is louder than (a). The *pressure changes* in (b) are greater. This means that they have a greater effect on the ear, so we sense it as a louder sound. The amount that the pressure changes above or below normal is called the **amplitude** of the change. The bigger the amplitude the louder the sound.

Graphs (c) and (d) show notes of the same loudness and pitch made on different instruments. The waves carry the same energy, and they would be 'in tune'. But their *shapes are different*, and we hear these differences as differences in **quality**. Activity B is about looking at sound waves made by different instruments.

Wave shape and pitch

The shape of the wave can also show the frequency of the sound. Graphs (e) and (f) show notes of the same loudness but of different frequency. Note (f) has a higher pitch than note (e). You can see that there are more waves

Picture 3 These graphs show how air pressure varies with different types of sound: (a) small amplitude, (b) Large amplitude, (sine wave, same frequency), (c) violin note, (d) clarinet note (same pitch and amplitude).

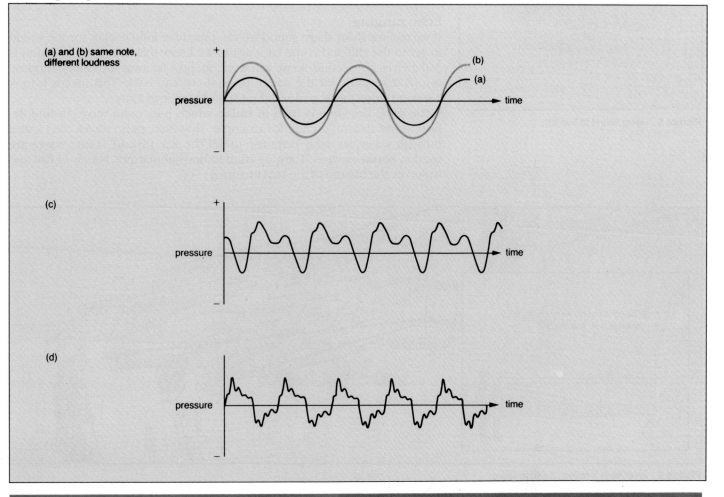

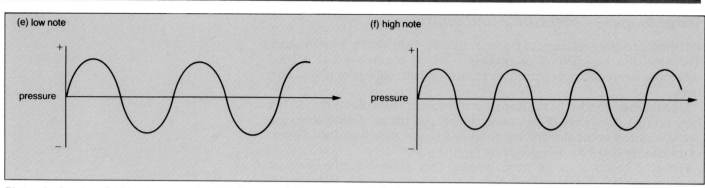

Picture 3 (continued) (e) low frequency, (f) higher frequency (sine wave, ame amplitude)

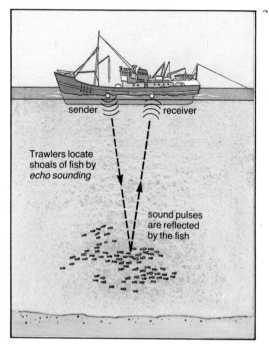

Picture 5 Using sound to find fish

arriving per second in (f) than in (e). The instrument must have been vibrating quicker. *The more waves there are per second (frequency) the higher the pitch.*

Echoes

Like other waves, sound can be reflected. We don't usually notice this, but we would find listening to music and speech in rooms and concert halls would be very strange if the walls didn't reflect sound.

Echoes are very obvious examples of sound being reflected. To hear a good echo, we must stand a few hundred metres from a cliff or hillside. When we shout, the sound travels to the reflector and bounces back (picture 4). If we are far enough away it means that you can finish a short sentence before the echo returns to us.

Echo ranging

If we make a short sharp sound we can time how long it takes for the sound to get to the cliff and come back again. We know that the speed of sound is 340 metres per second, so we can work out how far away the cliff is. Suppose it took 2 seconds for the sound to go there and back. This means it took 1 second to get there, so the cliff must be 340 metres away.

The same principle is used in **radar**, which uses radio waves to find the position of distant aircraft, for example. However, radio waves can't travel through water, so radar is no use under the sea. Instead, sound waves are used in **sonar** devices. They are used to find submarines, shoals of fish and to survey the bottom of the sea (picture 5).

Picture 4 An echo is a reflected sound wave

Activities

A High and low

For this activity you need a cathode ray oscilloscope (CRO), a sound generator and a loudspeaker or headphones. It will need to be set up by your teacher. You can use the controls of the sound generator to change the pitch and the loudness of the notes produced, as well as hearing them, you can see their wave patterns on the screen of the CRO.

Use this equipment to do the following.

1a See what happens when you make a given note louder and quieter.

Draw some typical results.

b See what happens when you change the pitch of the the note.

2 Measure the frequencies of the lowest pitch sounds and the highest pitch sounds that you can hear — i.e. the **frequency range** of your ears. You should start with the sound generator working at a very low frequency — this could be just one 'wave' per second and you should be able to hear the separate pulses. Increase the frequency slowly and note what happens. You will need to adjust the volume control as well!

a At what frequency do the pulses stop being heard separately and start to make a definite 'note'?

b At what frequency do you stop being able to hear the sound?

c Collect together the results of your class or group. Compare them with each other and with the results from your teacher and any other adults you can persuade to be tested. Comment on the results of this survey.

B Looking at sounds

For this activity you need a cathode ray oscilloscope (CRO), a microphone and as many musical instruments as you can borrow. Play each instrument into the microphone with a very steady note. You may need to adjust the CRO controls to get a good steady trace. Compare different instruments playing the same note. Compare different notes on the same instrument. See the difference between loud and soft notes played on the same instrument. Draw some of the traces that you see.

C Measuring the speed of sound in air

There are lots of ways of measuring the speed of sound. Here are some ideas:

1 Stand in front of a large wall (e.g the wall of the gym) and clap your hands. You should hear an echo. Now clap your hands at a steady rate. Time it so that the clap occurs at the same time as the echo reaches you. Get a friend to time 20 claps. Divide this time by 20. This is the time it takes for sound to go from you to the wall and back again.

Measure the distance from the wall to you and calculate the speed of sound:

$$speed = \frac{2 \times distance\ to\ wall}{time\ taken}$$

2 Get a friend to stand a long way from you, e.g. the far side of the playing fields or playground. Your friend should have two blocks of wood and hit them together to make a sharp sound. When you *see* the two blocks hit each other start a stop watch. Stop it when you hear the sound. This is the time it takes for the sound to travel the distance between you. Measure this distance and calculate the speed of sound from:

$$speed = \frac{2 \times distance\ to\ wall}{time\ taken}$$

3 Using electronic timing. Methods (a) and (b) are cheap but rely on human reactions in stopping and starting a watch. Electronic timers have quicker reactions and can measure much shorter times with good accuracy. They can be stopped and started when a sound pulse reaches two microphones just a few metres apart. In this way you can measure the time it takes sound to travel that distance.

Check with your teacher if you have such equipment available. If you have, plan an experiment to use it to measure the speed of sound in your laboratory.

Questions

1a What is the quietest sound you can hear?

b Why can very loud sounds damage your ear drum?

2 What are the three main ways in which sounds can be different from each other? What are the physical causes of these differences.

3 (For musicians) How are the three differences in question 2 shown on a musical score?

4a What is an echo? How is it produced?

b Echo sounders are used to find shoals of fish. How do you think they work?

5 A mountain walker notices that when she shouts she hears an echo from a distant cliff. She times it and finds that it takes 2.5 seconds for her shout to be returned as an echo. How far away is the cliff? (speed of sound in air: 340 m/s).

6 You can work out how far away you are from a thunderstorm by measuring the time between seeing the lightning flash and hearing the thunder it makes. This works because light travels so much more quickly than sound. A worried boy counted off 5 seconds (by saying 'alpha one, alpha two . . .') between the flash and the thunder. How far away was the thunderstorm?

7 The diagram below (picture 6) shows three traces of sounds, shown by a cathode ray oscilloscope, which of them: (a) would be the loudest, (b) would be the biggest in pitch, (c) is likely to be made by a flute?

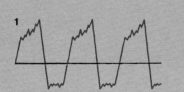

Picture 6

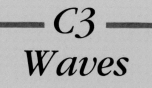

C3
Waves

Sound isn't the only thing that moves as waves. There are waves on water, light waves and earthquake waves.

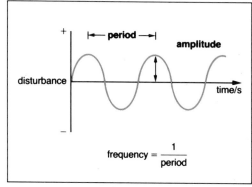

Picture 1 Amplitude and period — (disturbance *v* time)

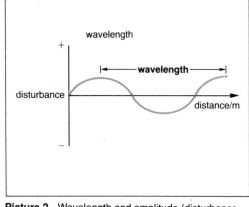

Picture 2 Wavelength and amplitude (disturbance *v* distance)

Waves in space

Light, and the whole electromagnetic spectrum of which it is a small part, travel through space as waves. Earthquakes set up waves which can cause great damage. They also travel right through the Earth, and scientists use them to find out what the inside of the Earth is like. Topics C9 and G2 deal more fully with these types of wave.

But whatever kind of waves they are, they are similar to each other. They all *move, carrying energy*. They all have a *pattern*, which *repeats itself*.

Picture 1 is a graph of a typical wave. It could be a water wave, a sound wave or a light wave. We can use it to explain the meaning of the key words we use to describe wave motion.

First, its **amplitude**. This is a measure of how much the wave vibrates the medium it passes through. It could be the height of a water wave, or the pressure of a sound wave.

Next we have the wave's **period**. This is the time it takes for the wave to repeat its pattern. It is the time for one **vibration** of the source of the wave.

The frequency of a wave is related to its period. The more vibrations are made in a second, the shorter is the time that each one takes. Thus, if a wave has a period of 1/100th of a second it repeats itself 100 times a second, so has a frequency of 100 Hz. That is:

$$\text{period} = 1/\text{frequency}$$

Wavelength

The graph in picture 1 was of the wave changes plotted against time. If we plot these against *distance* instead we get an idea of the size of the wave. This is done in picture 2. The marked distance is the **wavelength** — the distance between equivalent points on the wave pattern. This could be from peak to peak, or from trough to trough.

The wavelength of a typical sound (say middle C) is about 133 cm. The wavelength of light is very much smaller. For yellow light it is about 600 billionths of a metre. Long wave radio broadcasts use wavelength of over 1000 metres.

The wave speed formula

If a source of sound is vibrating 16 times a second it is producing 16 waves every second. At the end of that second the first wave has travelled 16 wavelengths away from the source. The wave speed is simply how far the waves move in a second. In this case it is obviously just 16 wavelengths. Picture 3 illustrates this.

If the sound source vibrated at 20 Hz, it would produce 20 wavelengths in a second. But sound travels at the same speed in air, whatever its frequency, so the waves are more squashed up — the wavelength is smaller. This is also shown in picture 3.

Thus, because the speed of sound is the same, however many waves are made each second they all have to fit into the same distance. This distance is 340 metres, in air. Looking at it mathematically:

$$\text{length of a sound wave} = \frac{\text{distance sound travels in a second}}{\text{number of waves made per second}}$$

$$\text{or: wavelength} = \frac{\text{wave speed}}{\text{frequency}}$$

This is usually written more neatly as:

$$\textbf{wave speed} = \textbf{frequency} \times \textbf{wavelength}$$
$$v = f\lambda$$

This formula applies to all waves.

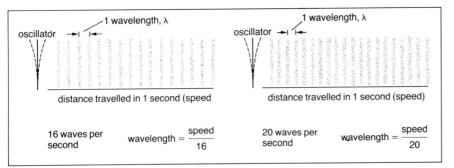

Picture 3 The speed stays the same, so wavelength gets less when frequency increases

To and fro, up and down

Sound waves compress and expand the air. The particles of air move backwards and forwards in line with the direction the sound travels. Waves on a large-coil spring can also do this. These kind of waves are called **longitudinal** waves.

When sea waves move, the water surface moves up and down as the wave moves along. Waves on a rope or a guitar string are also like this. Waves in which the carrier (medium) moves at right angles to the direction of wave movement are called **transverse** waves. Picture 4 shows these differences.

Waves through liquids

Sound waves travel well through water. Dolphins use sound waves as a sonar to hunt their prey. Sound travels through water because sound is a longitudinal wave. Transverse waves can travel along the surface of water but not *through* water. This is the case for all liquids. This fact has been used to prove that the core of the earth is liquid, because the transverse earthquakes waves don't get across to the opposite side of the earth. This is dealt with more fully in topic G2.

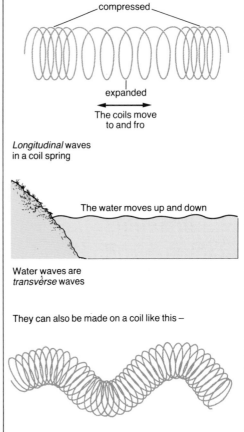

Picture 4 Longitudinal and transverse waves

Activities

You will need a strong, heavy rope and a long coil spring. You will also need a fairly long smooth floor, like a school corridor. You can make wave pulses in the rope by shaking it from side to side for a short time — making one or two 'wavelets'. These are transverse waves.

You can do the same with the coil, but you can also make longitudinal waves.

Investigate the properties of waves, and try answering these questions:

1 Can the waves be reflected?

2 What is the speed of the wave? Does the amplitude of the wave affect its speed? Does the frequency?

3 What happens when two waves pass through each other?

4 Can you alter the speed of waves in the rope? How?

5 Can you alter the speed of waves in the coil spring? How?

B Waves on water

Your school might have some ripple tanks which allow you to explore two-dimensional waves on water. There are lots of experiments that can be done. To begin with, try answering the questions 1, 2 and 3 in activity A for water waves. You can then go on to investigate:

4 What happens when waves are reflected by curved barriers?

5 What happens when waves go through a gap?

6 What happens when they go through two gaps, side by side?

Questions

1 Explain the difference between frequency and period for a wave.

2 Give a brief outline of an experiment you could do show:

a that sound waves can pass through each other,

b that water waves can pass through each other.

3 You should have learned that when light is reflected, 'the angle of incidence is equal to the angle of reflection'. Explain what this means. How could you test to see if sound waves obey the same rule?

4 Use the formula wavelength = speed/frequency, and the data in table 1 in topic C2 to calculate the following:

a the wavelength of a 500 Hz note in air,

b the wavelength of a 1000 Hz note in air,

c the speed of a note of 500 Hz which is measured to have a wavelength of 0.5 metres in carbon dioxide.

5 The speed of radio waves is 300 000 000 m/s. What is the frequency of the UK Radio 4 programme which has a wavelength of 1 500 metres?

C4 Making music

This topic is about how musical sounds are made and controlled.

Picture 1 Musical instruments

Labels: Sitar, Trumpet, Cello, Saxophone, Tabla, Pan pipes

The reed of a clarinet is a shaved-down piece of cane. It rests in the mouthpiece and vibrates when the player blows through it. These finger levels open and close the holes. This changes the length of the vibrating air column.

tip opening, tone chamber, reed, ligature

Picture 2

Like all sounds, music is made by vibrating objects. These may be strings, skins, pieces of wood or metal and even columns of air (see picture 1). Most of them have built-in 'amplifiers', which magnify the vibration of a small part of the instrument. For example, a guitar string doesn't make a very loud sound on its own, simply because it can't move very much air. The sound wave produced doesn't carry much energy.

Putting the string on a hollow wooden box makes the sound much louder. The box and the air inside it vibrate as well, in time with the string. Because they are bigger they give much more energy to the air.

When the guitar box vibrates in time with the string we say that it **resonates** with the string. The energy for this comes from the movement energy of the string. In a guitar the note soon dies away, but a violinist can make the note sound for longer. The player keeps giving energy to the string by drawing the bow across it.

Many musical instruments use resonance. In a clarinet the reed is made to vibrate and causes a length of air in the tube to resonate (picture 2). In a trumpet, the player's lips vibrate and make the air in the trumpet resonate. Some very simple instruments, like the triangle, just vibrate as a whole.

Getting the right note

The pitch of the note is decided by the frequency of the vibrator, as explained in topic C2. In a stringed instrument like a guitar this is decided by the string itself. For a particular string fixed at a certain tightness (tension), the shorter the string is, the more quickly it vibrates. The guitarist changes the pitch of the note by moving his fingers up and down the neck of the guitar. This alters the length of the string (picture 3). All stringed instruments use this principle.

To get a whole range of notes the instrument will have several strings. Some are thicker than others, and will be stretched to different tensions (see picture 3).

Making the string tighter makes the pitch higher. The strings can vibrate faster when they are tighter. Thicker strings are used to produce the lower notes. What counts is not so much the thickness as the mass per unit length of the string. The more mass there is the slower the string vibrates for a given length and tension.

Wind instruments

A wind instrument is worked by blowing into it. This makes a column of air vibrate. The pitch of the note is decided by how long this air column is. The shorter the column, the quicker it vibrates and the higher the pitch of the note. Picture 4 shows the pipes of a large organ. You can see the range of lengths needed to get the huge variety of notes this instrument can produce.

Fundamentals

In both stringed and wind instruments the vibrations can be simple or very complicated. This is partly due to the design of the instrument and partly to the skill of the player.

In a simple instrument, like a recorder or a guitar, the simplest way the string or air column vibrates is called its **fundamental mode**.

This is shown, for a string, in picture 5(a). This gives the lowest frequency that the string can produce, the fundamental note. But a string of the same length can vibrate in several other ways, as shown in picture 5(b). As you can see, the string is now vibrating not like one long string but like two or more shorter strings fixed together. The resulting vibrations are faster and the notes are higher in pitch. They are called **overtones**. The notes from most instruments will contain overtones as well as the fundamental note that the player is aiming for.

Some of these overtones make the note sound a lot more pleasant, others make it sound awful. A skilled player can make the nicest overtones stronger and cut down the unpleasant ones. This is the difference between a good violinist and a bad one — and between a good violin and a poor one.

More about resonance

Most objects can vibrate. Usually they will do so most easily in their fundamental mode. If you tap a glass with a spoon it will vibrate and give a sound. The basic note is its fundamental note and an expert musician can tell its pitch. If the musician then plays an instrument or sings at exactly the same note the glass will start to vibrate. It picks up energy from the vibrating air. Very loud singers can cause a glass to pick up so much energy that it shatters.

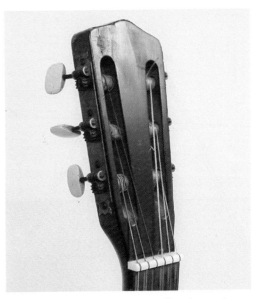

Picture 3 A guitar player has to adjust the tension in the strings

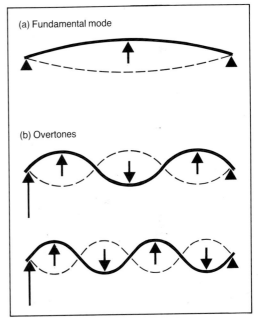

(a) Fundamental mode

(b) Overtones

Picture 5 A string can vibrate in many different ways

Picture 4 To get all the notes, an organ needs pipes

Picture 6 You have to push a swing at exactly the right time to make it swing really well

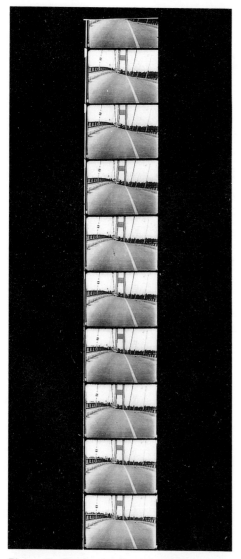

Picture 7 The Tacoma Narrows Bridge on the point of collapse

This is another example of resonance. It works just like someone being given a ride on a playground swing. To get the swing vibrating through a large distance (amplitude) you have to push it at just the right time. The frequency of your push has to be the same as the natural frequency of the swing (see picture 6).

An object which is designed to be good at vibrating, like the air column in a trumpet, doesn't need to be given the exact frequency at which it resonates. It will pick up energy from any sound as long it contains its natural frequency of vibration. Trumpeters purse their lips and blow hard. Their lips vibrate at a range of frequencies and the air in the trumpet will select its natural frequency and start to vibrate.

This vibration gives 'feedback' to the trumpeter's lips and soon all their energy of vibration is at that special frequency. The sound produced can be very loud indeed.

Resonance can cause damage. In an earthquake the earthquake waves may have a range of frequencies. Some of them may be the same as the natural frequency of a bridge or tall building. Just as a swing or a trumpet builds up a lot of energy from quite gentle pushes, so can a structure. The resulting vibration can be disastrous.

The most famous example of resonance damage is the destruction of the Tacoma Narrows Bridge, in Washington State, USA, in 1940. This bridge was able to resonate in time with regular gusts of wind passing over it. In a gale, the energy the bridge picked up was so great that it simply tore itself to pieces.

Picture 7 shows the final stages in the life of the bridge. Engineers and architects learned a good lesson from this, and all modern buildings are designed to have natural frequency well outside the range of any natural vibration that could occur.

Activities

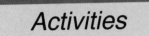

A Experimenting with musical sounds

1 Make a bottle-organ. When you blow gently across the neck of an open bottle the air inside resonates at its natural frequency.

Putting water in the bottle changes its frequency. Use eight bottles so that you make a musical scale when you blow over them.

2 A simple xylophone (wood-organ) can be made by laying blocks of wood across two long wooden strips. When you hit them with a small hammer they will make a dull but definite note. The pitch of the note depends on the size of the blocks of wood. Make a xylophone with eight pieces of wood that form a musical scale when you hit them.

3 Any hollow object can be blown into to make a note. Or you can blow across an open end. You can make it give out a set of musical notes if you blow harder or softer, purse your lips differently, and so on. But it takes patience to get the knack of doing this. People have made music from surprising objects such as lengths of hosepipe, bicycle frames and pieces of rubber tubing with a plastic funnel stuck in the end. Have a go!

B Investigating air columns

This invetigation is about how the length of an air column is related to the frequency of its natural (fundamental) mode of vibration. The diagram in picture 8 shows you a form of apparatus you can use.

You will need a source of standard frequencies. This could be a set of tuning forks or a calibrated variable frequency generator and loudspeaker.

Start with the tube full of water and let it out slowly as the highest frequency sound source is held above them open end. At a certain length the sound will get much louder — the air in the tube is resonating. Note the frequency and measure the length of the air column. Repeat this for five frequencies, ranging from about 200 to about 600 Hz. If you overshoot the length put some more water in the tube and try again.

1 Plot a graph of column length against frequency. What does your graph prove?

2 When it resonates there is a 'standing wave' in the air in the tube. This is caused by the sound wave going down the tube and reflecting off the water at the bottom. The tube length is actually a 'quarter wave' (see picture 8). Use this fact to calculate the speed of sound in (damp) air, using the formula $v = f\lambda$

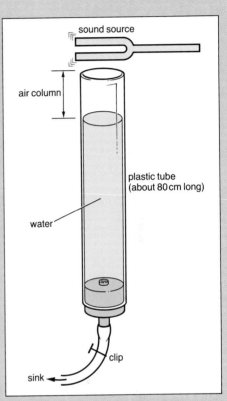

Picture 8 Resonance in a tube of air

Questions

1a Explain what *resonance* means.

b Give an example of resonance that you have seen or read about in everyday life.

2a What is a *fundamental* note?

b Explain how *overtones* are produced and what effect they have in music.

3 Describe any musical instrument that you know about or can look at. Explain:

a how it makes its sound,

b whether it uses the principle of resonance,

c how the player varies the notes it can produce.

4 Look at picture 1. For as many as you can of the instruments shown:

a say whether or not is uses resonance,

b say how the pitch of the note is changed.

C5
Record and playback

Most of the music we hear has been recorded. Even when it is 'live', we often hear it through electrical devices of some kind.

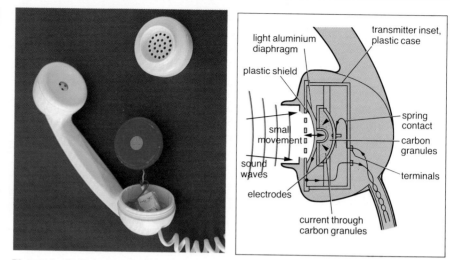

Picture 1   Carbon microphone

Both microphones and loudspeakers are **transducers**, as explained in topic C1. Transducers change information from one energy system to another.

A microphone has the job of changing sound waves into a varying electric current. The pattern of the changing current has to be a close copy of the pattern of the sound wave. The loudspeaker converts the changing current back into sound waves.

Microphones

Picture 1(a) shows a typical microphone — the **carbon microphone**. It is still used in some older telephones, but modern ones use **electret microphones** (picture 2).

The key component in the carbon microphone is the capsule with the grains of carbon in it. When sound waves reach the microphone they make a flexible piece of steel (the **diaphragm**) vibrate. As the diaphragm vibrates the grains of carbon are squashed closer together then moved further apart, in time with the vibrations.

Carbon is a conductor of electricity. When you pick up the telephone handset a current is switched on. This current flows through the grains of carbon. When the grains are squashed closer together they conduct better. But when they move apart a little their resistance increases.

The current through the carbon grains thus changes in the same pattern as the sound waves that affect the diaphragm. This current flows down the line to the handset of the person you are talking to. Picture 1(b) explains this.

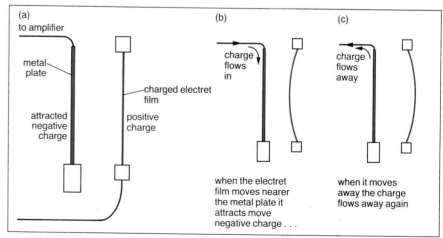

Picture 2   Electret microphone

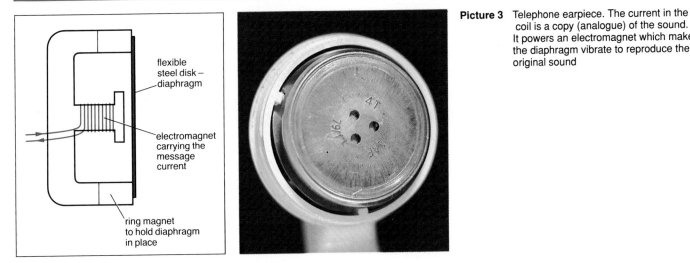

flexible
steel disk —
diaphragm

electromagnet
carrying the
message
current

ring magnet
to hold diaphragm
in place

Picture 3 Telephone earpiece. The current in the coil is a copy (analogue) of the sound. It powers an electromagnet which makes the diaphragm vibrate to reproduce the original sound

Most ordinary cassette recorders use electret microphones (picture 2). They are also used in some telephones. The key item in this type of microphone is a very thin piece of plastic film, 12 to 25 micrometres thick. One side of the film is metallised and is connected to an electric terminal. This is usually one side of an amplifier input.

The film is made of a special plastic that is a very good insulator. The plastic is electrically charged at the factory, and is such a good insulator that it can keep its charge for 200 years.

When the sound waves reach the microphone the plastic film vibrates in tune with it. The charged film moves nearer to or further away from a metal plate which is connected to the opposite terminal of the amplifier. The charge on the film attracts opposite charge to this metal plate. When the film gets nearer it attracts more charge. When it moves away it attracts less charge.

This makes a current which is a copy of the sound wave, which flows into or away from the metal plate. This is amplified for recording or for sending down the telephone line. Picture 2(b) shows how an electret microphone works.

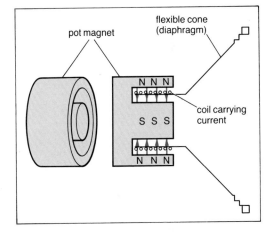

pot magnet

flexible cone
(diaphragm)

N N N
S S S
N N N

coil carrying
current

Loudspeakers

Pictures 3 and 4 show two everyday kinds of loudspeaker. The simplest is the type used in the telephone **earpiece**. The key parts are the electromagnet and the thin steel plate (diaphragm). As the size of the electric current changes in the coil it changes the pulling strength of the electromagnet. The plate moves to follow the changes in strength of the electromagnet. As it does so it moves the air in front of it — to produce sound. This is, of course, a copy of the sound wave going into the microphone at the other end of the line.

The telephone earpiece is good enough for conversations but cannot reproduce musical sounds accurately. A better way of doing this is to use a **moving coil** loudspeaker (picture 4). This uses the *motor effect*, which is explained in topic E8. As in the telephone earpiece, the changing current flows through a coil. This time the coil is inside a strong, steady magnetic field, produced by permanent magnets.

As a result of the motor effect, the coil moves. Its movements are an exact copy of the changing current. The coil is connected to a large flexible diaphragm, usually made of special paper or card. This has a large area and as it moves it produces sound waves in the air. A large loudspeaker can move a lot of air, and so produce very loud sounds.

Picture 4 Moving coil loudspeaker

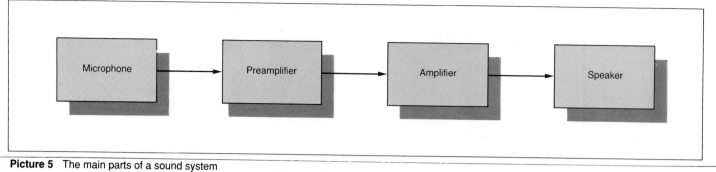

Picture 5 The main parts of a sound system

Headphones

These are made from two very small speakers and may be the moving coil type or may use an electret system (rather like the microphone, but in reverse). They don't need to produce sounds with a lot of energy, because they are so close to the eardrum. But they can produce enough energy to damage the ears. Doctors are worried that the use of 'personal stereos' is producing a generation of people who are all partly deaf. Headphones need to be used with great care. As a general rule, if other people can hear the music it is too loud for the wearer!

Amplifiers and electronics

The earliest 'sound systems' relied on the energy in the sound waves alone to make the electric current and even to make recordings.

Better results are produced when extra energy is injected into the system. This is what amplifiers do.

An amplifier can make a weak current produced by a microphone much stronger for making recordings. This is also better for carrying messages down a long wire.

At the loudspeaker end of the system the current can be amplified still more to power large, powerful loudspeakers. Most playback systems produce a very small current from the 'record' (disc, cassette, compact disc). This needs to be greatly amplified to make the speakers work.

Picture 5 shows the block diagram of a typical 'sound system'.

Electric sounds

Electric instruments — like electric pianos, violins and guitars — simply do away with mechanical resonators (see topic C3). Instead, the small, low energy vibrating source (e.g. a guitar string) is used to produce a small electric current which is fed to speakers via an amplifier (see picture 6).

Electronic instruments (see picture 7) don't need a mechanical input. An electric current with the pattern of a sound wave is made **electronically**. The mechanical vibrations (or oscillations) that produce the sound in an 'acoustic' instrument are replaced by electronic circuits (oscillators, wave shapers, etc). **Acoustic** is a word used to describe traditional instruments that produce sound using only mechanical means, like vibrating strings or air columns.

The vibrations in electronic instruments are made by circuits called **oscillators** which produce a voltage that can vary at any desired frequency. Other circuits combine the pure oscillations to make a copy of the complex waves that real instruments produce.

These signals are then amplified and played through a loudspeaker.

When you press a key on the keyboard you are simply operating a switch system which makes the circuits start working.

Picture 6 An electric guitar

Picture 7 Electronic keyboard

Storing music

Music is stored in three main ways:

- mechanically — on vinyl (LPs) or on 'compact' discs (CDs),
- magnetically — on magnetic tape,
- optically — on film sound tracks.

Mechanical systems record the music either in analogue or in digital form. In both, the **shape** of a material is altered. Vinyl records store the music as a long wavy groove that spirals around the disc. In a CD the information is stored in a long groove with 'holes' in it, as explained below.

The 'wave' in the groove of an LP is a copy of the sound wave of the original music (see picture 8). This is called an **analogue** of the original sound (see topic F4).

The master record is actually made on tape (see below), but it is copied onto the disc mechanically. A sharp pointed metal needle (a **stylus**) cuts the groove into soft plastic. This is then covered with a thin layer of metal to harden and preserve it. In this way a metal master is produced which is used to stamp out the discs we buy in the shops.

The music is read off the disc by another stylus which moves through the groove as the disc turns under it on a record player. The stylus is very light. It moves up and down and side to side as it follows the waves in the groove. The stylus is connected to a coil or a small magnet. The movement of these produces an electric current which is a copy of the original sound. This is then amplified and fed to the speakers.

Picture 8 The 'groove' in a vinyl record

The problem with vinyl discs is that the material has to be soft enough to be stamped, and so the grooves are easily damaged. Also, the groove has to be very long to get a reasonable amount of music stored in it, and this means that the disc has to be quite big. Thus it is easy for it to warp and distort the recording.

Digital recording is a great improvement. The information is converted to a set of numbers (digits), which are easy to store and quite hard to damage. They are binary numbers (see topic F4). The system relies on computer technology, which has always used binary numbers to store and control information.

Picture 7 in topic C1 shows how a sound wave can be converted to numbers which can be stored on a **compact** disc. It is called 'compact' because the information is stored in a small space. Two sides of an LP can be stored on one side of a much smaller compact disc.

The numbers are stored as a series of **pits** (small holes) in a strong plastic disc. (See picture 9 overleaf.)

The pits are very small — about half a micron (millionth of a metre) wide and a fifth of a micron deep. They are laid in a thin spiral, beginning at the

Picture 9 A compact disc

centre of the disc. The width of this spiral track is about one thirtieth the thickness of a human hair. It is this small size that allows so much information to be stored on a small disc.

The pattern of holes form a coded signal of '16-bit' binary numbers, each number standing for one feature of the sound wave. (A 16-bit number is a binary number between 0 and 65 536.) The sound wave is sampled and changed into numbers 44 thousand times a second. In music the sound wave changes quite rapidly, and if it was sampled less often some of the quality would be lost.

The groove with the holes in it does the same job as the wavy groove in a vinyl disc. But the pattern on a CD is far too small to be made or read mechanically, using a stylus. Instead **lasers** are used.

The series of pits is burnt into the plastic disc using a powerful, concentrated beam of light from a laser. The track is then silvered to make it a good reflector. In use the track is 'read' by another, weaker laser.

Laser light is a very accurate beam of light. When it is reflected from the disc it has gaps in it where the pits didn't reflect it properly. Thus the reflected beam is switched on and off in the same digital pattern as the pits on the disc.

The advantage of digital recording and playback is that it doesn't depend on the quality of the 'holes'. As long as the surface between the holes is smooth the numbers are the same and the result is exact.

Also, in a compact disc once the pattern is made the replay system doesn't touch it mechanically to wear it out.

Magnetic recording

Recording tape is a long piece of thin, strong plastic with a very thin layer of magnetic material on it. The magnetic material must be easy to magnetise — and demagnetise. The cheapest magnetic material is iron oxide (ferric tape), but you get better results from iron mixed with chrome or some other substance (ferrochrome and 'metal' tapes). The magnetising is done by a **recording head**, which is shown in picture 10. It is a very tiny electromagnet. The information to be recorded is sent as a changing electric current through its coil. This produces a small, varying magnetic field which magnetises the tape in the same pattern as the current in the coil.

The tape is read by the **playback head**, which is almost exactly the same as the recording head. As the tape moves under the head, it produces a changing magnetic field. This changing field is a copy of the current that made the recording.

This time, the changing magnetic field induces a current in the coil (see topic E8), which can be amplified to make a loudspeaker work.

Tape recording can be either digital or analogue. The large record companies use digital recording on special, very wide tape. As explained above, the original sound is converted electronically to a series of numbers before being recorded on the tape. Thus the tape recording consists of a series of simple on/off magnetic 'bits', which are a coded form of the sound.

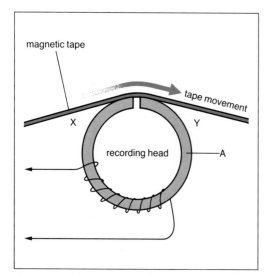

Picture 10 The recording head in a tape recorder

Sound films

Picture 11 shows the 'sound track' of a film. It is an analogue recording. The density (lightness or darkness) of the image on the film corresponds to the pattern of the sound that made it. This pattern is read by a **photocell** as the film is projected. The photocell is a device that gives a smaller or larger current of electricity, depending on how much light reaches it. Thus the current from the cell is a copy of the 'sound track' and the sound that made it.

Picture 11 A film sound track. The sound track runs down the side.

Activities

A Recording

Making a tape recording is a useful skill, and it isn't as easy as it looks. Use a cassette recorder to record speech and music and investigate what decides the quality of the result. You could try:

1 recording in different rooms and in the open air,

2 based on the results of (i), making a simple 'studio',

3 using different types of microphone,

4 using different types of tape,

Write about what you find out. Keep it simple and write it in two parts:

1 A report of your investigations,

2 How to make a good recording.

B Making a decision

You have a birthday coming up and your rich uncle has promised to buy you a good hi-fi system. Use magazines or books to find out what you can about these, and to decide which one you would advise your uncle to buy for you.
(If you'd rather have a portable cassette player make out a case for that, instead!)

C Looking at transducers

You might be able to get an old cassette player, microphone, loudspeaker or telephone, etc. Pull it apart (as carefully as possible) and identify the key parts of the transducers described in this topic.

Questions

1 The list below contains a number of different recording/playback devices. Sort them into two groups, headed *analogue* and *digital*:

human brain, cassette recorder, compact disc, LP record, sound film track, book, computer ROM chip, photograph, video cassette.

If you don't know, guess!

2 Give *two* reasons why compact discs are better than the older vinyl LPs. Are there any ways in which they are worse?

3 Describe what you would imagine to be the perfect 'sound system' — one that you might get for your birthday in the year 2010.

4 Explain the meaning of: analogue, digital, binary, bit, distort, transducer.

5 'There ought to be a law against playing music loudly'. Do you agree with this? Give arguments for and aginst this suggestion.

6 Explain either: (a) how a moving coil loudspeaker works, or
(b) how a carbon microphone works.

7 List *five* ways of storing information. For each way, state how that information is *retrieved* again.

8 'In a modern democracy, a supply of information is as important as a supply of energy.' Obviously, *energy* is vital for life. Is *information* really that important? Write a short essay (or a long one if it interests you!) giving your opinion and justifying it.

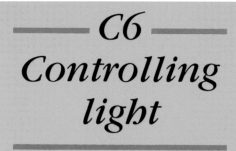

C6
Controlling light

*Light is very useful!
This topic is about
how we can control it.*

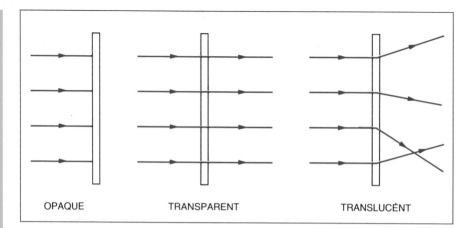

Picture 1 What happens to light in opaque, transparent and translucent materials

OPAQUE TRANSPARENT TRANSLUCENT

Keeping light in — or out

There are some materials that light cannot pass through. They are **opaque**. We stop light getting into a camera by using a metal or plastic material that is opaque to light.

Curtains and blinds do the same thing. A material that lets light go through it is either **transparent** or **translucent** (see picture 1). A transparent material, like glass, air, water and some plastics, lets light through it in straight lines. We can see things clearly through them.

A translucent material breaks up the light so that we don't get a clear picture of the object sending out the light. Finely scratched glass, or 'ground' glass is like this. We use translucent materials in some light bulbs and in bathroom windows.

White painted materials can reflect light so that more of it goes where we want it to, but without a glare. The inside of lampshades are sometimes painted white for this reason (see picture 2).

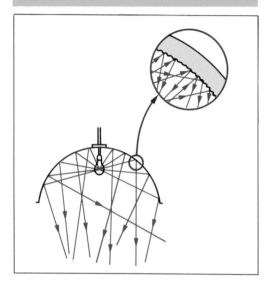

Picture 2 Diffuse reflection — light goes off in all directions

Mirrors

Mirrors reflect light in a regular way (see picture 3 and compare it with picture 2). The light is reflected according to the rule:

the angle of incidence equals the angle of reflection.

In other words, the light rays make the same angle to the mirror going in as they do when they come away from it. We normally measure these angles between the light rays and a line at right angles to the mirror (called the **normal** line).

This means that flat (**plane**) mirrors reflect light to give a clear image. But it is a 'mirror image', in which left hands turn into right hands, and vice versa (picture 4).

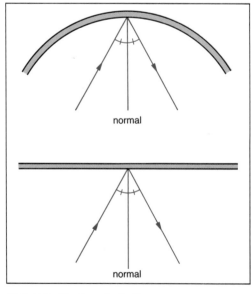

normal

normal

Picture 3 The law of reflection

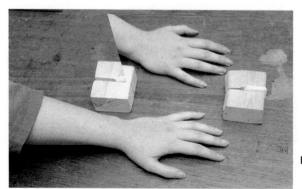

Picture 4 An image in a plane mirror. A left hand becomes a right hand

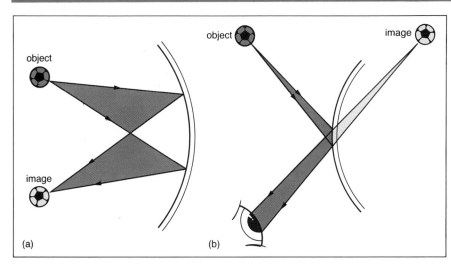

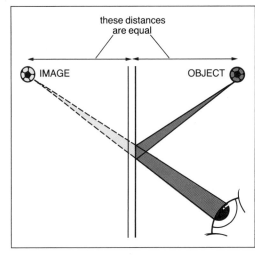

Picture 6 Curved mirrors and their effects on light

Picture 5 The image that appears in a flat mirror seems as far behind it as the object is in front

We can work all this out using the idea that light travels in straight lines. Picture 5 shows how this idea explains why an object in front of a mirror produces an image that is as far behind the mirror as the object is in front of it.

Curved mirrors

Curved mirrors produce interesting effects. They can magnify and make smaller, depending on which way they curve and how far away the object is (see picture 6). If the shiny (reflecting) side curves inward, it is a **concave** mirror. If the reflecting surface curves outwards, it is a **convex** mirror.

Concave mirrors are easy to find in everyday life. They are most often used just to straighten up beams of light, as shown in picture 7. Torches, searchlights and car headlights use mirrors in this way.

They work as they do because of the basic law of reflection, given above. The curved shape of the concave mirror changes the direction of the rays in such a way that a beam of parallel light rays are all reflected to pass through the same point. This is the **principal focus**. Light leaving the focus retraces the same path and leaves the mirror as a parallel beam. This is shown in picture 7, where the lamp filament is placed at the focus of the mirror.

You will often find convex mirrors in shops, or at the corner of the stairs in a double-decker bus. They are used to give the shopkeeper or bus conductor a **wider field of view**. This helps to stop theft or fare-dodging. This time the shape of the mirror allows light to be collected from a wide angle to be reflected towards the observer (see picture 8).

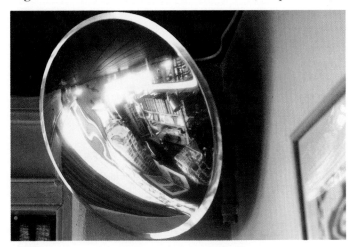

Picture 8 Convex mirrors give a wide field of view

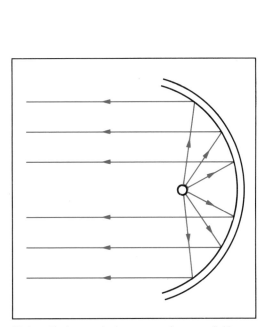

Picture 7 A curved mirror can make a parallel beam of light — as in a searchlight

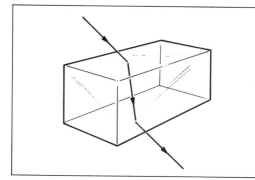

Picture 9 Refraction in a glass block

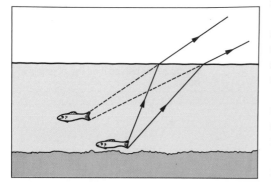

Picture 10 Why pools look shallower than they really are

Picture 11 A straight stick looks broken in water. Why?

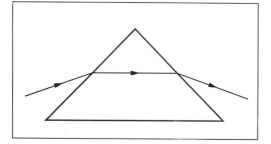

Picture 12 Light passing through a prism

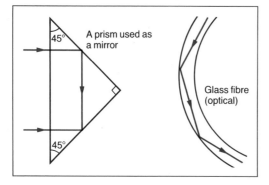

Picture 14 Total internal reflection in action

Using transparent materials

When light passes at an angle from one material into another it changes direction. This is called **refraction**, and is caused by the fact that light travels at different speeds in the different materials. This effect is used in lenses and optical fibres. Topic C7 deals with how these components are used in different kinds of practical devices, like cameras, microscopes and binoculars.

Refraction

Picture 9 shows what happens to a ray of light as it goes from air into glass, and from glass back into air. As it goes into the denser material, it changes direction to make a bigger angle with the surface. The opposite happens on the way out.

Refraction can cause some optical illusions. For example, a pond or swimming pool always looks a lot shallower than it really is. Picture 10 shows why. The light from a pebble on the bottom is refracted, and the light **appears** to come from somewhere else, nearer the surface. This is why a straight stick looks broken when you put it in water (picture 11).

A **prism** is a block of glass or plastic with straight sides. It is usually triangular in shape. Light that enters the prism at right angles to a surface carries on unchanged. But if it goes in at an angle it changes direction due to refraction (picture 12).

But prisms are not used to change the direction of light by refraction. The reason is that white light would come out coloured. Prisms produce a **spectrum** of the light. How they do this is explained in topic C8.

Trapped light — total internal reflection

Picture 13 shows light travelling out of water. See what happens as the angle the light makes with the surface is reduced.

At a certain angle the light doesn't get out at all. It is trapped inside, or **totally internally reflected**. The same thing can happen with glass.

This effect is used, in prisms, to make very good mirrors (see **binoculars**, topic C7). It is also the effect used to send light down long thin fibres of glass — **optical fibres**. Picture 14 shows total internal reflection in action in a prism and a glass fibre — see topic C7 for descriptions of how they are used in practice.

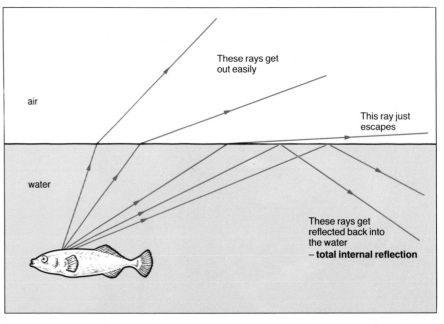

Picture 13 How total internal reflection happens

Focusing and making images

One of the most useful applications of refraction is in **lenses**. Lenses are curved pieces of very clear glass. When a straight beam of light reaches them the outer part of the beam is bent inwards by refraction at both surfaces of the lens (picture 15).

This happens because the light reaches the glass surfaces at an angle. The further out from the centre of the lens the bigger the angle is. So the outer part of the beam is bent inwards more than the inner parts. The beam comes to a point (converges) before spreading out again. This effect is called **focusing**.

The lens in picture 15 is a **positive** or **converging** lens. A lens shaped 'the other way', as in picture 16, makes the light beam spread out (diverge). It is a **negative** or **diverging** lens.

The more curved the lens surfaces are, the more powerful is the lens. Picture 17 shows this.

Lenses are used to make images in various kinds of optical instruments. How they do this is described in the next topic.

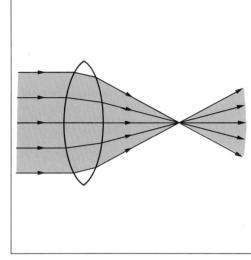

Picture 15 What a positive lens does to light

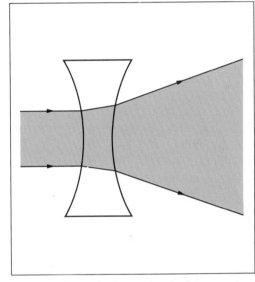

Picture 16 A negative lens makes the light spread out — or diverge

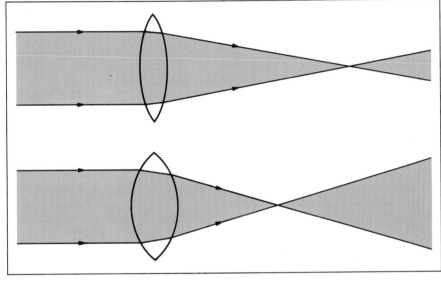

Picture 17 The more curved the lens the more powerful it is

Activities

A Using mirrors

Devise a way of getting a beam of light (from a torch or ray box) from one side of the room to the other, starting with the beam at floor level and finishing with the beam 2 metres above ground.

Rules: 1 You aren't allowed to move any furniture
2 The beam has to change direction 5 times.

B Where are mirrors used?

Find as many everyday applications of mirrors as you can. You should find at least five. List them all and state whether they use plane or curved mirrors. Describe as exactly as you can how one of them works.

C Making images

Use a small lamp and a positive lens. You will need a card screen — and maybe a white wall will come in handy. By trial and error you should be able to get a clear image of the lamp filament on the screen.

1 Investigate:
a where the lamp and lens has to be to get an image *smaller* than the lamp filament,
b how to get a *magnified* image,
c how to get a really *huge* image.
2 Take measurements to test the prediction that:

size of image × distance of lamp from lens = size of filament × distance of image screen from lens

3 Design and carry out experiments to find out if:

a the rule of reflection (page 78) applies to curved mirrors as well as plane ones,

b the image in a plane mirror is really as far behind the mirror as the object is in front of it.

D Reflecting safety

Get some **reflective paint** or some material that cyclists use on a reflective belt. How do these substances work? (Look at a sample under a microscope. Investigate what it does to light.)

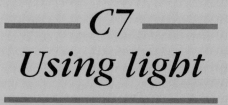

C7
Using light

This topic explains how light is used and controlled in some everyday devices.

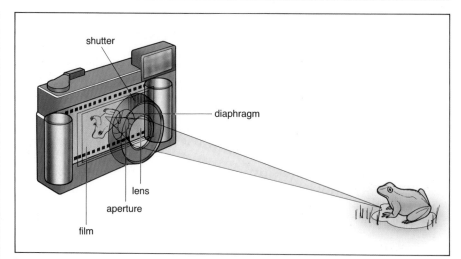

Picture 1   How a camera works

Cameras

Cameras use positive lenses to make a clear image on a light-sensitive screen. Picture 1 shows how this is done in a simple camera. A single lens of the kind that you might work with in school doesn't make clear enough pictures. In real cameras more complicated lenses have to be used.

In an ordinary camera the screen is a piece of thin plastic — the **film**. The light produces a chemical change in light-sensitive chemicals held by this film. Later, the film is **developed**, which means that another chemical is used to make the changed chemicals visible. They turn black (in a black-and-white film) or to a particular colour in a colour film.

In a **TV camera** the screen is made up of very small light-sensitive electronic devices (see picture 2). When light reaches these they produce an electrical signal. The camera is programmed so that each picture cell (pixel) is looked at in turn, or **scanned**. The signal from each pixel is sent down a cable one after the other to make a long coded message, which represents the picture on the screen. The screen is scanned completely 25 times a second. The coded signal is then recorded or broadcast directly.

Picture 2   The screen of a TV camera senses light

The eye

The eye is like a TV camera. The screen at the back (the **retina**) is made of light-sensitive cells. The 'cable' is the optical nerve which carries the coded message to the brain (see picture 3). The human eye is described more fully in *The Living World*, topic D4.

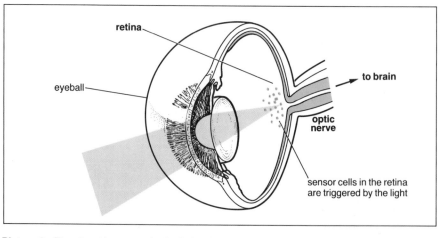

Picture 3   The signal is sent to the brain by the optic nerve

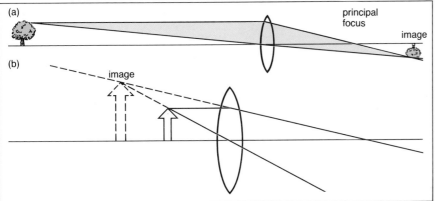

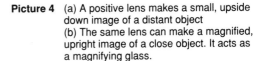

Picture 4 (a) A positive lens makes a small, upside down image of a distant object
(b) The same lens can make a magnified, upright image of a close object. It acts as a magnifying glass.

Seeing things bigger

When you look through a positive lens at an object some distance away it looks smaller, and the image is upside down. If you bring the lens closer and closer to the object you will reach a position where it makes an enlarged image of the object. This image is the right way up. The lens is now acting as a **magnifying glass**. Pictures 4(a) and 4(b) show what the light is doing to produce these two different kinds of image.

Two positive lenses can be used to make a simple **microscope**. The first lens, placed near the object being looked at, is very powerful. It is called the **objective** lens, and it makes an upside down image of the object. This image is then viewed through another lens — the **eyepiece** lens. The eyepiece lens acts as a magnifying glass to make this first image look even bigger. Picture 5 shows what happens.

Seeing further — telescopes and binoculars

A **telescope** works much like a microscope, but the eyepiece is now the most powerful lens. The first, weaker lens again makes an upside down image. The eyepiece lens is more powerful. Again, you use it as a magnifying glass to look at the image made by the first lens (the objective lens). Picture 6 shows how this is done.

The first lens makes an upside down image. The second one (the **eyepiece**) works just like a magnifying glass, and so the image stays upside down. This kind of telescope is used by astronomers to look at stars and

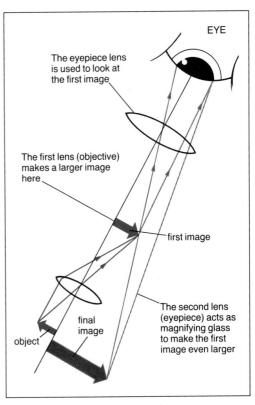

Picture 5 How a microscope works

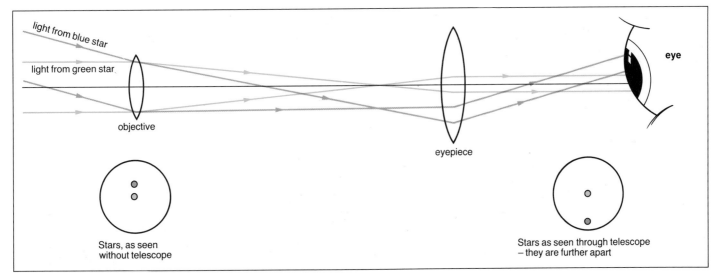

Picture 6 What a simple telescope does to light

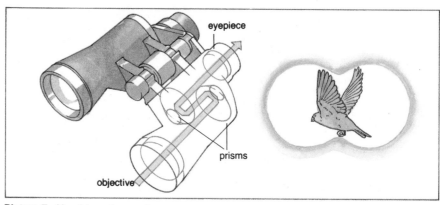

Picture 7 How binoculars work

planets. They don't seem to mind that they see the universe upside down. In most astronomy books the pictures of the moon are actually *printed* upside down, so that they don't get confused.

Sailors don't like seeing distant ships and landfalls upside down. They may prefer to use **binoculars**. A pair of binoculars is simply two telescopes fixed side-by-side, one for each eye. They are shorter than astronomical telescopes. They also make an image which is the right way up. Both of these effects are produced by using **prisms**.

Picture 7 shows a pair of binoculars. The two prisms make the light go up and down the tube three times. This means that the binoculars need only be one-third the length of a telescope. At the same time the mirror faces of the prism swop the light rays around so that the final image is the right way up.

Seeing inside

Thin glass fibres can be used to carry a beam of light so that the light cannot escape. This is due to **total internal reflection** (see topic C6). A bundle of very thin fibres is used in medical research and in hospitals to look deep inside the human body. Picture 8 shows how the fibres are arranged.

One set of fibres carry light down into the body. The others carry reflected light back — the image of the part of the body that is illuminated.

This useful device is called an **endoscope**. They can be made small enough to slip down a vein and bring pictures back from inside a living heart or other organ. Picture 9 shows the inside of the human gut.

Picture 8   Fibres in an endoscope

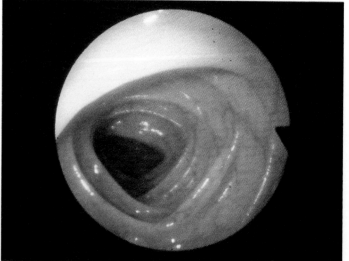

Picture 9   The human gut seen through an endoscope

Communications

As explained in topic C1, thin glass fibres carrying 'light' signals are much better for carrying information than almost any other medium. There is more about these optical fibres in topic F3.

Activities

A Using lenses

You need a thin, weak lens, a medium strength lens and a small, powerful lens. They should all be positive.

1 Which lens makes the best magnifying glass?

2 Picture 6 shows the principle of a telescope. Use two lenses to make one. Hold the lenses in your hands while you are experimenting. When you get it right it is better to use some kind of holder for the lenses.

3 Telescopes are for looking at distant objects.

Experiment with the lenses to see if you can find a pair of lenses that you can use together to magnify *near* things,

like the printing in this book. If you succeed you have made a **microscope**.

B Making a periscope

A periscope allows you to see around corners or over other people's heads in a crowd. A simple periscope uses two plane (flat) mirrors held in a tube. Design a periscope. Get your design checked by your teacher, then make it.

Questions

1 Copy the following diagrams (picture 10) and complete them to show what happens to the rays or beam of light, for:

(a) a strong lens, (b) a weak lens, (c) a glass block (d) a curved mirror.

2 A positive lens can be used to start a fire.

a Explain how it can do this.

b Why is it dangerous to leave empty glass bottles in a wood?

3 'When you look into a river or lake and see a fish, it isn't where you think it is'. Draw a diagram to justify this statement (with you standing on the river bank and ray of light coming to you from a fish in the water).

4 Explain what is meant by *total internal reflection*.

5 What is the difference between a *transparent* material and a *translucent* material?

6 Why are 'fat' lenses more powerful than 'thin' ones?

7 Use the idea that light travels in straight lines to explain how eclipses of the sun or of the moon occur.

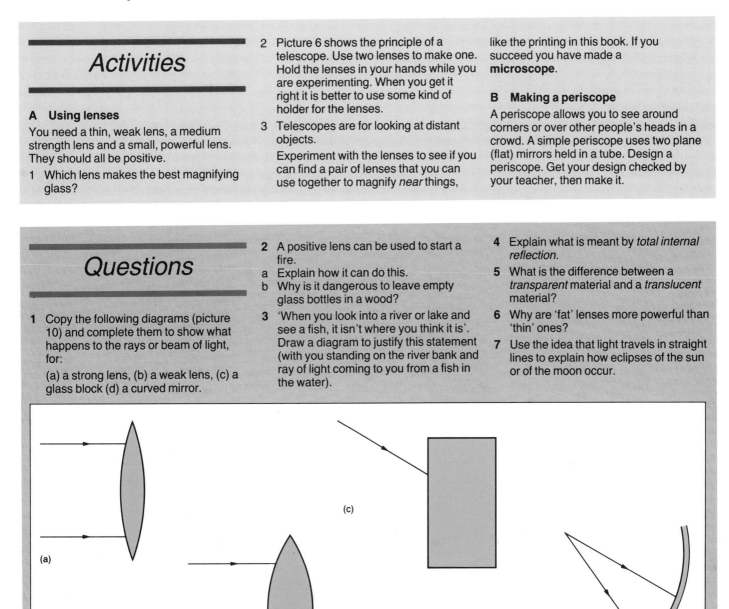

Picture 10

8 Name three things that use lenses.

9 Compare the human eye with a simple camera. In what ways are they similar, and in what ways are they different?

10 Give three uses for total internal reflection. Describe how one of them works.

11 Name an optical device that can be used to:

a make things appear smaller,

b carry messages long distances,

c turn light through 180 degrees,

d make an upside-down image on a screen.

12 A TV camera takes 25 pictures every second. When we see them, one after the other, we get the impression of movement. What do you think the difference would be if the pictures were taken just ten times a second?

C8
What is light?

We take light, and the fact that we can see it, for granted. But light has strange properties....

Light is just the *visible* part of a whole family of 'radiations' that we call the **electromagnetic spectrum**. These radiations all travel through space at the same speed, 300 million metres per second (3×10^8 m/s).

The speed of light

This was first measured by a tidy-minded Danish astronomer, Olaf Römer, as long ago as 1676. He noticed that the moons of Jupiter were sometimes a few minutes late in disappearing behind the planet. The moons of Jupiter orbit at a very constant rate, and it was easy to calculate when this disappearance should take place. Romer explained why the moons were late by saying that light took time to travel.

This might be obvious to us, but at that time many scientists believed that light took no time at all to go from one place to another. But Romer said that this was not so: light had a definite speed. He said that light took longer to get to the Earth from Jupiter when the two planets were further apart.

As picture 1 shows, at its furthest point from Earth the light from Jupiter has to cross an extra distance equal to the diameter of the Earth's orbit. He measured the time difference this extra distance caused. The size of the Earth's orbit around the Sun was known fairly accurately in 1676, and so Römer was able to calculate a value for the speed of light.

His measurements of the times were not very accurate, however. His measurements gave the speed of light as only two-thirds of the modern value. But it was a start. The speed of light is now very accurately measured, and is so reliable that we use it to measure distance. Accurate surveying is done by measuring the time it takes for laser beams to travel a particular distance. The times are then converted to distances:

$$\text{distance} = \text{light speed} \times \text{time of travel.}$$

All electromagnetic waves travel at the same speed in a vacuum. Short radio waves are used in **radar** systems to measure the distances of aircraft in air traffic control.

Radar has been used to measure the distances of planets from Earth. This has given us an accurate measurement of the scale of the Solar System (see topic I2). The distance of the Earth from the Sun is the base line used to measure the distances of stars. The distances of stars and galaxies are so great that we measure them in **light years** — the distance travelled by light in a year.

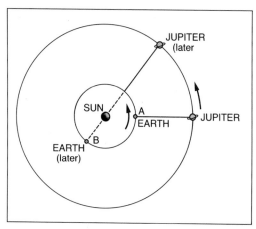

Picture 1 How Romer measured the speed of light

Picture 2 Modern laser surveying instrument in use

Making light

Light usually comes from very hot objects — flames, the Sun, hot filaments. But it can also come from insects (fireflies, glow worms), from the fluorescent paint in TV tubes and some lamps, and from the glowing gases in advertising lights ('neon lights'). But all these sources produce light in the same basic way — by giving energy to atoms to make them unstable.

Picture 3 reminds you what an atom is like — a positive nucleus with some electrons orbiting around it (see *The Material World*, topic J3) . If an atom is given the right amount of energy, the outermost electron jumps up to a slightly higher energy level. It stays there for a while and then falls back to its normal level. When it falls back it gives back the extra energy as light.

The atoms of different elements have their electrons in different levels. Thus the light they give out is different in colour. This means we can tell what kind of element it is from its spectrum (see *Signals from Space*, page 100).

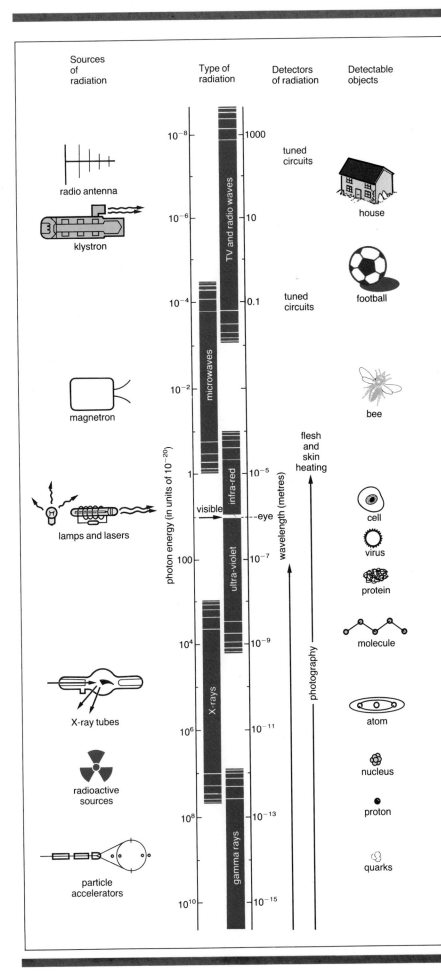

Picture 4 The main parts of the electromagnetic spectrum

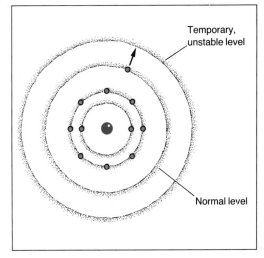

Picture 3 A simple model of the atom

The electromagnetic spectrum

Picture 4 shows the main parts of the electromagnetic spectrum. It also summarises how the radiations are made and how they are detected. In one way or another, all ways of making the radiations involve the movement of charged particles.

Radio waves are made by making electrons move rapidly up and down a wire — called the **aerial**. In other words, the aerial has an alternating current in it. The frequency of this alternating current is the same as the frequency of the broadcast. For a VHF (very high frequency) radio broadcast it is about 90 MHz (90 million waves a second).

The **microwaves** in a microwave oven are made by electrons which move round and round in a small metal box with no air in it (a magnetron). This device was first used in **radar** in World War II, to detect aircraft at a distance.

Infra red is produced by making whole atoms vibrate inside their molecules. This happens when materials are heated, so we associate infra red with hot objects.

As we go along the spectrum towards the **gamma ray** end, the radiations are more and more dangerous to life. This is because of the large packets of energy they carry (see page 94). The changes in the atom that produce these radiations are due to electrons involved in large energy changes, or changes inside the nucleus itself (see topic D6).

Electromagnetic waves

Electromagnetic radiations travel as **waves**. They are called electromagnetic because when electric charges move (i.e., there is an electric **current**) it always produces a magnetic field. This is covered in topic E8. An electron is a charged particle. Thus it has an electric field surrounding it.

When electrons move to and fro or go around in circles they produce changing electric and magnetic fields. This is what the **electromagnetic radiations** are — a set of constantly changing, combined electric and magnetic fields that travel through space at 300 million metres a second. This is shown in picture 5.

This also means that they don't need anything to 'carry' them. 'Normal' waves, like water waves or sound waves, need a **medium**. Water waves need water!. Sound needs air, or a liquid or solid. Sound cannot travel through a vacuum because there is nothing there 'to be waved'.

Electromagnetic waves carry their own 'waviness' with them, so they can travel through the vacuum of empty space

The range of electromagnetic waves is shown in picture 4.

The discovery of radio waves

The electromagnetic theory of light was first put forward by a great Scots physicist, James Clerk Maxwell, in 1873. It was a good theory because it not only explained what scientists knew about light, it also made **predictions**.

Picture 6 One of the earliest 'Marconi transmitters', used by the Royal Navy

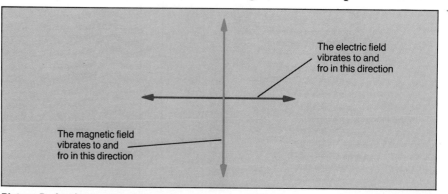

The electric field vibrates to and fro in this direction

The magnetic field vibrates to and fro in this direction

Picture 5 An electromagnetic wave has two moving fields

Maxwell predicted that there ought to be other radiations which travelled through space at the same speed as light. These waves were not discovered until 15 years later, in 1888. They were discovered by the German scientist Heinrich Hertz, who detected what we now call radio waves. Although they had been predicted, they were discovered by accident! Radio waves can be made by electric sparks. You may have noticed this when you play a radio close to a car engine.

Hertz was working with a machine used for making static electricity. He noticed that whenever the static machine made a spark, so did a small coil of wire on the other side of the room. Energy was travelling from one side of the room to the other as radio waves. We name the unit of frequency the hertz after Heinrich Hertz.

The whole of the radio and television industry is based on these discoveries. The first radio transmitters were 'spark transmitters' and they were in use, in ships at sea, by 1902 (see picture 6).

The next topic deals with the evidence for the wave nature of light. Topic C10 then tells about some astonishing discoveries that showed, once again, that the world of physics was not as simple as it seemed.

Activities

A   The uses of electromagnetic waves

The chart of the electromagnetic spectrum (picture 4) shows the main uses of the different parts of the electromagnetic spectrum. Use reference books to find **other** uses for any section of the spectrum that interests you.

B   Make a radio!

Picture 7 shows the components you need to make a simple radio. You can connect them together using crocodile clips. To make sure it works you need to have a good long aerial wire and a good connection to earth.

Use a book on radio to find out how the circuit works.

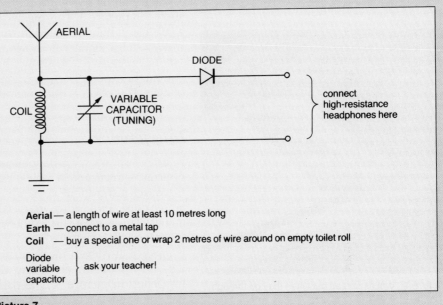

Aerial — a length of wire at least 10 metres long
Earth — connect to a metal tap
Coil — buy a special one or wrap 2 metres of wire around on empty toilet roll

Diode
variable } ask your teacher!
capacitor

Picture 7

Questions

1   The famous scientist Galileo tried to measure the speed of light using two people with lanterns, standing about a mile apart from each other. One person was supposed to send a light signal back when he saw the light sent to him by the first person. Galileo tried to measure the time this took. The experiment was a complete failure. Suggest one or two reasons why it failed.

2   Design an experiment to show that infra red (heating rays) travel at the same speed as visible light.

3   Which parts of the electromagnetic spectrum,
a   can be used for heating?
b   cause skin tanning?
c   can pass through flesh but are partly stopped by bones?
d   can cause cancer?
e   are stopped by the ozone layer in the atmosphere?
f   are used to find the positions of distant aircraft?

4   The speed of light is 300 000 000 metres per second (3×10^8 m/s). Use the wave formula **speed = frequency × wavelength** to calculate the following.
a   the wavelength of MW radio waves that are broadcast at 1500 kHz
b   the frequency of BBC Radio 2 which is broadcast on a 'long wave' of 1500 metres.
c   The wavelength of the waves in a microwave oven which works on a frequency of 2450 000 000 Hz (2.45×10^9 Hz)

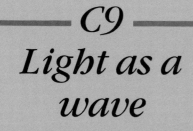

C9
Light as a wave

How do we know that light is a wave? How can we use its 'waviness'?

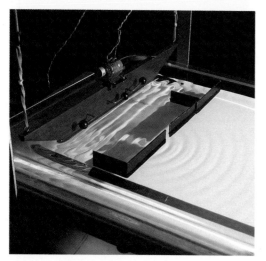

Picture 1 Waves spread out when they go through a gap

Waves

Topic C3 deals with some of the main properties of waves. Water waves and waves on ropes and springs are easy to see. We can show sound waves on an oscilloscope, and investigate how the **frequency** and **amplitude** of the waves change when we alter pitch and loudness.

Light waves are harder to investigate. Their wavelength is so small. But all waves behave in much the same way, in that they can show **diffraction** and **interference** effects. We use these effects to investigate light. They are explained below.

Diffraction

Picture 1 shows water waves moving through a gap in a barrier. When they reach the gap the waves spread out. This effect is called **diffraction**. Picture 2 shows what would happen if particles, like bullets, were fired at a barrier with a gap in it. The bullets that go through the gap carry on in a straight line.

When light is shone through a narrow gap it actually spreads out. This is shown in picture 3. This means that light is behaving like a wave, and not like a particle. If light travelled in straight lines, like a steam of particles, it would make a sharp shadow as shown in picture 2.

But as picture 3 also shows, light makes a more complicated pattern than you might expect. There are zones of light and dark outside the main spread of light. This effect is due to another property of waves — **interference**.

Interference

Picture 4 shows what happens when waves meet. Two pulses travelling in opposite directions on a rope can pass through each other. When they meet the waves just 'add up'. Two 'up' parts of a wave (the crests) add up to make a larger crest. Two 'down' parts (the troughs) add up to make an even bigger trough. This is what you would expect.

But it might be surprising to see that when a crest meets a trough the result is a 'zero'. It is even more surprising that this also happens with light. Two light waves can meet — and the result is darkness! But this is what has happened in picture 3. The dark places are where two sets of light waves have met and cancelled each other out.

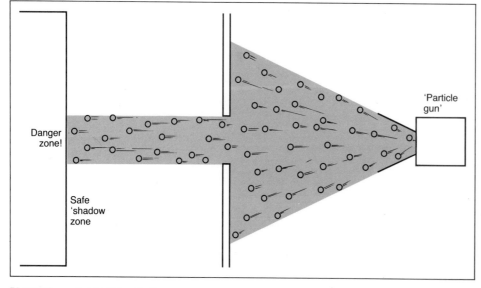

Picture 2 .....but particles don't

Picture 3 Light spreads out — like a wave

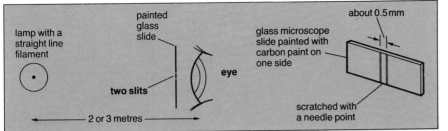

Picture 5   How to see 'Young's fringes'

We say that when two waves meet they **interfere**. They can add up to make a bigger wave — or cancel each other out (picture 4).

The Two-slit Experiment

This experiment was first done by a doctor, Thomas Young, in the early part of the 19th century. It was the first clear proof that light travelled as a wave. At that time scientists believed that light was made of particles, like tiny bullets. They believed this so strongly that Young's work was accused of being 'absurd and illogical'. It was completely ignored for twenty years.

Picture 5 shows how you could set it up to do it yourself. When you look through the two narrow slits on the painted glass slide you will see an **interference pattern** like the one shown in picture 6. Picture 7 shows how the waves from the two slits combine. At some places they meet so that crests always meet with crests. The following troughs always meet with troughs. The result is a light patch.

At other places, the crest of a wave from one slit always meets with a trough from the other slit. The result is darkness. The activities at the end of the topic are about investigating these effects with light and radio waves.

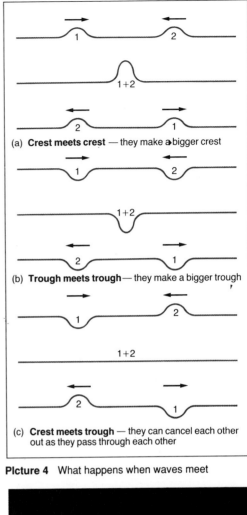

(a) **Crest meets crest** — they make a bigger crest

(b) **Trough meets trough** — they make a bigger trough

(c) **Crest meets trough** — they can cancel each other out as they pass through each other

Picture 4   What happens when waves meet

Picture 6   This is what we see in Young's experiment

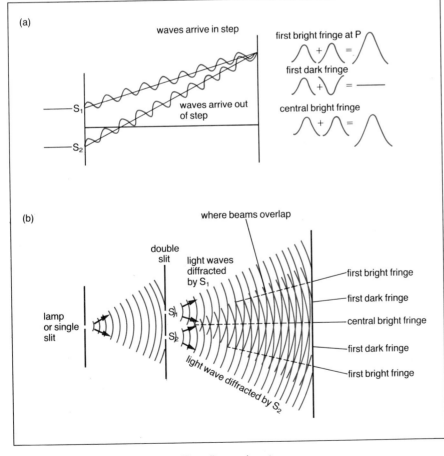

Picture 7   What happens in Young's Two-slit experiment

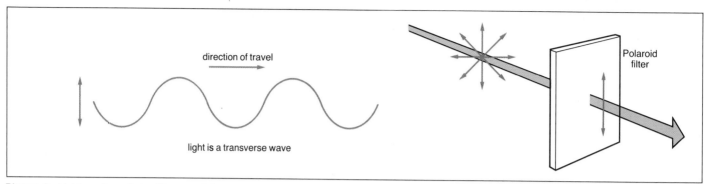

direction of travel

light is a transverse wave

Polaroid filter

Picture 8 Light can be polarised because it is a tranverse wave

Polarised light

Light is a **transverse** wave, which means that it vibrates at right angles to the direction in which it travels (see page 63). Normally, a ray of light is a mixture of waves, all vibrating at different angles to each other, as shown in picture 8. A filter made from a special substance called **polaroid** only lets through the waves which vibrate in one direction. The light that comes through is said to be **polarised**.

If you hold another piece of polaroid film in the path of the polarised light it will let it through — but only if the film is aligned the right way. If you turn it through 90 degrees it cuts out all the light. This effect is shown in picture 9.

Polaroid **sunglasses** work because light reflected from shiny surfaces, like water or glass, is partly polarised. The glasses are made of polaroid film. They cut out the polarised light. Thus they cut down 'glare' from shiny surfaces — which in summer is usually strong reflected sunlight (picture 9).

Picture 9 Polaroid filters can cut out glare. When they are 'crossed' they cut out all the light

Activities

A Looking through holes

These are experiments about the wave nature of light that you can do at home.

You will need a distant street lamp, or at least a bright torch placed about 50 metres away. You can also do this in the laboratory, by looking at a small hole lit by a bright lamp.

1 Face the spot of light and look at through the gap between two fingers (e.g. the first and second fingers of one hand). Gradually bring the fingers closer together to make an ever narrowing slit. Just before the light from the lamp disappears you should see it broaden out. This is **diffraction**.

2 Get a piece of kitchen foil about 10 cm long by 5 cm wide. The exact size doesn't matter. With a sharp pin or needle, make a small hole in the foil. When you look through the pinhole at the light source you will also see

diffraction, but this time there should be a more definite pattern to it. This is really worth seing.

The difficult bit about this experiment is finding the small pinhole in the dark. Mark it before you go out by putting a small piece of sticky paper next to it. You could go on to try the effect of making pinholes of different diameter.

3 Now make two pinholes in the foil, about a millimetre or so apart. You should be able to see through both holes together. When you look through

the pair of holes at the light source you will not only see the spreading out effect (diffraction) but also an **interference** pattern. This is caused by light from the two separate holes combining ('interfering') with each other.

B Radio waves

The wavelength of light is very small. This is why we don't notice the waviness of light very often. VHF radio and UHF TV have much longer waves — about 3 metres long for radio and about half of that for TV. You can show interference between radio waves using sheet of metal and a small radio with a telescopic aerial. The metal sheet could be a length of kitchen foil pinned to some cardboard. It needs to be a about a metre square for the radio experiment, about half that will do for the TV.

1 Tune the radio to a 'weak' VHF station, or make it just off-tune for a strong one. The aerial should be horizontal, as shown in picture 10. You can tell it is weak or mistuned by the fact that the sound gets distorted now and again, especially when the speech or music is loud.

2 Hold the sheet of metal upright about 2 metres from the aerial. Walk towards or away from the aerial, moving the metal sheet closer or further from it. As you do this, you will find places where the signal is made stronger, and places where it is made weaker.

For this experiment to work the radio or TV will have to be in line between you and the station broadcasting the signal. If you don't know where the station is you will have to find the best place by trial and error.

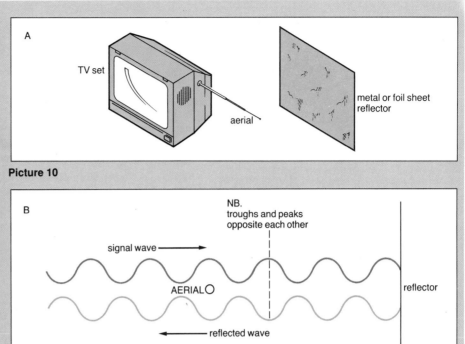

Picture 10

Picture 11

Picture 11 shows what is happening. The reflected wave is sometimes reinforcing the direct wave. In other positions the reflected wave cancels out the direct wave, making reception worse. The distance between successive 'cancel points' or 'reinforcement points' is half the wavelength of the radio broadcast.

C Water waves

You can experiment with water waves using a ripple tank. This is a shallow plastic tray with water in it. You can do similar experiments in a bath or washing-up bowl at home.

You can make circular pulses by touching the water surface with the end of a pencil. You can make straight waves by touching the surface of the water with a piece of round wooden rod held level.

Put objects of different shapes into the tank and see what effect they have on waves. Investigate reflection and diffraction. What happens if you have curved barriers? What happens when waves pass through gaps of different widths?

Your teacher may be able to give you extra apparatus to take these investigations further.

Questions

1a Explain the difference between *diffraction* and *interference*.

b Why don't we usually notice these effects with light in everyday life?

2 Design an experiment to show (a) diffraction (b) interference using sound waves.

3 Name — or describe very briefly — two scientific discoveries which were ignored or said to be wrong when they were first made.

4 Describe an experiment you have seen or done to show that (a) water waves spread out (diffract) when they pass through a narrow opening (b) when water waves from two sources pass through each other, the 'up' part of one wave can be cancelled by the 'down' part of the other wave.

5 Explain what *polarised* light is. Why do 'polaroid' sunglasses cut down glare on bright sunny days?

6 What evidence is there that light is a wave? Describe some experiments and show how they support this idea.

7 You are given a box which gives out some mysterious, unknown 'rays' when you press a switch on it. The rays make a certain kind of paint glow. How could you test if these rays were waves? (They might be particles)

C10
Light as a particle

Sometimes light behaves as a wave, but sometimes it's more like a particle

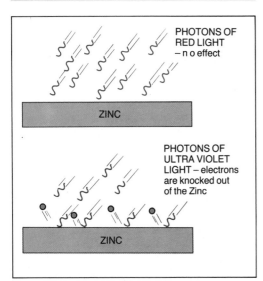

Picture 1 Light arrives in packets of different energies

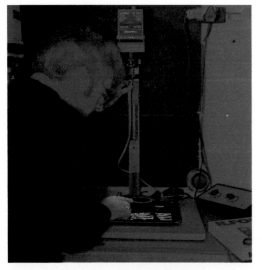

Picture 2 Pictures can be developed in red light but not in blue light. Why?

A beautiful theory spoiled by an awkward fact

Maxwell's theory (page 88) joined together light, electricity and magnetism. It explained why elecromagnetic radiations travelled through space as waves, with a speed of 300 million metres a second. As a scientific theory it was one of the great successes of 19th century science.

It was also of immense practical importance. It dealt not only with light, but with new discoveries like radio and X-rays. But we now know that it doesn't tell us everything about electromagnetic waves.

In 1905 the certainties of 19th century physics were shattered. Albert Einstein came up with some surprising additions to the theory of light. One was his **theory of relativity** (see topic B11). The other was the idea that light might **travel as a wave**, but when it met matter it seemed to behave like a **particle**!

Particles of light?

Isaac Newton believed that light was made up of particles of different colours — but he was wrong. As time went on, his theory didn't match the evidence. Experiments carried out in 1801 by Thomas Young proved without any doubt that light was a wave (see topic C9).

The photoelectric effect

But in 1905 Einstein asked physicists to think about a simple experiment that could be done with a zinc plate and some ultra-violet light.

When you shine UV light on the zinc, it gives off electrons. This is the principle of the **photocell**. Photocells are used to detect light and to read the signals coded on to sound films in cinemas.

The strange thing about the experiment with zinc is that it doesn't work when you shine *visible* light on to the zinc plate. However bright you make the light, no electrons are produced. But the very weakest trace of UV light will produce electrons, which shoot out of the metal at speed.
Einstein argued that this simple effect could only be explained by a particle theory of light, as follows.

Einstein's explanation
Light hits the metal as small particles, which are now called **photons**. Each photon carries some energy. If it carries enough energy, it can knock an electron out of the metal. It takes a definite amount of energy to do this, depending on the metal.

If the photon doesn't carry enough energy then the electron stays in the metal. Each photon of UV light **does** carry enough energy to knock an electron out of zinc. Photons of visible light don't carry enough energy — they are too feeble, and however many you throw at the metal they produce no effect. (See picture 1.)

It is rather like a coconut shy. To get a coconut out of its cup needs some energy. A lightweight ball travelling at a certain speed simply doesn't have enough energy. A faster ball, or a more massive ball travelling at the same speed would. Slow throwers could throw millions of light balls at the coconuts and not win one. One heavy ball travelling at the same speed would get a result — if it hit the coconut, of course. This is illustrated in picture 3.

But this analogy is slightly misleading. Wooden balls can travel at different speeds and have different masses. All photons travel at the same speed — the speed of light. They do not carry any mass. The difference between photons of different colour is simply how much **electromagnetic energy** they carry.

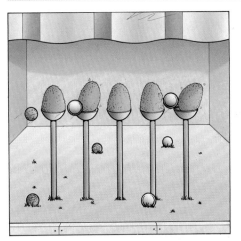

Picture 3 The coconut analogy

Einstein worked out that the energy, **E**, carried by a photon of light depended on its frequency, **f**.

photon energy = a constant × frequency
E = hf

The constant, **h**, is called the **Planck constant** after the German physicist Max Planck. You will notice that the particle (photon) theory of light has to use the wave theory as well. It needs to use the **frequency** of the wave.

The higher the frequency of a wave the shorter is its wavelength. Blue light has a shorter wavelength than red light, and each photon of blue light carries more energy than a photon of red light. Photons of short-wave ultra violet rays and X-rays carry even more energy. This is why they so dangerous to living material.

All this should not really be a surprise to you, living at the end of the 20th century. But it is hard to imagine something that is both a wave (spread out over space) and a particle (all in one small space). Yet both models of how light behaves are useful. You will find that science magazines and TV programmes use both theories of light, as they see fit. They use whichever one seems to give the best explanation for what they are talking about.

Waves or particles?

Photography, sun-tanning, the retina of the eye and the physics of the ozone layer (see *The Material World*) all need the particle theory to explain what is going on. In each of these cases, a little packet of energy is delivered to make something happen. If the packet isn't big enough, nothing happens. This is why you can develop a black-and-white photograph in red light. The photons of red light do not carry enough energy to affect the light-sensitive chemicals in the photographic paper (see picture 2).

Radio broadcasts, aerials and the action of lenses all need the wave theory to explain how they work. You will learn more about this if you take more advanced courses in physics.

Both theories are 'right', because they are backed up by experiments. They also make good predictions about new effects. Scientists have stopped being worried by the fact that they seem to contradict each other.

The wave-particle duality

Just to make things fair, we now know that 'real' particles, like electrons and alpha particles, can also behave like waves. Electrons can diffract and interfere, just like light waves. The **electron microscope** uses electron waves to see very small objects, just as ordinary microscopes use light waves.

Electron waves are not electromagnetic waves. They do not travel at the speed of light, for example. They are **matter waves**.

C11 Light and colour

The world is full of colour. What makes colours? How can we see them?

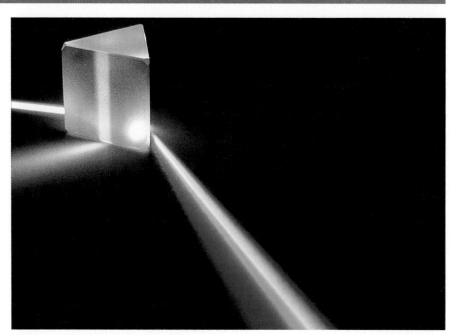

Picture 1 The spectrum of white light

Splitting up light

White light is a mixture of colours. When we pass white light through a prism the colours become separated out into a **spectrum** (see topics C8 and C10). Picture 1 shows a prism splitting up white light. The same effect can be produced by tiny drops of water, and causes the **rainbow**.

Keen-eyed people say thay can detect seven colours in this spectrum. In order, they are red, orange, yellow, green, blue, indigo and violet. You can remember this by the sentence 'Richard Of York Gave Battle In Vain'. Light from the sun, or any white hot object, also includes invisible radiations, like ultra violet and infra red. See topic C8.

How prisms separate colours

When light goes from air into glass it slows down. This is why it changes direction — it is **refracted** (see topic C6). All electromagnetic waves travel at the same speed in a vacuum. They slow down when they enter a transparent medium, like glass. This change of speed causes refraction.

But the different colours of the spectrum travel at **different** speeds in glass. For example, violet light travels more slowly than red light. This means that it is refracted more, so that its direction is changed more than red light. The other colours fit in between, depending on their speeds in glass. Light of different colours is thus spread out into the spectrum. This spreading out is called **dispersion** (picture 2).

Putting white light back together again

If you collected all the seven colours of the spectrum and joined them together again you would once more see white light. You could try this, using one prism to separate the colours and another to put them back again.

Primary colours

This effect may not be a surprise to you, but what might be surprising is that you can recreate white light by using just three colours — **red, green** and **blue**. Picture 3 shows three beams of light shining onto a white screen.

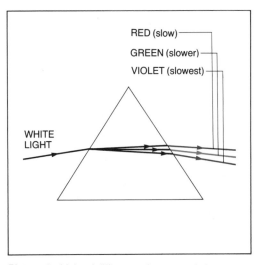

Picture 2 Light of different colours travels in different paths

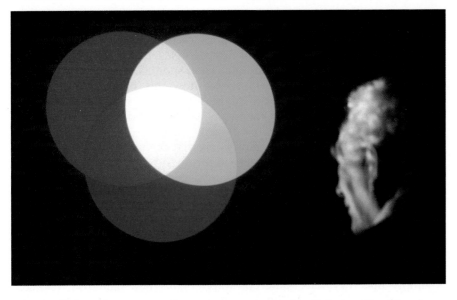

Where all three beams overlap you can see that the *light is white*. The three colours that do this are called the **primary colours**.

At other places, just two colours overlap. They combine to form other colours:

- Red and green combine to make **yellow**,
- Red and blue combine to make **magenta**,
- Green and blue combine to make **cyan**.

These 'double' colours are called the **secondary colours**.
You get these effects when the three primary colours are balanced so that they are equally effective. By using different brightnesses you can create all the colours of the rainbow, and many other colours that go to make up the everyday world.

Colour television uses this property of primary colours. In the TV screen there are just three kinds of paints which glow when electrons hit them (**phosphors**). The phosphors glow red, green or blue. The screen contains thousands of small 'picture elements' (pixels) each containing a set of the three phosphors. By making these glow at different brightnesses we can get the whole range of colours that we see on a TV screen. Colour films and colour printing work in much the same way, using just three basic colours. (See picture 3.)

Why is a red book red?

Paints and dyes work by **reflecting** light. When white light shines on a red surface, the dye absorbs all colours except red, which it reflects. If you shine blue light, for example, on the surface it looks dark, because there is no red light present to be reflected. The same happens when you shine green light on to the red surface.

But if you put the red surface in **yellow** light it will look red. This is because yellow light is a mixture of green and red light, as explained above. The red surface absorbs the green light but can reflect the red. This is shown in picture 4.

Activity B is about experimenting with different coloured light and different coloured surfaces. The fact is that the colour of an object depends not only on its own 'colour' but also on the colour of the light you view it with. The 'white' of artificial light is different from the 'pure' white of sunlight. They contain a slightly different mix of colours. This is why clothes bought in a shop under artificial lighting may look quite different when seen out of doors in sunlight.

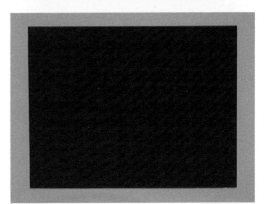

Picture 4 A red object can't be seen in blue light. Why?

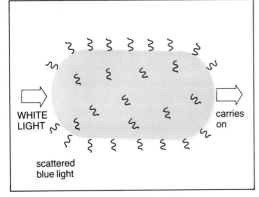

Picture 5 Atmospheric scattering

Why is the sky blue?

The upper atmosphere contains millions of tiny particles, ranging from molecules of various gases to small dust particles, ice crystals, etc. Many of these are good at absorbing blue light. Then soon after absorbing the blue light they reradiate it. Picture 5 shows how this happens; it is called **scattering**. Thus the blue light from the sunlight passing high above us in the atmosphere is first trapped, then some of it is sent down to Earth. So the sky looks blue.

This effect also explains why sunsets are red or orange. Red light is not absorbed in this way and so carries on through the atmospheric particles. When we look at the setting sun we see this reddish light. Most of the blue light has been absorbed and then sent out sideways (see picture 6)

Colour, energy and wavelength

Light travels as a wave. The difference between red light and, say, blue light is simply that they have different wavelengths. Red light has a longer wavelength than blue light. The diagram of the electromagnetic spectrum in picture 4 of topic C8 shows this (page 87)

The eye and brain work together when we see light. The effect of light of one wavelength makes us see 'red', other wavelengths trigger the sensation of green or blue.

One theory of vision says that our eyes work like the TV screens described above. Some cells in the retina at the back of the eye detect the red light. Others detect only blue and a third kind detect green light.

When all three types of cell are triggered we see the light as white. Combinations of cells create different colours in the brain in much the same way as combinations of red, green and blue in the pixels of a TV screen create different colours.

Picture 7 shows some of these cells in a human retina. See *The Living World*, topic D4 for more on how the eye works.

The triggering effect of light on cells in the eye is caused by the energy carried by the light. The effect can only be explained by using the model of light that says it is made of particles called **photons** (see topic C10). The blue photons carry more energy than the green ones, and the red photons carry the least energy.

Picture 7 The retina has cells which can detect colour, (cones) but most of the (rod-shaped) cells can't

Picture 6 Sunsets are red and yellow because the blue light has been scattered

Infra red photons don't carry enough energy to trigger any of the cells. Ultra violet photons are invisible because they carry too much energy, and actually damage the eye cells. Our eyelids screen this dangerous radiation coming from above. But if it comes up into the eye by reflection from water or snow we can be harmed. Skiers, Eskimos and polar explorers guard against this 'snow blindness' by using goggles whenever the sun shines brightly.

Activities

A Producing a spectrum

1 Let nature do it for you! Keep an eye out for **rainbows**. Where must the sun be, compared to the raindrops, for a rainbow to be made?

2 Make your own rainbow. You need a sunny day and a garden hose with a fine spray. Wear a bathing costume or raincoat, depending on the air temperature. Where do you have to stand to see the rainbow effect?

3 In the laboratory you can set up a raybox to produce a beam of white light. Pass it through a prism and view the spectrum on a white sheet of card. Picture 8 shows how to do this.

4 Repeat method 3 using a **diffraction grating** instead of a prism. A diffraction grating uses the interference effect of light. You can see similar effects in a compact disc, and the colours of butterfly wings are made using the same principle.

B Mixing light

You will need three ray boxes (or the equivalent), a white screen and some coloured filters: red, green, blue, yellow, cyan and magenta. You should have some simple way of holding the filters in the paths of the beams. The best results are obtained if the room is darkened.

Here are just a few of the investigations you can do. You should be able to think of some more.

Picture 8

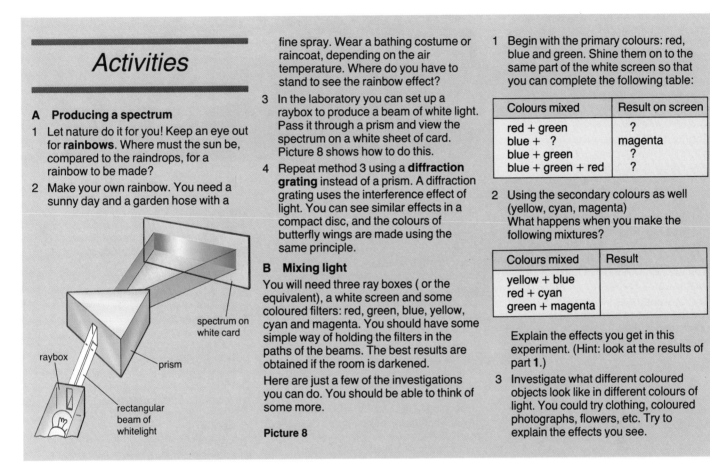

raybox

prism

spectrum on white card

rectangular beam of whitelight

1 Begin with the primary colours: red, blue and green. Shine them on to the same part of the white screen so that you can complete the following table:

Colours mixed	Result on screen
red + green	?
blue + ?	magenta
blue + green	?
blue + green + red	?

2 Using the secondary colours as well (yellow, cyan, magenta) What happens when you make the following mixtures?

Colours mixed	Result
yellow + blue	
red + cyan	
green + magenta	

Explain the effects you get in this experiment. (Hint: look at the results of part **1**.)

3 Investigate what different coloured objects look like in different colours of light. You could try clothing, coloured photographs, flowers, etc. Try to explain the effects you see.

Questions

1 Broxton United wear white shirts with blue shorts. Wexley Wanderers wear yellow shirts with black shorts. When they turn out for a floodlit match in the park the referee sends one team back to change their strip. Explain why the referee had to do this. (Hint: There was nothing wrong with the ref's eyesight. They were playing under yellow sodium lamps.

2 It is said that bees can see ultra violet light. Design an experiment to test this statement.

3a Suggest a reason why ordinary electric filament lamps produce a 'white' light that is different from sunlight.

b Explain why the difference between sunlight and lamp light should affect the colour of clothing.

4 Explain clearly the difference between the **refraction** and the **dispersion** of light by a prism.

5 Fluorescent light tubes bought in England usually give a slightly yellow light. But fluorescent tubes bought in Saudi Arabia give a slightly bluish light. Suggest why they are made with these differences.

6 Yellow is sometimes called 'minus blue'. Explain why.

7 Why do both the human eye and a colour TV camera have just **three** kinds of colour-sensitive cell?

8 A red rose and a yellow rose are passed through the spectrum from a bright lamp. What colour would each appear to be in:

a the blue light,

b the green light,

c the yellow light,

d the red light?

Signals from space

The only way we can learn about the universe is from the signals it sends us. For millions of years the only signals from space that humans could detect were carried by light. The ancient astronomers of Egypt, Babylon and Greece observed the Sun, the Moon and the stars. They saw how they changed and plotted their movements through the heavens. They produced the first theories about what the universe was like, as shown in picture 1.

The Ancient Romans weren't very interested in astronomy, and when Italy and Western Europe were overrun by 'barbarians' from the steppes of Asia in around 400 AD the old knowledge of astronomy was almost completely lost. But the study of astronomy was carried on by the Arab Muslims who conquered the Middle East and parts of Europe in the years 700 to 1500 AD. If you look at a good sky map you will find that many of the star names are in Arabic.

The telescope was invented in 1610 and over the next two centuries telescopes got bigger and better. Fainter and more distant stars could be seen. Knowledge about the universe increased, but these instruments still used ordinary, visible light.

A breakthrough was to come with the marriage of two old ideas — the **spectrum** of light and the **telescope** — with a new technique — **photography**. But like many scientific discoveries, it was a long time before what had been discovered made any sense.

Newton had explained the 'colours of the rainbow' back in 1666, and had investigated the spectrum of white light (see topic C8). Then in 1802 an English scientist, William Wollaston, noticed that the coloured spectrum of sunlight was crossed with a number of dark lines. He did not know what caused them, and it was 40 years before the mystery was solved.

High technology 1857: the Bunsen Burner

Robert Bunsen invented the bunsen burner to investigate spectra. He looked at the coloured light given out by elements heated in his clear, colourless gas flame. He discovered that each element had its own spectrum, different from all the others. When heated, it gave out light energy in definite wavelengths and in its own pattern. It could be used as a 'fingerprint', to detect the very tiniest traces of any element (see picture 2).

Picture 1 The Egyptian universe

His fellow worker Gustav Kirchhoff made the key connection. The dark lines in the spectrum of sunlight were caused by elements in the Sun that **absorbed** light energy at their own special wavelengths. Immediately, astronomers were able to work out what the Sun was made of! In fact, one *new* element was discovered, up until that time unknown on Earth. It was named **helium**, after the Greek Sun-god Helios.

Photography was discovered in about 1800, and the solar spectrum was first photographed in 1842. Since then, millions of photographs of the spectra of stars, planets, galaxies and comets have been taken. Photographs are needed because some of these objects are very faint. A photograph can collect light for many hours, and so build up its image until it is clear enough to be developed and measured. The details of the spectrum can tell us what elements there are in the star, its temperature, and even whether it is moving or not.

Then, in 1931, a new radiation was observed coming from outer space: **radio waves**. This led to the development of **radio astronomy**, and the discovery of radio stars and galaxies, pulsars and quasars (see topic I4 and picture 3).

Since then, astronomers have been able to use nearly every part of the whole electromagnetic spectrum, from gamma rays at one end to long radio waves at the other (see topic C8). They have found that the **pattern** of the spectrum of a given element is the same all over the Universe. But they also noticed that the actual wavelengths were sometimes different. This effect was caused by the movement of the star.

The American Edwin Hubble noticed that this difference was greater, the further away the stars or galaxies were. This could only be explained by the theory that these objects were moving away from us. This led to the theory of the expanding universe — and of the Big Bang that started it. The new 'space telescope' is

Picture 2 The helium spectrum

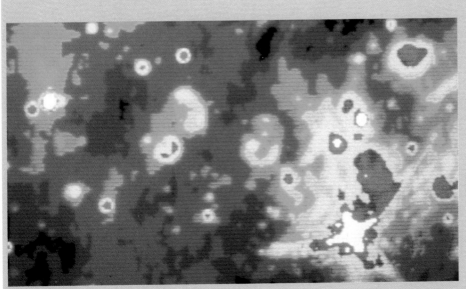

Picture 3   This image was produced using a radio telescope

named after this great astronomer. Answer the following questions:

1   Suggest why so much early astronomy was done by people living in the desert areas of the Middle East (compared, for example, with people living in Britain).

2   Why are radio telescopes so much bigger (maybe 30 metres or more across) than light telescopes (up to 2 or 3 metres across)?

3   Suggest a reason why it took nearly 200 years for Newton's discovery of the spectrum to be useful in astronomy.

4   Why are light, radio waves and X-rays all thought to be part of the same type of radiation? (Check with topic C8 if you are not sure.)

5   Radio waves don't affect photographic plates. How are these radiations detected and recorded?

Seeing with sound

Modern science uses all kinds of radiation to 'look' at things which are invisible, or to see through materials that light cannot penetrate. X-rays can see through flesh to spot broken bones or faulty hearts or lungs. Even the tiny magnetic fields of hydrogen atoms in our bodies can be used to give us a picture of what is going on deep inside our bodies.

One of the most useful body probes is the *ultrasound scanner.* This produces pictures like the one shown in picture 1, which shows a scan of an unborn baby.

Ordinary sound, which we can hear, has frequencies in the range of 20 Hz to 20 000 Hz. Ultrasound has frequencies

well beyond the upper limit that we can hear. It uses sounds at frequencies between 1 and 15 **million** hertz. This is higher than any animal ear can sense.

At this high frequency, the sound can travel through most materials. But some of it is always reflected back when it goes from one type of material into another. How much of it is sent back depends on the material in the way. In fact it behaves very much like light does in going through materials of slightly different transparency.

Also, the speed at which the ultrasound travels in the material also depends on the material it is travelling in. This means that it can be focused. This is done in the same way as glass lenses do for light by using an ultrasound transparent material with curved surfaces.

The advantage of using sound is that it doesn't harm the living cells, as X-rays may do. But if it is to see fine detail, the sound waves must be very small. They have to be slightly smaller than the small parts (e.g. blood vessels) of the object being looked at (picture 2).

This is another reason for using such very high frequencies. The higher the frequency, the smaller the wavelength. This is because of the wave formula:

$$\text{wavelength} = \frac{\text{speed}}{\text{frequency}}$$

The speed of sound in the human body is about the same as it is in salt water — about 1500 metres per second. If we want to see detail to about 1 mm, the wavelength has to be no more than this length. This means a frequency of 1.5

million herz (1 MHz). In hospitals, ultrasound scanners use frequencies between 1 MHz and 15 MHz. Answer the following questions.

1   Why can't we hear ultrasound?

2   Doctors prefer to use ultrasound for looking at babies in the mother's womb, even if the images produced aren't quite as clear as they could get using X-rays. Why is ultrasound preferred to X-rays?

3   Use the formula given above to calculate the size of the smallest object you could 'see' using ultrasound at 15 MHz

4   Bats find their way around at night, and see their insect prey, using ultrasound at about 50 kHz (50 000 Hz). Suggest why they don't need to use frequencies a lot higher than this, as in ultrasound scanners.

5   Draw a diagram showing what an ultrasound scanner might 'see' if it looked at an orange.

6   Another use for ultrasound is for cleaning things. When ultrasound is beamed at dirty fabrics the particles of dirt fall off the fibres. Suggest why: (a) the particles fall off, (b) this method is used for cleaning very old or very expensive materials.

Picture 2   High frequency sound waves

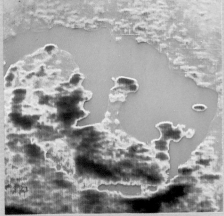

Picture 1   Ultra sound scan of baby in the womb

D1
Where does energy come from?

All life on Earth needs energy; we can control it, but cannot create it. Are we using it wisely? Is the world going to run out of energy?

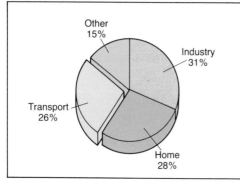

Picture 1   What do we use energy for?

Picture 2   Nuclear fusion changes mass to energy

We use energy to make things work. We use it to cook with, and the food we eat is our personal supply of energy. Without this energy, we would die in a matter of weeks. Long before then, we would begin to feel ill and very weak. *The Living World*, (topics C1–C4) deals in detail with how our bodies make use of energy.

This topic is about where the energy we use actually comes from. We usually take it for granted. Just press the switch and electrical machines start to work, or the heating comes on. We stop at a garage and fill up with petrol; someone comes and delivers oil, coal or coke to our homes. The gas supply is always there when we want it.

The pie chart (picture 1) shows where we use energy. Most of it is fairly evenly shared out between home, industry and moving things about. A smaller amount is used for what are called 'services', which means things like schools, hospitals, town halls, shops, etc.

How much energy do you use?

One way or another every person in this country uses, each year, the energy that could be got from burning 3.5 tonnes of oil. North Americans use more than twice as much, and the average for the world is 1.5 tonnes per person. Topic D2 is about measuring energy, and it shows you how to measure the amount of energy you use, per day.

Changing energy to suit our needs

Of course, the energy you use is not all supplied from oil. One of the great things about energy is that it can be changed into various 'forms', and can be moved about in so many different ways. What is really happening is that the energy is being transferred from one 'system' to another.

A system can be something quite simple, like a spinning wheel. It can also be something very complicated, like a human cell or a power station. Let's take a look at how the energy you get from an electric fire actually got there. It took a longer time than you might think!

Moving along the energy trail

Think of sitting in front of an electric fire. The energy has moved along quite a long trail before it reached you.

The start of the trail — the Sun

The Sun is millions of kilometres away, and the energy from the fire that warms you started just there, millions of years ago. Nuclear fusion (see topic D7) makes the sun very hot (picture 2). This makes the sun send out energy as various kinds of **radiation** (see topic C8). A very small fraction of this gets to the earth and keeps the earth warm.

Some of the radiation energy, visible light, is used by plants to make the chemicals they need to grow. *The Living World*, topic C13 describes in detail how this process, called **photosynthesis**, actually works. In photosynthesis, the simple molecules of carbon dioxide and water are turned into new, more complicated molecules of sugar and starches. These are **carbohydrates**.

Forming fossil fuels

Over millions of years, the carbohydrates in ancient plants have been converted to fossil fuels: oil, coal and natural gas. *The Material World*, topic I3 tells you more about the special conditions which formed fossil fuels when Britain had a tropical climate.

Fossil fuels contain many different chemical compounds, but the most important ones are **hydrocarbons**. There is more about these compounds in *The Material World*, topics H1 and I3.

Energy from fossil fuels

Fossil fuels are now our major source of energy.

The fuel-oxygen system

The hydrocarbon chemicals in fossil fuels can only release their energy when they change into other chemicals. The most common way is by combustion (burning). They combine with oxygen and heat up their surroundings, producing the waste gases steam and carbon dioxide. The energy cycle has brought us back to where it started all those millions of years ago, as shown in picture 3.

From coal to electric fire

Most of our electricity is generated in power stations that burn coal. (See topic E10). As it flows through the resistance wire in the bars of our electric fire, it delivers energy and the wire glows red hot. It is not as hot as the sun, but it gives out energy in the same way, as radiation. Our skins absorb this radiation, warming up as they do so. It has been a long trail from the sun to the electric fire, and picture 4 sums up the changes.

The problem with fossil fuels

The main problem with fossil fuels is that we are using them up far more quickly than they were made. For example, by the year 2040 there will be no more oil left. Britain's North Sea oil wells will be pumped dry by the time you are settling down to raise a family, but coal will last longer because there is much more of it. *The Material World* topic I3 tells you more about how long we expect the reserves of fossil fuels to last. They could last longer, or be used up sooner. This depends how sensibly we use them.

A great deal of the energy we obtain is wasted because we don't manage it sensibly — see topic D4.

We might also begin to make use of alternative sources of energy. These could be the renewable sources such as wind, sun, tides and waves. We can make use of the energy of 'biomass' — from plants and from animal wastes. These sources of energy are dealt with in topic E10. We might also make more use of nuclear energy — see topic D7.

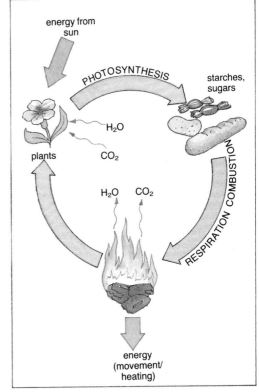

Picture 3 Water and carbon dioxide are recycled in the Earth–Sun energy system

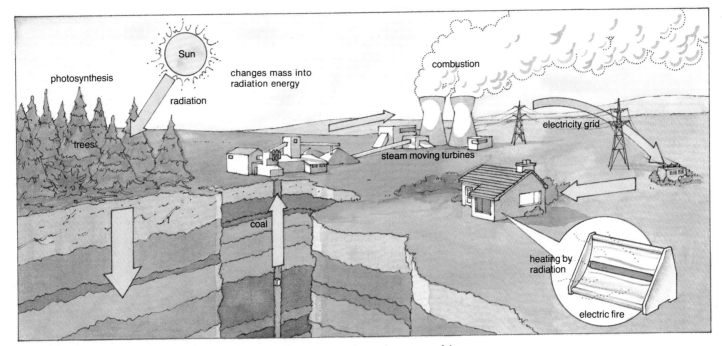

Picture 4 The Sun is our main energy source — but it might take a long time to become useful

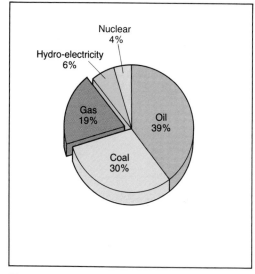

Picture 5 The world's main energy sources

Energy supplies

Most of the energy that we use on earth comes from fossil fuels — at present this is almost 90% of the total. The pie chart in picture 5 shows the main sources of energy that the people of the world use. But it does not show all of it. It shows only the 'artificial' energy that we buy in the form of fuels and electricity.

The chart does not show the energy that we get 'free', for example, the energy comimg from the sun.

All the crops on earth rely on this solar energy to keep them warm enough to grow, and to give the light they need to photosynthesise. It warms up sea and land, and keeps people warm in summer, so that they don't need to buy so much fuel. We should remember that for most of human history, energy from the sun was the only energy we were able to use.

Neither does the chart show the energy used by millions of people all over the world that they collect for themselves. It might surprise you to learn that most people on earth rely on **wood** for their heating and cooking. These people live in Africa, and much of Asia. Picture 6 shows the simple type of wood-burning stove that is widely used. The advantage of wood is that it is a **renewable** energy source — provided people keep on planting trees to replace the ones they cut down for fuel.

What is energy?

We can usually recognise energy when we come across it, but it is hard to say what it is. The simplest way to describe it is to say what it can do. The main things that it can do are:

- **work**, by means of applying forces to move things,
- **heat**, making things get hotter, melt, boil or evaporate.

Energy is not a 'stuff', like air or atoms, or even like electric charge. It is to do with the way things are arranged.

When an elastic band is stretched, the molecules are rearranged. The forces between them can pull the rubber back to its original shape. When you stretch the band you have done some work — by applying a force and moving it. When the band goes back it can apply a force — perhaps to a stone in a catapult. It does work on the stone. The stone moves and can hit a window, and apply a force to break it. Each time the force is applied to move something it does work — and energy is transferred to another system.

Picture 6 Most people on Earth use wood as their main fuel

Potential energy

A stretched elastic band looks innocent enough. If you had never seen or used one you wouldn't guess what might happen if it was let go. But it is able to give energy — dangerously — to a stone (picture 7). We use a special term to describe the energy 'hidden' in the stretched elastic band. We call it **potential energy.**

A stone at the top of a cliff can fall and do some damage. Water in a dam high in the mountains can flow downhill and make a hydroelectric power station produce electricity. These are another two examples of potential energy. This time it is due to the way 'things are arranged' in a gravity field. We call it **gravitational potential energy**. Gravitational energy is dealt with further in topic B5.

Kinetic energy

It takes **work** to make something move. Things will only start to move if a force acts on them, like gravity, or the force of a stretched piece of rubber. Work is a way of transferring energy, and the moving object picks up the

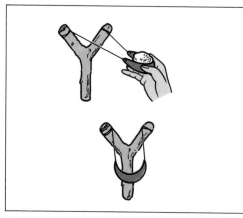

Picture 7 The stretched elastic stores potential energy, then gives it to the stone as kinetic energy

energy from whatever system provides the force. The energy it carries can be used to do more work. Because we make so much use of moving objects we have a special name for the energy they carry. It is called **kinetic energy**.

Are we using too much energy?

Fifty years ago most people travelled long distances by bus or train. Now we tend to go by car. The, people washed clothes by hand, instead of using a washing machine. Men shaved with a sharp piece of metal that they worked themselves, instead of using an electric razor. Not many people had central heating. They just wore more clothes in winter!

We now expect to find fruit and vegetables in the shops at all times of the year, not just in their 'natural season'. This is because we can freeze them and keep them for years, if necessary.

All this makes life a lot more comfortable, but it does take a lot of energy. There was a time when people were very worried that the world was 'running out of energy'. There was an 'energy crisis'. The price of energy went up. Oil, coal and electricity cost more. This made people try to use less energy.

In the 1990s energy has become slightly cheaper. In the UK we have been able to use gas and oil from the North Sea. People have become more careless in their energy use. Now, the problem that is worrying people is the *polluting effect* of energy use. Burning fuels produce carbon dioxide, which increases the greenhouse effect (see *The Material World*, topic I4). They often give off gases which make rain more acid.

Do we need to use as much energy as we do? Do North Americans need to use twice as much energy as Britons? What changes would there be in your life if you had to cut your personal energy use by half? And what will happen when everybody in the world is able to use energy at the rate that Western countries do?

Of course, fossil fuels will not last for ever. British oil and gas will run out by early next century. Would it be wise to make it last longer by using it more carefully?

Topic D4 tells you more about energy, and in particular about the laws of energy that allow us to control it. Knowing more about what energy is should help us use it more wisely.

Activities

A   Where does your energy come from?
Make a list of the ways in which energy gets into your home. Find out from your parents which source of energy costs you the most per month. It is likely that the one that costs you the most is the kind that you use most. Try making a pie chart of your results.

B   The geography of energy
Use a geography book or atlas to find out where the main sources of energy are in the United Kingdom. Explain why some parts of the country are (a) sources of coal, (b) good for making hydroelectricity, (c) used as the sites for nuclear power stations.

Questions

1 Where does the energy we use come from? Name three of the main energy sources used in this country.
2 Why are coal and oil called 'fossil fuels'? Name one other fossil fuel.
3 Name three devices (or things) that can store energy so that it can be used later. Describe how one of them works.

4a What is the energy source most commonly used in transportation (moving people and goods about)?
b Suggest two ways in which the UK could cut down on the amount of energy it uses for transportation. For each way, write a sentence or two explaining whether or not you think people would readily accept your suggestion.
5 Think about the energy involved when you ride a bicycle up a hill. Write down an 'energy trail' (as in picture 4) showing the various systems the energy goes through. Start with the energy in the food you ate before you rode the bike.

Make a guess as to where all the original energy has gone by the time you are standing on the top of the hill.
6 Think through a typical day, from getting up in the morning to going to bed at night. Make a list of as many as possible of the energy-using things you have used during the day, and say what energy source it tapped. One kind is given for you to start with.

Device	Source
1 bedside lamp	electrical power station

D2
Measuring energy

Wherever energy comes from we can measure it in the same unit — the joule.

Picture 1 Using energy

NUTRITION INFORMATION

Typical composition by weight

	per 100g	per 45g serving
Energy	1410kJ 330kcal	635kJ 150kcal
Protein	9.8g	4.4g
Carbohydrate	72.8g	32.8g
Fat	2.1g	1.0g
Dietary Fibre	10.5g	4.7g

Picture 2 Breakfast energy!

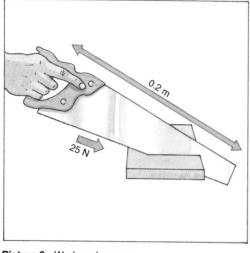

Picture 3 Work and energy

Look at the packet

Energy is measured in **joules.** Picture 2 shows the label on a packet of cereal. Amongst other things it tells you that when you eat one serving (45g) it could supply 635kJ (635 000 joules) of energy. This is a lot of joules — nearly half a million! It is roughly the same as the amount of movement energy you would gain if you fell off a cliff 1000 metres high. It is twenty times the energy carried by a high-velocity rifle bullet.

How much energy do you need to live?

The average 16-year old girl needs to take in 9 million joules of energy a day. Boys of the same age need about 12 million joules a day.

12 million joules is about enough to lift a 60kg person a height of 20km. Mount Everest is less than 9km high! Obviously you do more with your energy intake than just **move**.

So what do you actually do with all this energy? See *The Living World*, Section C for an answer to this important question. It seems that just keeping alive takes a lot of energy!

Working

One way that scientists measure energy is by seeing how much work it can do. This is quite simple, and is dealt with in topic B4, where we look at energy and force in a gravity field. Forces do work when they move something. First we need to measure the size of the force (in newtons). Then we measure the distance (in metres) that the object moves in the direction the force is acting.

Then we can calculate:

Energy used = work done

= force × distance moved in the direction of the force

E joules = ***F*** newtons × ***d*** metres

Suppose it takes a force of 25N to move a saw when you cut a piece of wood, and you move it 0.2 metre each time. Then one cutting movement needs 25N x 0.2m = 5J of energy. If it takes 20 sawcuts to get through the wood, you need to supply 100J (picture 3).

James Joule — heating with energy

James Joule lived over 150 years ago. He was one of the first scientists to do experiments which actually measured how much energy was needed to **heat** something.

It seems obvious now that it takes energy to make things hot — just think of the electricity and gas bills. But in those days scientists thought that heating was done by a mysterious gas that seeped in and out of things to make them hot or cold.

James Joule did experiments to show that when you do *work*, by moving a paddle wheel against the friction of water in a container, the water gets hot (picture 4). More importantly, he showed that the more work he did the hotter the water got. He was changing *energy of movement* into *heating energy*.

We now know that energy can appear in all kinds of different ways, doing different jobs. We have also discovered the strange fact that we can only use the energy when *something changes*.

Energy and change

Norway gets most of its electrical energy from hydroelectric power stations. In Britain we get most of ours from power stations which run off coal. In both cases, changes have to occur before we get the electricity we need.

In Norway, water stored in dams high in the mountains has to run downhill. The water has to *change its position* in a gravity field (picture 5). In an ordinary power station, coal or oil has to be burned. This is a *chemical change*, in which hydrocarbon molecules combine with oxygen in the air.

In both examples there are other changes as well. The falling water gains movement energy (kinetic energy) and gives it to the turbines which drive the generators. In a coal-fired power station the chemical reaction heats up water, which turns to steam. The steam applies force to turbines, and they do work in turning the generators.

Systems

A useful idea is to think of the energy as being in a **system**. The water in a dam is able to do work because of the pull of gravity on it. This pull is caused by the Earth, so it makes sense to think of the *water-earth system* as having the energy. The water wouldn't be much use on its own.

In a coal-fired power station the fuel will only heat water if it burns. This needs oxygen. So here we have a *fuel-oxygen system*. When we use energy we are always moving it from one system to another — or to several others. But there is a very important law of physics that says that the total quantity of energy always stays the same. The trouble is that most of the time it doesn't all go from the starting system to the one we want it to go to! See topic D4 for more about the laws of energy.

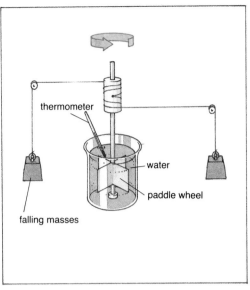

Picture 4   Joule's Paddle Wheel Apparatus showed that 'heat is a form of energy'

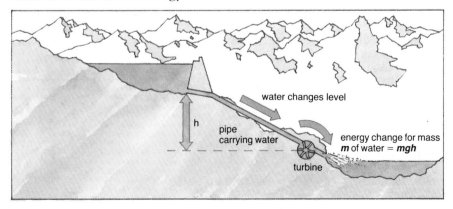

Picture 5   Energy transfers in a hydroelectric power station

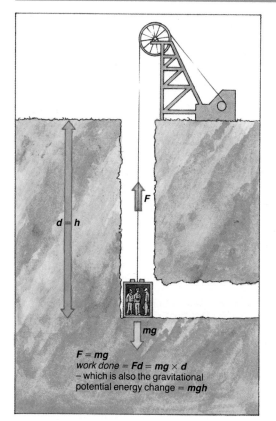

Picture 6 Doing work against gravity: *E* = *Fd*

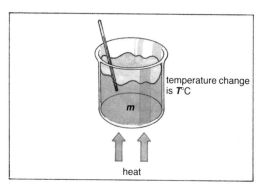

Picture 7 Heating water *E* = *msT*

A joule is a joule is a joule

We can measure energy using a lot of different techniques once we realise that moving energy from one system to another doesn't alter the fact that it is still **energy**. We can lift a can of beans up on to a shelf in all sorts of ways. But whatever source the energy comes from, it always takes the same amount of work to do the lifting. So the same quantity of energy has to be transferred, whether it is done by an electric robot, a conveyer belt or a human being.

Similarly, it takes a standard quantity of energy to warm a kilogram of water by 1°C. This is the principle behind the ways of measuring energy described below.

How to measure energy

This is a useful reference section: use it when you need to measure energy for an investigation or when dealing with energy in another topic.

1 Doing work.   You need to measure the force (F in newtons) involved and the distance (d in metres) moved. Then use the formula:

$$\text{work done} = \text{energy } \boldsymbol{E} \text{ transferred} = \text{force} \times \text{distance}$$

$$\text{or} \qquad \boldsymbol{E} = \boldsymbol{Fd}.$$

2 Measuring potential energy   When you do work against the force of gravity you always increase the potential energy of the object being moved. If it has a mass **m** it will have a gravity force on it of **mg**. This is its **weight**. When you move it a through a height **h** you do work and transfer energy to the object as potential energy. As above we can calculate this as:

$$\text{work done} = \text{force} \times \text{distance}$$

$$\text{or energy transferred as potential energy, } \boldsymbol{E} = \boldsymbol{mgh}$$

3 Using temperature measurements.   You need to measure the mass (**m** in kg) of material being heated and the temperature change (**t** in °C) produced. Then look up in tables the amount of energy (**s**) needed to heat 1 kg of the material through each celsius degree. This is the **specific thermal capacity, s**. In the case of water, s = 4200 J/kg/°C. Finally, you can use the formula:

$$\boldsymbol{E} = \boldsymbol{msT}$$

to find the energy **E** transferred by heating.

4 Using electrical measurements.   The energy delivered by an electrical current to a device is measured by using the formula

$$\boldsymbol{E} = \boldsymbol{VIt}$$

(**V** is voltage, **I** is current in amps, **t** is time in seconds; see picture 7 and topic E6)

5 Kinetic energy.   The energy, **E**, possessed by a moving object is calculated using the formula $\boldsymbol{E} = \frac{1}{2}\boldsymbol{mv}^2$, where **m** is the mass of the object in kilograms and **v** is its speed in metres per second.

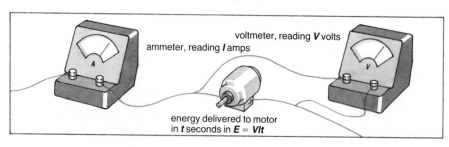

Picture 8 Measuring electrical energy changes

Activities

A Self-energised heating

1 Rub your hands together quickly. Describe any energy changes that you notice.
2 How does your body use muscles to keep itself warm without moving your arms and legs?

B How much work do you do . . .

1 When you climb upstairs?
2 When you lift a stool and put in on a table or workbench?
3 When you open a door?
4 When you move a laboratory trolley across the room?
5 When you lift this book from the floor to the top of your workbench or desk.

To answer these questions, make measurements of force and distance. You will need a newton meter, and a set of bathroom scales to measure your own weight. Assume that a mass of 1 kg has a weight of 10 newtons. Use the formula for doing work, $E = Fd$.

C Investigating the heating effect of electrical energy

1 Set up the circuit shown in picture 9. Your task is to find out how much energy is delivered to the heater by the electric current in five minutes, and what effect this has on the material X. X may be a liquid or a solid block.

CARE! Do not let the heater cool in a liquid — switch off, take the heater out, allow it to drip and cool in air.

You will need to measure:

i) the temperature of X before you start and after the current is switched off,
ii) current (**I**), voltage (**V**) and time of flow (**t** seconds). Use the formula **E** = **VIt** to calculate how much energy was supplied. 'X' will be given to you by your teacher. You may be given more than one material, or different groups may investigate different materials.

2 Collect together the results for different materials and put them into a table. Write a note describing how the different materials were affected by the energy supplied to them.

3 Can you suggest why materials behave differently?

D Measuring the effective energy output of a bunsen burner?

1 Get a large tin or beaker and put 0.5 kg (0.5 litres) of cold water into it.
2 Put the container with water on a tripod and place it on a heating mat.
3 Measure and record the temperature of the water.
4 Light the bunsen burner and put it under the water container. Start a stopwatch as you do so.
5 Stir the water with the stirring thermometer and when the temperature reaches about 50°C turn off the bunsen burner and stop the stopwatch. Keep stirring the water until its temperature becomes steady. Note down this final temperature and the time the bunsen was used for.
6 Calculations.
   i) Work out the rise in temperature of the water, **T**.
   ii) Calculate the energy given to the water by using the formula **E** = 0.5 × 4.2 × **T** kilojoule
   iii) Work out how much energy was supplied to the water per second. This is the effective power output of the bunsen burner.
   iv) Was this a fair test of the energy that a bunsen burner can provide? Give reasons for your answer.

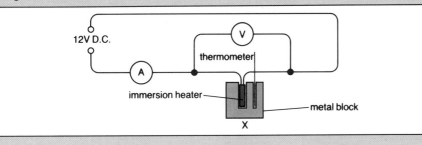

12V D.C.
thermometer
A
V
immersion heater
metal block
X

Picture 9

Questions

Use the formulae given in this chapter to answer the following.

1 It takes a force of 30 newtons to push a box along the floor. How much work do you do when you move it 5 metres?
2 How much energy is released in a 12 V electric lamp when a current of 0.5 A flows in it for 1 minute?
3 A kilogram of copper needs 400 J to raise its temperature by 1°C. How many joules would it take to:
a warm 5 kg of copper by 1°C?
b warm 6 kg of copper by 20°C?

4 Mount Everest is 8848 metres above sea level.
a Calculate how much energy it would take to lift someone weighing 600 N straight up through this height.
b In fact, a climber would use much more energy than you have calculated in (a) to climb Mount Everest. Give at least three reasons for this.
5 Design an experiment to find out one of the following.
Say what you would need to measure and give an idea of how it might be done.
a How much energy can you get by burning 1 kg of wood?
b How much energy does a gerbil need in a day?
c How much energy does the Sun deliver on a square metre of ground per minute on a sunny day?

d How much energy does it take to do the washing up?
6 How much does the temperature of a 2 kg mass of water rise when it is given 21 kJ of heating energy? Use the fact that it takes 4.2 kJ to warm 1 kg of water by 1°C.
7 A rock weighs 2 kg and is just on the edge of a cliff 12 m above the beach.
a What is the potential energy of the rock? (Assume **g** = 10 N/kg).
b The rock falls off the cliff. How much kinetic energy will it have just before it hits the beach?
c Use a calculator to work out the speed of the rock just before it lifts the beach. How much will air resistance affect the speed of the rock?

D3
Using energy

We use energy — and misuse it. Studying this section and the next could save you a lot of money . . . .

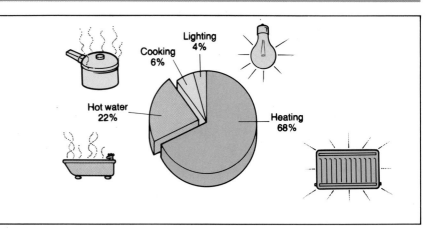

Picture 1   How we use energy in the home

The costs of energy

Energy is expensive. You have to pay for producing it. You have to pay for moving it to where you want to use it. *The Material World* (topic I3) deals with the problems of getting fossil fuels out of the ground. 'Human fuel' — food — is even more expensive to produce. The way we get energy from food is dealt with in *The Living World* (topic C4).

This section deals with the way we use energy in the home, and how we might save money by using less, and wasting less. Look at the pie chart in picture 1. It shows what we use energy for at home.

But there are other costs when we use energy. We get most of our energy from fossil fuels. The unwanted by-products are carbon dioxide and some other gases which cause pollution in one way or another. We are only now waking up to the fact that the bills for the greenhouse effect and acid rain and other kinds of pollution have still to come in.

Home heating

Most of the energy we use at home is used to keep ourselves and our surroundings warm. Nearly all of this *heating energy* comes from the combustion of fossil fuels, such as coal, oil or gas. We can use the fuels directly or indirectly (see table 1).

Electricity is not a fuel: it is just a very good way of moving energy from one place to another. But the energy it carries is produced in power stations and most of these use fossil fuels.

Heating costs

Table 2 shows how much it costs to provide heating using different sources of energy. The most expensive is 'daytime' electricity. Prices will change, of course. This may be due to inflation and changes in world production of oil, for example. But the prices tend to stay much the same in comparison with each other.

Energy is measured in joules. One joule is a very small quantity of energy, it is about how much you would use to lift this book from a chair to a table. It usually takes a lot of energy to heat things up. For example, to boil a kettle holding 1 kg of water needs about a third of a million joules (0.33 MJ). If this energy were used in working instead of heating, it could raise the kilogram of water higher than the top of Mount Everest! Heating things uses up a lot of energy, and it's by far the biggest part of your household energy use. A 1-bar electric fire is rated at 1 kilowatt (see below). This fire delivers a thousand joules *every second*. In an hour it would deliver 3 600 000 joules (3.6 MJ). This amount of energy is also called a **kilowatt-hour**.

Table 1   Direct and indirect use of fuels

Directly
coal, oil or **gas** in fires, stoves, room heaters, central heating systems
Indirectly
electricity in fires, convector heaters, storage heaters

Table 2   Costs of producing energy

Energy source	cost of producing	
	Yearly cost for typical house/£	1 megajoule p
solid fuel/coal	360	0.02
oil	250	0.015
gas	270	0.016
bottled gas (propane)	455	0.027
Electricity		
—daytime	1000	1.7
—night-time	335	0.6

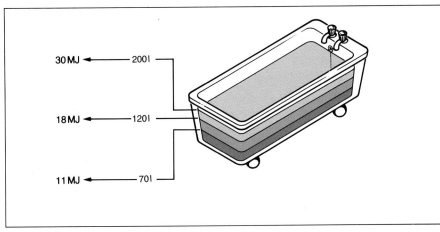

Picture 2 The energy needed to have a bath

Cleaning, washing and bathing

How much does it cost to have a bath? This depends on how hot you like it, and how much water you use. A standard bath, just about as full as you can have without it spilling, will take 200 litres of water (picture 2). A litre of water has a mass of 1 kg, so you would use 200 kg of water. Tap water is at an average temperature of 10°C in winter, a hot bath is about 45°C.

We can use the formula $E = mst$ (see topic D2) to calculate how much heating energy this needs:

$$\text{energy needed, } E = 200\text{g} \times 4200\text{J/kg°C} \times 35\text{°C}$$

$$= 29.4\text{MJ}$$

You can use the data in table 2 to work out how much this would cost using different energy sources. Using the most expensive, electricity, it would cost about 50p. It should be cheaper to take a shower — activity B asks you to estimate how much taking a shower costs (picture 3).

Table 3 shows how much energy is needed for different cleaning tasks.

Table 3 Energy needed for cleaning tasks

Washing up (using typical quantities of water)		
in a sink (8 litres)	2 MJ (very hot)	1.3 MJ (hot)
in a plastic bowl (5 litres)	1.3 MJ (very hot)	0.8 MJ (hot)
Washing machine (for clothes)		
using 17 litres of water	heating 5 MJ pump, motor 1 MJ	
Vacuum cleaning		
for 1 hour	3.6 MJ	

Picture 3 Taking a shower

Cooking

Cooking means heating food to a temperature high enough to change the tough fibres of meat and vegetables into a softer form that we can digest. It also kills any bacteria that might be in the food. Cooking takes time, to make sure that the changes, which are in fact chemical reactions, take place.

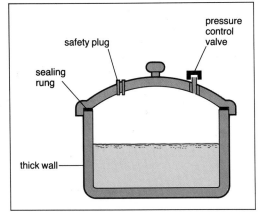

Picture 4 Pressure cookers save energy. How?

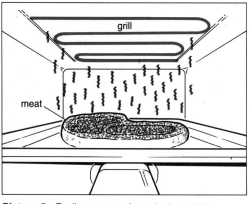

Picture 5 Radiant energy from the hot grill is absorbed in the surface of the food

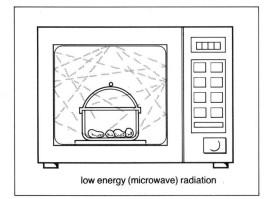

low energy (microwave) radiation

Picture 6 Low energy microwave radiation can pass through glass and plastic and carry heat energy deep into food, where it is absorbed easily by water and fat

The higher the temperature the quicker the changes take place (see *The Material World* topic C1). Frying foods is quicker than boiling them. Frying and roasting temperatures are high, from 150 to 240°C, compared with the boiling water temperature of 100°C. Salt water boils at a higher temperature than pure water, but only by a degree or two.

Pressure cooking

Water under pressure needs an even higher temperature before it boils. This effect is used in pressure cookers (picture 4). In these cookers, water is kept boiling very gently (just *simmering*) at about 105°C. Even this small rise in temperature reduces cooking times by over a half, thus saving quite a lot of energy.

Cooking with radiation

You do this every time you grill some food. A grill is simply a piece of metal, heated by gas or electricity. Red hot metal gives out not only energy we can see (as red light) but also energy radiation that is invisible — *infra red* radiation (see topic C8). In fact, most of the radiant energy in a grill is carried by the infra red radiation (picture 5). This is absorbed by molecules in the food and the food gets hot, so getting cooked.

Microwaves are like infra red radiation, but the energy is carried by longer waves (picture 6). The advantage of microwaves is that the waves travel deeper into the food before getting absorbed. This means that the food is cooked on the inside as well as the outside.

In a grill — and with frying and boiling — the energy has to travel in from the surface. This is slow. We can try to speed it up, by making the fat or the grill hotter. But if we do, the outside may burn while the inside is still raw. Another advantage of microwave cooking is that the microwave radiation is not trapped by glass or plastic. It just affects the food material, and particularly the water in it. This saves energy because only the food is heated up, not the food container.

Lighting

Good lighting is important. We all spend a great part of our lives under 'artificial' light, and poor lighting causes eyestrain and headaches. Well designed lighting also makes places pleasant to work and live in. The physics of lighting involves:

- how light is produced
- the colour and colour balance of light
- how light is reflected or absorbed by objects
- the energy costs and efficiency of lighting.

Topics C6 and C11 deal with these ideas.

Energywise

Most of the energy we use in the home is for heating. We need to keep warm, and we need hot water to keep things clean. We can cut down the costs of this in all sorts of ways. Most of them mean *insulating* our homes better. This includes keeping cold air out as far as possible — but no so much that we then suffocate to death!

Picture 7 shows the ways the average home could save on energy loss. The 'payback time' is the time it would take for energy savings to pay for the improvement that causes them.

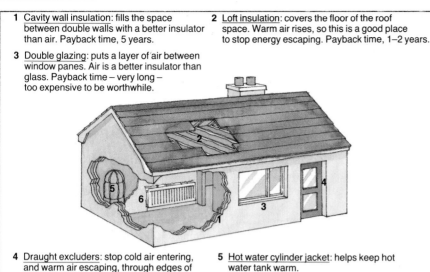

1 Cavity wall insulation: fills the space between double walls with a better insulator than air. Payback time, 5 years.

2 Loft insulation: covers the floor of the roof space. Warm air rises, so this is a good place to stop energy escaping. Payback time, 1–2 years.

3 Double glazing: puts a layer of air between window panes. Air is a better insulator than glass. Payback time – very long – too expensive to be worthwhile.

4 Draught excluders: stop cold air entering, and warm air escaping, through edges of doors and windows. Payback time, 2–4 years.

5 Hot water cylinder jacket: helps keep hot water tank warm. Payback time, less than one year.

6 Radiator foil: a shiny material fitted behind radiators on outside walls stops energy escaping as it reflects radiation back into the room. Payback time, 1–2 years.

Picture 7 How to save energy at home

Activities

A Energy and you

What energy do you use at home during a day? Keep a diary for one day and note down what energy-using devices you use, and how long you use them for. Don't forget to add in your share of the family energy use, for cooking, watching TV etc. Make a list like this:

Device   time used (hours)
         power rating (kilowatts)
         energy used (time x power rating)

Hints.

1 Don't worry about accuracy too much. A rough idea is all you need.

2 Electrical devices have a label on them giving the power rating. If it is in watts, divide the number by 1000 to get kilowatts.

3 You may have to make reasonable guesses about non-electrical devices; ask for help from your teacher if you need to.

B Take a shower!

The calculation on page 111 tells you how much it costs to take a bath. Is showering cheaper? How much energy does it use? To find out you will need to measure:

1 How much water the shower gives per minute.

2 How long you take to shower.

3 How much hotter the shower makes the cold water (the temperature rise).

(*Note:* it takes 4.2 kJ to heat 1 kg of water by just 1°C.

C Who left that light on?

1 Count up how many electric lights you have in your home or the school science department. What (on average) is their wattage (this is written on the bulb)? How many watts are used when they are all switched on at the same time?

2 How many watts is a typical electric heater rated at? (If you have one, look at its label to find out.) Write a comment about the energy cost of heating compared with the energy cost of lighting.

Questions

1 Put the following energy users into an ordered list, biggest users first: vacuum cleaner, electric kettle, torch bulb, electric cooker oven, room lamp, central heating system, microwave cooker.

If you find this difficult, the table on page 155 of topic E6 may help you.

2 Electrical energy costs 6p a 'unit'. An electric fire uses 2 units of electrical energy an hour. How much does it cost to use the fire for 6 hours a day, 7 days a week?

3 We spend a lot of money heating our homes in the winter. This energy comes into the home in coal, oil, gas or electricity. Where does energy go to in the end?

4 The pie chart in picture 1 on page 110 shows what the average home does with the energy that comes into it. Can you think of any other uses of energy that are *not* shown? Suggest why they don't appear in the chart.

5 Useful energy devices change one kind of energy into another, or move it from one place to another. What kind of energy changes or moves are involved in the following?

a An electric drill.

b A TV set.

c Using a hand saw to cut through a piece of wood.

d Using a telephone.

6 A full bath of water might need 30 MJ of energy to get the water to a nice comfortable temperature. Use table 1 to check that this would cost about 50p using electricity to heat the water. How much would it cost if the water was heated by an oil burner?

D4
Saving and wasting — the laws of energy

Energy cannot be created, and it cannot be destroyed. But it can be wasted.

Picture 1 Working against gravity

Whenever we use energy we are moving it from one system to another. The first three topics in this section give many examples of this **transfer** of energy. We are using the energy to **do work**. When you lift a weight you are transferring energy from your body system (sugar and oxygen in the blood) to do work against the force of gravity. If the weight is large you know that you are doing work! (see picture 1). After you have lifted it the weight is higher above the ground and so has more **gravitational potential energy**. (See topics B5 and D1.)

There are very many other examples of energy transfers like this. The fact that energy can move from one system to another is the basis of life on earth. It is the basis of all movement, and of all the industries that we rely on in the modern world.

In studying energy changes scientists have discovered the laws which describe how energy behaves. They are called the *Laws of Thermodynamics*, and this topic is about the two most important of these.

The First Law — you can't win

This law is very simple. It says that whenever any changes take place there is just as much energy at the end as there was in the beginning. Like many of the most important ideas in science, it is impossible to prove this! But the law has been tested, time and time again.

Very careful measurements in all kinds of experiments have shown that in every single case **energy has neither been lost nor created**. Of course, one day someone might do a test in which energy is lost or created, and this would disprove the law.

This first law is the law of the **conservation** of energy. Many people have tried to disprove this law, and all of them have failed. It would be so nice to be able to create energy from 'nothing'. Life on Earth would be very convenient if someone could design an engine or a machine which gave out more energy than you put into it.

Picture 2 shows one design for such a machine. Look at it. Can you see how it is supposed to work? Can you see why it won't work?

What the law means in practice is that with any machine or engine the best we can do is get out as much energy as we put into it. But even this is optimistic. For real machines we always get *less* energy out than we put in. This is because of the *Second Law*.

The Second Law — you always lose!

Whenever we use energy to do a job of work some of the energy we put in seems to escape. It just doesn't go where we want it to.

When you lift a weight you give it some potential energy. But this is less than the energy produced in your muscles by the chemical reaction (respiration) between sugar and oxygen. Some of that energy goes to make your muscles warmer (see picture 3).

When you do a lot of work you have to sweat a lot to get rid of this 'waste energy'. It ends up warming your surroundings. The molecules in the air move a little faster as a result. They gain some kinetic energy.

The Second Law simply says that when work is done there is always some energy that will somehow escape and spread itself out. It usually ends up as 'heat', which in basic energy terms means that millions upon millions of molecules move a little faster.

Can we get this 'waste' energy back?

Yes — we **can** get this energy back. There is no law of physics that stops us. All we have to do is slow down all those molecules and do some useful work

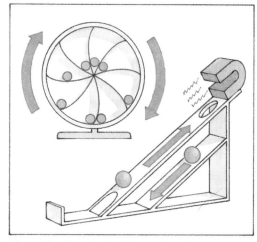

Picture 2 Perpetual motion machines. Will they work?

with the energy they lose. Or we could concentrate the energy in one place to warm something up. For example, we could put this energy into water in a central heating system.

The problem is, how can we concentrate this spread-out energy in one place? It is not easy to do, particularly if the energy has spread out a long way over millions of different particles.

Making spread-out energy concentrated

If you have a refrigerator at home this is exactly what it is doing. It is taking movement energy from the molecules of the foods inside it. They cool down as a result, so that the food stays fresh longer. The energy they lose is given to the air. You can check this effect for yourself. At the back of the refrigerator you will find a set of tubes which feel warm because of the energy that the food has lost (picture 4).

But this energy is still wasted. After all, we don't normally buy refrigerators to use as house-warmers. But what if we used the same idea, say, to take energy *out of the ground* and feed it into our homes? This is simply using a 'refrigerator' in reverse! Such heaters do exist, and they are called **heat pumps**. They take energy out of the air or the ground instead of out of food, and 'pump' it *into* the house to warm it.

Picture 5 shows a commercial heat pump. It is used in large buildings. These need ventilating, which means that warmed, smelly air has to be taken out of the building. This has to be replaced by fresh air — which is cold in winter. One end of the heat pump is placed in the warm 'exhaust air' and takes the heat energy out of it. This energy is pumped to the cold air coming into the building, so that it is 'prewarmed'. This cuts down the heating bills.

Something for nothing?

But this effect doesn't happen all by itself. Refrigerators and heat pumps contain a liquid which has to be pumped around the system. The pumps are usually driven by an electric motor and this needs energy to run it. When all the energy sums are done we find that we still get less energy out than goes in, and that *overall* the energy still gets more spread out!

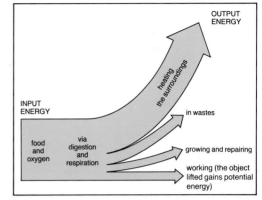

Picture 3 What the body does with the energy it gets — only a small amount can be used to do work

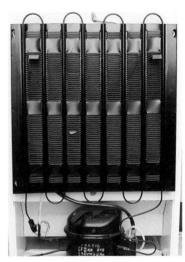

Picture 4 In a fridge the energy taken out of the cooling food warms the air

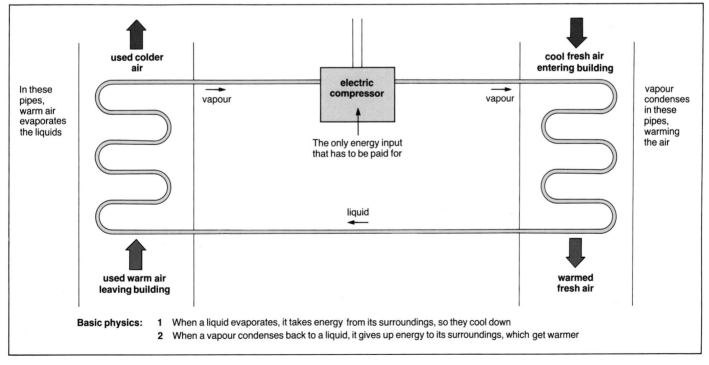

Basic physics: 1 When a liquid evaporates, it takes energy from its surroundings, so they cool down
2 When a vapour condenses back to a liquid, it gives up energy to its surroundings, which get warmer

Picture 5 How a heat pump works

Efficiency

Being *efficient* means being effective at doing a job. The **efficiency** of a machine or engine measures how good it is, in terms of how much of the energy put into it is used where you want it. Efficiency is measured by how much useful energy you get out compared with what is put in. It usually converted to a percentage:

$$\text{efficiency} = \frac{\text{useful energy output}}{\text{energy input}} \times 100\%$$

Table 1 shows the typical efficiency of some energy devices. Some of them, like engines and some machines, use energy from a fuel system to do work. Cars, bicycles, electric drills and trains do this. Other devices are used to make energy more 'usable'. They might do this by taking energy from a fuel-oxygen system and putting it into an electrical system. This is what power stations do. Electricity is very useful because it makes it so easy to move energy from one place to another.

Table 1   Typical efficiency of energy devices

	engines and machines producing movement or doing work %	energy 'movers' %
Train: diesel engine	36	
Car: petrol engine	15	
Train: steam engine	15	
Train: Electric motor	90	
Car gears		93
Bicycle		90
Car jack		15
Electric power station	35	
Transformer (electrical)		97
Human muscle	40	

Gears and *transmission systems* in cars, trains and bicycles are the mechanical versions of electric transmission. They carry energy from one part of a **mechanical** system to another. They also allow the force that eventually does the work to be made small or large, according to what is needed.

Useful laws

The laws of energy were discovered in the 19th century. They were also very useful in helping to design engines. It was scientists' understanding of these laws that helped improve steam engines and led to the development of car engines (internal combustion engines).

Some puzzling questions — and a surprising answer!

But there were very important questions about energy that these laws couldn't explain. Two of these questions were to do with the Sun and the Earth. We know that the Sun is sending out huge quantities of energy. It has been doing this for thousands of millions of years. What is not so well-known is the fact that the Earth is also giving out energy (see topic G1). This energy is seeping up from the centre of the Earth, causing volcanoes and earth-quakes, warming the atmosphere and, eventually, being lost to space. *Where does this energy actually come from? Why does the Sun keep on shining? If the Earth is losing so much heat, why doesn't it cool down?*

The answer to these questions was given by the physicist Albert Einstein. He was able to prove that **matter** — the stuff that makes ordinary atoms and molecules — can be converted to energy. The Sun keeps on shining by actually losing mass — at the rate of four million tonnes a second.

Most people have heard of Einstein's famous equation that calculates the amount of energy E that can be obtained by destroying a quantity of matter, m:

$$E = mc^2$$

where c is the speed of light.

This discovery meant that scientists had to think again about the First Law, the Law of Conservation of Energy. *It would only work if mass was thought of as a kind of energy!* This was a very strange idea. This story continues in topic D5.

Activities

A   Talking about biology and energy

Read about — or remind yourself about — the energy flow through an 'ecosystem'. Discuss, and come to an agreement about the following questions:

- What is the source of the energy that all the organisms make use of?
- Where does it all go in the end?
- Which organisms are best (most efficient) at using energy?
- In an ecosystem, materials are often recycled. Is energy recycled?

Write down, briefly, your agreed answers to these questions.

B   Moving forever

Design a 'perpetual motion' machine — a machine that goes on moving by itself, forever, without needing any fuel supply or energy input. Draw a picture of your design, with a few words of explanation.

C   The efficiency of a machine

Design an experiment to measure the efficency of a machine. This could be a pulley system, a car jack, a wheel-and-axle, etc. When you have designed your experiment, get it checked by your teacher before you do it.

Hints and tips

You will need to measure the forces involved — the **effort** and the **load**. You will also need to measure the **distances** that these forces move.

The input energy is the work done by the effort. The output energy is measured by what happens to the load.

In both cases the same formula applies:
work = force × distance moved
(measured in the direction the force acts). If you have forgotten about machines look up topic A5.

D   The efficiency of an electric motor

Use a low voltage electric motor. Make it lift a load and measure how much work it does in doing this. The formula is given in activity C.

You can measure the electrical energy supplied to the motor in one of two ways. The easy way is to use a *joulemeter* which is a special electrical energy meter.

Or you can use an ammeter, a voltmeter and a stopwatch. The energy supplied electrically is then given by the formula:

energy = voltage × current × time in seconds.

Questions

1   A car stands in front of the owner's house. It has a full tank of petrol. After a long drive the car comes back and stops in front of the house. The tank is now empty. What has happened to the energy that was stored in the petrol? (There is a very short answer to this question!)

2   The Moon is the main cause of the tides in the sea. Careful measurements show that the Moon is actually slowing down as it goes around the Earth. Use the Law of Conservation of Energy to explain this.

3   A small hoist (pulley system) lifts a load and gives it 3000 J of potential energy. The person who used the hoist supplied it with 5000 J of work. What is the efficiency of the hoist?

4   The efficiency of a type of car jack is 25%. It takes 1000 joules of energy to lift a car high enough to replace a wheel. How much work must be done by the worker who used the jack?

5   Why do you think engines (petrol, steam) are so much less efficient than gears or bicycles?

6   Picture 6 shows the energy system of a light bulb. It is called a 'Sankey Diagram'. It shows that energy is delivered electrically, and that most of this energy is wasted. Only a small percentage is delivered as useful light. Draw a similar diagram for one of the following:

a   Boiling a kettle of water on a gas ring.

b   Sawing through a log of wood.

c   Just sitting on a chair.

d   Winding up an old-fashioned (spring) clock.

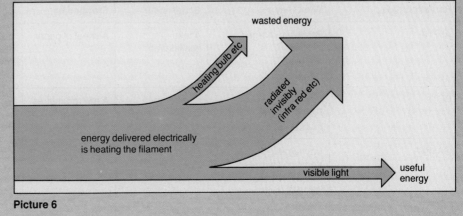

Picture 6

D5 Radioactivity

One of the great discoveries of the 20th century was a new energy source. Matter itself could be transformed into energy.

Picture 1 Not everyone wants to use nuclear energy

Picture 1 shows the nuclear power station at Sizewell in Suffolk. It also shows a group of people who are protesting against the proposal to build another one there.

We live in a world which is using up its fossil fuels many times more quickly than they can possibly be replaced. When they are burnt they release carbon dioxide which will very likely change the Earth's climate, by the greenhouse effect (see *The Material World*, topic I4).

Nuclear energy seems an ideal alternative to fossil fuels. The materials are quite cheap. When they are used they do not make any polluting gases. In coal-fired power stations there are transport problems in getting hundreds of tonnes of coal to the site every day. Nuclear power stations do not have this problem.

These are the kind of reasons which have led many countries to build nuclear power stations. In France, for example, more than half of their electricity is generated by nuclear power. In the UK we generate just under 20% of our electricity in this way.

But there are also many objections to using nuclear power. Many countries have decided not to build any more nuclear power stations. This topic and the next deal with the physics of nuclear energy, its benefits and dangers.

Radioactivity

Radioactivity was discovered in 1895 by a French scientist called Henri Becquerel. A **radioactive** substance is one that sends out very energetic rays. There are three types of ray. All of them can damage living things, but they are also very useful. When they were first discovered nobody knew exactly what they were. So they simply named them **alpha, beta** and **gamma** rays, after the first three letters of the Greek alphabet.

Alpha, beta, gamma

Table 1 summarises the properties of these rays. It took scientists 20 years after they were first discovered to work out what the radiations actually were, and that they were coming from the **nucleus** of the atom.

The Material World, topic J3 describes the structure of the atom, and this is summarised in picture 2.

One property that these radiations possess is their ability to **ionise** atoms and molecules. The radiations knocked electrons out of the atoms to turn them into **ions** (picture 3). The atoms and molecules become positively charged. Thus the radiations are sometimes called **ionising radiations**.

Table 1 Nuclear radiations

Radiation type	What they are	Range in air	Stopped by	Comments
ALPHA	Positively charged. NUCLEI of Helium ($^4_2$He)	A few centimetres	A sheet of paper	Because they are so massive and carry a double positive charge they easily affect atoms. They make lots of ions and don't travel far. Alpha emitters are quite safe unless they get into the body.
BETA	Negatively charged. Fast moving ELECTRONS	A few tens of centimetres	A few millimetres of aluminium	Electrons are small and so can travel further than alpha particles, as they don't collide as often. Dangerous, but easily stopped.
GAMMA	Uncharged. Very short wavelength ELECTROMAGNETIC WAVES (high energy photons)	They go on indefinitely	A metre or two of concrete	These are genuine 'rays'. They don't ionise atoms very easily and so travel a long way. They travel at the speed of light. Dangerous because they are hard to shield against.

The reason they are dangerous to life is that they ionise atoms in living cells, which can kill the cells (see topic D8).

The radiations travel at high speeds, as if they are shot out of the nucleus like a bullet from a gun. What was puzzling to the scientists who first worked with these radiations was where they got the energy from to do this. We now know that it comes from the conversion of matter into energy, and this is explained more fully in topic D7.

Detecting the radiations

We detect the rays from atoms using the fact that the rays ionise other atoms. Picture 4 shows the tracks of alpha particles travelling through damp air. The water vapour in the air condenses as droplets on the **ions** made from the air by the alpha particles. This leaves a 'vapour trail' like the ones made by high flying aircraft. This picture was taken as the alpha particles travelled through a special 'cloud chamber'. This contains cool damp air, which is good at making clouds and vapour trails.

Picture 5 shows another detector which makes use of ionisation.

It is a GM tube, or **Geiger-Muller tube**. It contains a gas at a low pressure. The inner wire is at a high voltage. When some ionising radiation goes into the tube it makes ions in the gas. The wire attracts them and so there is a short pulse of current when the ions hit the wire. This is amplified and can be counted by an electronic **pulse counter**. Thus each time a 'ray' gets into the tube it produces a pulse which can be counted.

Geiger counters like this are the most common way of finding out how much radioactivity is present.

Alpha particles

Alpha particles are very good at ionising other atoms. This suggests that they carry a lot of energy. This is movement energy (kinetic energy).

But they don't travel very far in air. They barge through the air like a bull in a china shop, hitting lots of air molecules and knocking electrons off them (picture 4). As they do this they lose energy quite rapidly, so they soon slow down.

If we put a piece of paper in the way the alpha particles can't get through it! This is because they hit a lot of atoms in the paper.

Alpha particles are electrically charged, carrying 2 units of positive charge. This is twice as much as the (negative) charge carried by an electron.

Because alpha particles are positively charged they are affected by both electric and magnetic fields. Scientists have worked out exactly what alpha particles are by measuring how strongly they are affected by these fields.

These measurements show that alpha particles are in fact **helium nuclei**, which are four times as heavy as hydrogen nuclei. They have, therefore, an atomic mass of 4. They are made up of two protons (positive particles) and two neutrons, joined very firmly together (picture 6). See *The Material World*, topic J3 for more about protons and neutrons.

Beta particles

Beta particles do not cause as much ionisation as alpha particles. Also, they travel much further in air before running out of energy. This suggests that

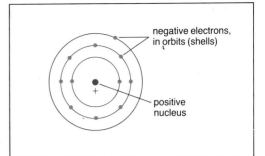

Picture 2 Rutherford's model: a nuclear atom

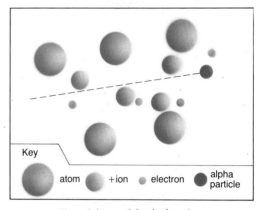

Picture 3 How alpha particles ionise atoms

Picture 4 The tracks of alpha particles in a 'cloud chamber'

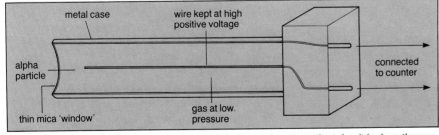

Picture 5 A GM tube. When an alpha particle, for example, enters the tube, it ionises the gas inside. This triggers off a sudden flow of freed electrons to the wire. This registers as a 'count' or a 'click'.

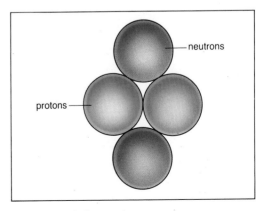

Picture 6 A helium nucleus

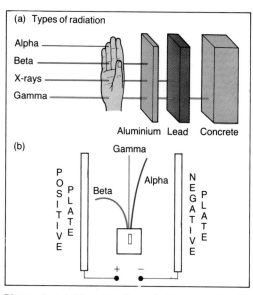

(a) Types of radiation

Alpha
Beta
X-rays
Gamma

Aluminium   Lead   Concrete

(b)

Gamma

P O S I T I V E   P L A T E

Beta

Alpha

N E G A T I V E   P L A T E

+   −

Picture 7   (a) What stops the radiations — or doesn't! (X-rays are shown for a comparison)
(b) What happens to the radiations in an electric field

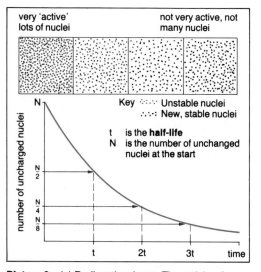

very 'active' lots of nuclei

not very active, not many nuclei

N

Key   Unstable nuclei
      New, stable nuclei

t   is the **half-life**
N   is the number of unchanged nuclei at the start

number of uncharged nuclei

N/2

N/4
N/8

t   2t   3t   time

Picture 8   (a) Radioactive decay. The activity of a radioactive substance gets less all the time. This is because the number of unstable, active nuclei gets less. But there is a catch — the new nuclei may also be unstable!
(b) A half-life graph

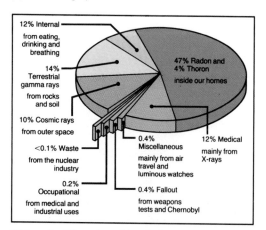

12% Internal
from eating, drinking and breathing

14% Terrestrial gamma rays from rocks and soil

10% Cosmic rays from outer space

<0.1% Waste from the nuclear industry

0.2% Occupational from medical and industrial uses

47% Radon and 4% Thoron inside our homes

0.4% Miscellaneous mainly from air travel and luminous watches

12% Medical mainly from X-rays

0.4% Fallout from weapons tests and Chernobyl

Picture 9   Where harmful radiation comes from

they are lighter and smaller than alpha particles. They carry a single unit of negative charge. Measurements using electric and magnetic fields show that they are in fact **high-speed electrons**. An alpha particle is 7333 times as massive as a beta particle.

Because they are so small, beta particles can travel through matter more easily than alpha particles can. But a few millimetres of aluminium will stop them.

Gamma rays

Gamma rays do not cause much ionisation and they travel quite freely, through air. They carry no charge and so are not at all affected by electric or magnetic fields. This means that they are very **penetrating**. They can travel through thick blocks of concrete, and it takes up to a centimetre or so thickness of a heavy metal like lead to cut down the radiation by a half.

Gamma rays are different from the alpha and beta particles because they have no mass. They travel at the speed of light, and are a type of **electromagnetic radiation**, like light and X-rays. But they carry more energy than X-rays do.

Picture 7 shows how the three kinds of radiation compare in their ability to pass through matter, and what happens to them in an electric field.

Half-life

Radiations are given out by decaying atoms. As the atom decays to something else, it 'spits out' the radiation. In a sample of uranium the nuclei don't decay all at once. In fact, only a very tiny fraction of the nuclei in the sample decay each day. It will take four and a half thousand million years for just half of the nuclei in a lump of uranium (U) to decay.

On the other hand, it will take only 52 seconds for half of the nuclei in a sample of radon gas (Rn-220) to decay. These times are called **half-lives**, and we can measure them very accurately.

Picture 8(a) shows how the number of undecayed nuclei left in a sample of a radioactive element changes with time. Picture 8(b) shows this in the form of a graph. The **count rate** is a member of the number of rays given out per second.

As time goes on this **count rate** will go down, as more and more of the nuclei have decayed. What happens to the nucleus when it decays is dealt with in the next topic.

Your teacher should be able to show you an experiment about how the rate of radioactive decay gets less as time goes on. You should be able to get results to allow you to measure the half-life of a radioactive material. Some sample results are given in question 3 on page 121.

In activity A you can simulate radioactive decay.

Background radiation

Nuclear radiation is all around us. It comes from the rocks and the soil, it comes from plants and even animals. This is because they all contain tiny amounts of radioactive materials that are naturally present on Earth. Some comes from elements made radioactive by radiation from outer space, some from atomic weapon testing thirty or forty years ago. A small amount comes from the operation of nuclear power stations. Also, the nuclear accident in Russia (at Chernobyl) in 1986 produced radioactive 'fall out'.

Picture 9 summarises where this background radiation comes from. It is too low to have a serious effect on health, but it does have some effect. Doctors estimate that background radiation causes 1200 deaths from cancer per year in Britain. It is also likely that this radiation affects the genes in sex cells, so causing slight changes from one generation to the next. This may be one of the main causes of biological **variation**, which is necessary for evolution to occur.

Activity

Half-life

The idea of *half-life* is very important in nuclear physics. It is an example of how *random* events can produce a predictable result. A random event is one that we cannot predict — like the number you get when you throw a dice. Many calculators and computers will also be able to generate 'random numbers'. Radioactive decay is an example of a random event because we cannot predict when any given atom will in fact decay and give off radiation. Here are two ways of investigating such random numbers to help understand the idea of half life.

1 **Using dice**

Six-sided wooden cubes can be used instead of real dice, which are too expensive. You will need 500 of them! Each cube has a spot painted on one face. When you throw one cube, there is a one-in-six chance of the spot side facing upwards. The cubes represent atoms, and if they fall spot side up they have 'decayed'.

This activity is best done by a team, otherwise it takes quite a long time. Choose a clean floor, preferably free of furniture.

a **Guess!** How many cubes will have the spot side up when you throw all the dice?

b **Now try this**. Put all the cubes in a large box and shake them a lot. Pour them out on to the floor — taking care not to lose any. Pick out all the cubes that have fallen spot-side up and put them in aside. Count them. Was your prediction reasonably accurate?

c Collect all the dice that didn't 'decay' on this first throw. Now repeat stage (b) again, using these undecayed cubes. Again, collect and count the number that fell spot side up, and take them out of use. Repeat the throwing of the undecayed cubes another four times, so that you have six 'counts' altogether.

d Plot a graph of the number of cubes that decayed (the 'count') at each throw against the throw number (1 to 6). Does this look like the half life graph of radioactive decay in picture 8(b)? Explain why they are similar, and why they are different. Hint: a millionth of a millionth of a gram of radon gas contains over 2 thousand million atoms.

e Use the graph to calculate the 'half life' of the cubes — the number of throws needed for half of the cubes to 'decay'.

2 **Using a computer**

This simulation uses the ability of a computer to generate random numbers. It is written in BASIC and should run on all computers. You don't need to type in the REM statements.

```
10 CLS: REM . . .. a simulation of
radioactive decay
20 N = 2000: T = 0: remainder = N:
REM . . . N is number of nuclei at start,
when time T = 0, remainder are the
nuclei left unchanged
30 DIM nuclei(N): REM .. sets up an
array to store results
40 FOR T = 1 TO 10
50 D = 0: REM .. D is the number of
nuclei that will decay in each loop
60 FOR P = 0 TO remainder
70 LET nuclei(P) = RND(6): REM this
generates random numbers between 1
and 6
80 IF nuclei(P) = 1 THEN N = N – 1: D
= D + 1: REM ..if the random number
is 1 then a nucleus decays
90 remainder = N
100 NEXT P
110 PRINT T, D: REM. . . D is the
number of nuclei that decay in each
loop of the program printed next to the
'time' T
120 NEXT T
130 END
```
(NOTE: some BASIC needs to change line 70 to
```
70 LET nuclei(P) =INT(RND*5)
```

a Type in the program and RUN it. Copy down (or print out) the list of numbers and plot them against T.

b Does this graph look like the half life graph of radioactive decay in picture 8(b)? Explain why they are similar, and why they are different. Hint: a millionth of a millionth of a gram of radon gas contains over 2 thousand million atoms.

c Use your graph to work out the half life of the nuclei in this simulation.

d If you have time, investigate the effect of changing N and of changing the random numbers allowed (e.g. from 1 in 6 to 1 in 4).

Questions

1 Consider alpha, beta and gamma radiations. Which:

a travels at the speed of light?

b causes the most ionisation?

c has the same mass as an electron?

d is stopped by a few millimetres of aluminium?

e is the most massive?

2 Why aren't gamma rays affected by an electric field?

3 Use the figures in the table to work out the half life of the radioactive isotope (Polonium-218). There's a quick way and a long way – use the quick way!

Time/minutes	0	1	2	3	4	5	6	7
count rate (counts/second)	260	205	160	129	104	82	64	51

4 A sample of a radioactive substance was sending out 4000 alpha particles a second at the start of an experiment. Ten minutes later it was sending out 2000 particles per second.

a What is the half life of the substance?

b How long after the start would you expect to measure a count of 500 particles a second?

c How much activity (in counts per second) would you expect after 10 half lives?

d The natural 'background' count of radiation comes from radioactive materials in the ground or in the air. It is about two counts per second in most places. How long would it take for the radiation from the radioactive waste to be just less than this background count?

D6

What happens when atoms decay?

They make **new** atoms!

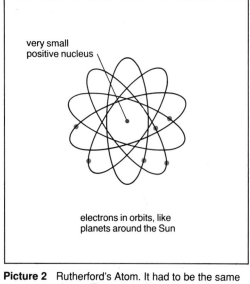

Picture 1 Thomson's Atom — the Plum Pudding Model

electrons (the plums)

— embedded in positive matter (the pudding)

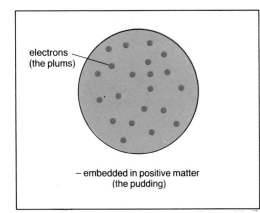

very small positive nucleus

electrons in orbits, like planets around the Sun

Picture 2 Rutherford's Atom. It had to be the same size as Thomson's Atom. But it was mostly empty space

The nuclear atom

We now know that atoms are made up of a small, massive **nucleus**, which is electrically charged (positive). This is surrounded by a cloud of very tiny negative particles — **electrons**.

Electrons had already been discovered in 1895 before Henri Becquerel discovered radioactivity. At that time, physicists thought that the electrons were negative particles somehow embedded in a kind of positive blob, like bits of chocolate in a chocolate chip cookie (picture 1).

A team of physicists led by the New Zealander Ernest Rutherford showed that the atom was not like a cookie at all. It was more like a tiny solar system, with electrons orbiting a large positive nucleus like planets orbit the Sun. Picture 2 shows Rutherford's model of the atom. There is more about the structure of the atom in *The Material World*, topic J3.

The nucleus is itself made up of simpler particles — **protons** and **neutrons**. These particles are very nearly equal in size and mass, but protons are positively charged and neutrons are uncharged (neutral).

What causes radioactivity?

The nuclei of some elements are so large that they are unstable. The uranium nucleus is like this. As time goes on, sooner or later, the nucleus will break apart. This is called **nuclear decay**. As the nucleus breaks apart it gets rid of some energy. The energy is carried away by the radiation, the alpha, beta or gamma rays. For example, a nucleus of uranium 238 decays by sending out an alpha particle at a very high speed. But the nucleus is still unstable, and will decay once more. This time it sends out a beta particle.

Even after this, the nucleus is still unstable. In fact, these changes will keep happening for quite a long time. For most of the changes the nucleus sends out a gamma ray as well as some mass in the form of a particle. The original uranium nucleus goes through 14 changes before it becomes stable, and so non-radioactive. U-238 has become a lead nucleus, Pb-206.

What happens to the nucleus when it changes?

The nucleus of uranium 238 is the largest nucleus of any natural element. The '238' refers to the fact that it has 238 particles in it. It contains 92 protons and 146 neutrons.

When uranium 238 decays it sends out an **alpha particle**. An alpha particle has two protons and two neutrons. The nucleus is now lighter (by 4 particles) and has less charge (by 2 units). *This means that it has become the nucleus of a different element.* In fact it has become a nucleus of thorium — thorium 234. One of the neutrons inside the nucleus sends out an electron, as a beta particle, and turns into a proton. But the thorium nucleus is also unstable.

The *mass* of the thorium nucleus stays much the same. After all, it has only lost an electron, which doesn't weigh very much at all. But it has now one more proton. This means that it behaves chemically like a different element. It has become an element called protactinium.

The decay chain doesn't stop there, however. Protactinium is also radioactive. It too decays by sending out a beta particle. This means that it too becomes a different element. Yet more changes have to take place before, finally, a stable nucleus is formed. This will be the element **lead**.

Some of these changes are shown in picture 3.

In many of these changes the nucleus also gets rid of some energy by emitting gamma radiation. This has no mass or charge, so the nucleus does not change into a different one.

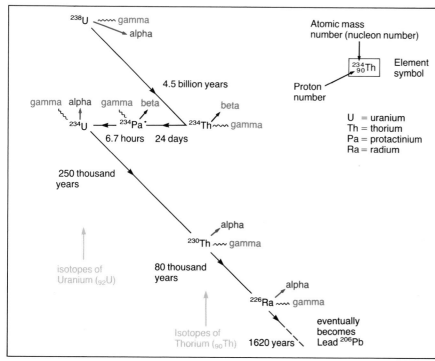

Picture 3 Radioactive decay — how one nucleus changes into another

Isotopes

As the nuclei change, they may become isotopes. A nucleus can lose two positive charges by emitting an alpha particle, then two negative charges by emitting beta particles. The charge on the nucleus is back to what it was originally. But the nucleus has lost mass. Nuclei with the same charge but with different masses are called isotopes.

Picture 3 shows some examples of this.

Why don't nuclei decay all at once?

No one knows the answer to this question. All we know is that some nuclei tend to decay very quickly and others very slowly. Also, no one can predict when any given nucleus will decay. It all seems to be a matter of chance, or 'luck'.

It may seem strange that what happens in such a precise subject as physics depends on 'luck'. But the laws of 'luck' are quite well known. They deal with the **probability** of some event actually occuring. Scientists have known for a long time that many everyday things are decided by the laws of probability as much as by the precise rules of Newton, for example. Many things seem to happen, or to move, in quite a **random** way. Just think of the weather, or the way molecules move in a gas.

But we can still make predictions about gases and about the weather. They might not always be accurate predictions, especially when it comes to weather forecasting. But the 'gas laws' (see *The Material World*, topics C1 and C2) are quite accurate, even though we can't predict how any single molecule in the gas is going to move. The laws are accurate because there are *so many* molecules in any reasonable quantity of gas. The different speeds and energies of the molecules average out to something quite predictable.

In much the same way we can predict quite accurately the 'activity' of a collection of radioactive nuclei. Again, this is because there are so many of them. In just one hundredth of a gram of uranium, for example, there are 25 billion billion nuclei! We cannot predict what any one nucleus is going to do, but on average we can say quite accurately how many of them will decay in the next hour.

Questions

1a Why does an atom normally have just as many protons as electrons?

b What happens to an atom when it loses an electron?

c Give an example of what might make an atom lose an electron.

2 What is the difference between a proton and a neutron?

3 In the simple model of an atom, we believe that electrons go around the nucleus in 'orbits'.

a Explain what an *orbit* is.

b What might happen if the electrons stopped moving?

4 'The radioactive decay of an atom is a random event'. Make up two other sentences in which the word 'random' is correctly used.

5a The probability of getting a 'six' in a game of dice is 1 in 6. Why?

b What is the probability of choosing a Queen of Hearts in a card game with a well shuffled pack (52 cards)?

c The probability of a certain type of nucleus decaying in the next 10 seconds is 1 in 5 million. How many would you expect to decay in the next ten seconds in a sample of 100 million nuclei?

D7 Nuclear energy

Nuclear energy powers the movement of continents. We can also use nuclear energy directly — but can we do this safely?

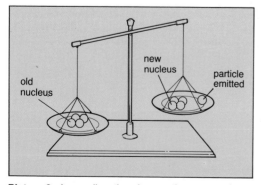

Picture 1 A heart pacemaker: it uses a 'nuclear battery'

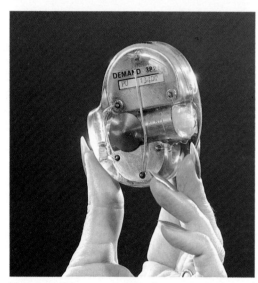

Picture 2 In a radioactive change, the new products weigh less than the original nucleus

'Moonshine!' says top nuclear scientist

Ernest Rutherford was the physicist who found out what alpha particles were, and who discovered the nucleus of the atom. *There is enough energy in a gram of uranium to send a liner across the Atlantic!* he said in a newspaper interview, 60 years ago. *But the idea that you can actually use it is moonshine. It will take over 4 billion years to get just a half of it out!*.

He was thinking of radioactivity, not nuclear power stations. When a natural radioactive substance decays, it gives out energy — but only slowly.

Nowadays we have artificial isotopes that do give out their energy in a shorter time. They are used to power instruments in Earth satellites. They are also used in heart pacemakers.

Even so, the power they can generate is very small and it is very expensive. Nuclear power stations do not rely on 'ordinary radioactivity' to produce energy. But where does the energy come from?

$E = mc^2$

When atoms decay, the energy they release appears as movement energy (kinetic energy) of the high-speed particles they send out. It was Albert Einstein (see topic B10) who came up with the surprising theory that this energy was produced by changing some of the mass of the nucleus into energy. This theory was proved when careful measurements were made. The mass of the emitted particle plus the mass of the nucleus left behind was *less than the mass of the original nucleus* (picture 2).

Energy had appeared. Mass had disappeared. These two facts contradicted the laws of physics as understood in 1905. These laws said that:

- energy could not be lost or created (Law of Conservation of Energy),
- mass could not be made or destroyed (Law of Conservation of Matter).

Einstein would say: *If we say that matter is really a kind of stored energy, then both laws can be correct. In fact, my Theory of Relativity predicts that this should happen. The 'energy value' of any piece of matter is given by my formula* $E = mc^2$. *No problem.*

The missing mass, *m*, was converted to kinetic energy, *E*. The quantity *c* is the speed of light. This is a very large number — 300 000 000 m/s. It is the *square* of this number which multiplies with the mass to give the value of the energy. This means that it doesn't take much mass to produce a lot of energy.

But the main problem still remained. There was no way to speed up the rate at which ordinary, naturally radioactive materials decayed and produced their energy. The energy was there, but dammed up so well that it could only trickle out at a uselessly small rate.

Nuclear fission

The breakthrough into 'atomic energy' came in 1938. Physicists working in Berlin proved that some of the unstable uranium nuclei don't just decay by giving out a small particle or some gamma radiation, as described in topic D5. Instead, they split up into two nearly equal parts.

But just as with radioactive changes, mass was lost and converted to energy. What was more important, *the splitting could be controlled*. Then, in 1939, the Second World War began. It was clear to some physicists that the immense store of energy in a lump of uranium could be released very quickly. The result would be a huge explosion — a **nuclear bomb**.

The bomb was built — it took five years to do this — and two 'atomic bombs' were dropped on Japan by the USA in the summer of 1945. The nuclear age had begun.

The process of splitting nuclei to give energy is called nuclear **fission**. The same process is used in a **nuclear reactor**, but of course it is controlled so that it happens much more slowly than in a bomb. To explain how it works we need to remember what the nucleus of an atom is like.

Atomic nuclei

Atoms contain negative electrons moving around a positive core — the nucleus. But a nucleus is not just a blob of positively charged matter. The main parts of a nucleus are the protons and the neutrons (see picture 3). Protons are positively charged and neutrons do not carry an electric charge.

Some nuclei have too many protons and neutrons and tend to be unstable. This causes radioactivity (See topics D5 and D6). But some very large nuclei may split into two parts, instead of undergoing ordinary radioactive decay.

The chain reaction

This splitting is what goes on in a nuclear bomb or reactor. When the nucleus splits in two main parts it also shoots out one or more spare neutrons. These can fly into another nearby nucleus quite easily — and make that nucleus split. In turn, the new neutrons may shoot off into other nuclei and cause them to split.

This builds up into a **chain reaction**, with nucleus after nucleus splitting and triggering off others. Each time a nucleus splits it gives out energy. This process is shown in picture 4. You can compare it to an avalanche on a mountain slope. When someone throws just one stone it can cause them all to cascade down the slope.

The result of an uncontrolled chain reaction is an explosion. This is what happens in a nuclear bomb. In a nuclear reactor the reaction is controlled. The uranium is spread out, as thin rods (**fuel rods**). In between the rods are other rods, made of a material that absorbs neutrons. If all the neutrons are absorbed no further reactions are possible and the reactor stops giving out energy.

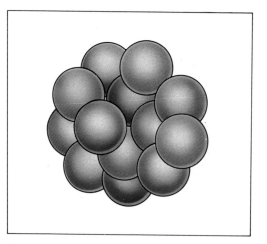

Picture 3 An oxygen nucleus. It has 8 protons and 8 neutrons

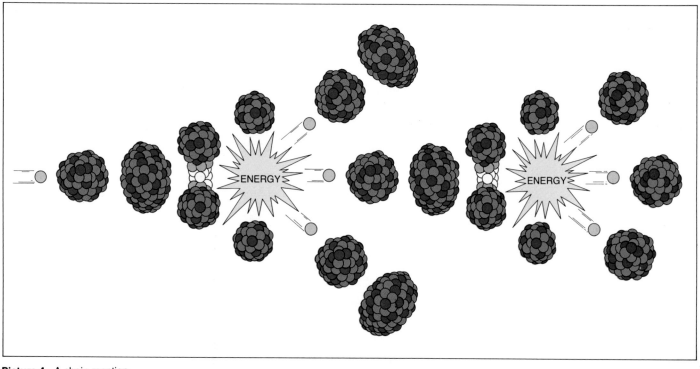

Picture 4 A chain reaction

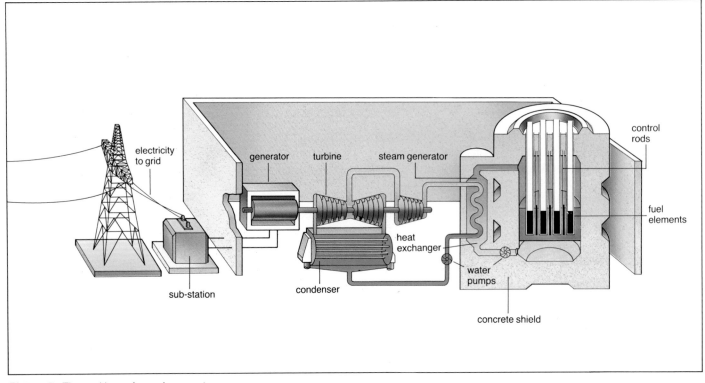

Picture 5 The workings of a nuclear reactor

In a typical reactor, the absorbing rods (**control rods**) are moved in and out of the fuel rods. This controls the rate of the fission reactions and the amount of energy released. Picture 5 shows the main parts of a nuclear reactor. Other rods (**moderators**) slow down the neutrons so that they are better at causing fission.

The energy released when the nuclei split up heats the fuel rods. The hot 'core' of the reactor then heats water or some other fluid. This hot fluid may become radioactive and has to be kept inside the reactor. The fluid moves to carry its energy to the heat exchanger where it is transferred to boil water. The steam produced drives an ordinary steam turbine.

The turbines turn generators, as in an ordinary coal-fired power station.

Nuclear waste

When the uranium nuclei in the fuel rods split they form two smaller nuclei. These nuclei are radioactive. Eventually the fuel rods are used up and have to be replaced. The old rods are now very radioactive.

Also, the neutrons that are not used in the fission process are absorbed by control rods and other parts of the reactor. This makes them become radioactive as well.

After anything from 20 to 50 years, the working parts of the reactor are worn out, and have also become highly radioactive. When the reactor is dismantled these materials have somehow to be disposed of. This process is called **decommissioning**. It now seems that this could be very expensive. This means that the total costs of nuclear power are much higher than was thought when the first nuclear power stations were built.

Nuclear fusion

But there is yet another way to get energy from nuclei. The Sun and the stars get their energy from the opposite process to nuclear fission. They use the fact that when light nuclei join together to make heavier ones, energy can be released. Once again, the energy comes from lost mass.

In the Sun, the process involves the nuclei of hydrogen isotopes (protons and deuterons) joining together to form helium nuclei. This process is called nuclear **fusion**.

Could we use fusion to produce energy on Earth? The This energy would come from a very cheap material — hydrogen. The process would produce less radioactive waste. Research has been going on for nearly fifty years to produce **controlled fusion**. So far this research has been unsuccessful, the only 'practical' device to have been developed is the hydrogen bomb (picture 6). Although the latest experiments have shown some success, this source of energy is not likely to be available until well into the 21st century.

Picture 6 A hydrogen bomb. So far, we can't control nuclear fusion to give a steady supply of energy

Activities

A Find out about nuclear power stations

Read all you can about nuclear power stations. People disagree about whether nuclear power is a good idea or not.

Make a list of the points that could be made both *for* and *against* using nuclear power to generate electricity. You should consider: *safety, cost per unit of electricity, pollution, convenience of use, storing nuclear waste, the Greenhouse Effect, etc.*

Some of the points will be 'scientific', but some may be important but not based on scientific evidence. Mark the ones you think are scientific with a capital S.

B The case for and against nuclear energy

Use a library to find out about the following. Make short notes on each one, then combine your thoughts and knowledge into a poster which could be entitled either:

1 Ban the bomb! or
2 Nuclear weapons have kept the Free World Free!

Topics:
1 The effects of nuclear radiation
2 The hydrogen bomb
3 Hiroshima
4 The end of the 'Cold War'
5 ICBMs
6 The nuclear winter

Try to get a good balance of fact and opinion.

Questions

1 Explain briefly why a nuclear chain reaction is like an avalanche of snow or rocks on a steep mountainside.
2 Nuclear 'fuel' — uranium — is quite cheap. Give two reasons why it is still expensive to produce electricity from nuclear power stations.
3 Explain what the *control rods* do in a nuclear reactor.

4 Some revision from your chemistry studies:
a Describe an atomic nucleus.
b 'Carbon 14 is an *isotope* of carbon 12'. Explain what an isotope is, in terms of the difference between these two types of carbon.
5 People argue a great deal about whether nuclear power stations should be built. Give three reasons in each case:
a in favour of nuclear power,
b against nuclear power.

6a Use the formula $E = mc^2$ to calculate how much energy in joules could be obtained from 1 kg of matter if all of it could be turned into energy. The speed of light, c, is 300 000 000 m/s.
b The total energy we can actually get from 1 kg of uranium is very much less than the answer you should have got for part (a). Why is this?

D8 Radiation and life

Ionising radiations are all around us. Only a small amount is due to human activities.

Picture 1 Granite outcrop in Derbyshire — granite often contains radioactive elements

Picture 2 Nuclear waste transporter

Why is radioactivity dangerous?

The alpha, beta and gamma radiations from a radioactive material travel at high speeds. Also, alpha and beta particles are electrically charged. When they go into living cells they can kill or damage them. A large 'dose' of radiation kills so many cells that the effect is like being burned by fire.

Alpha particles are easily stopped. Even a sheet of paper is enough. But beta and gamma radiations can get deep into the body so that cells in internal organs are damaged (see topic D5).

For low levels of radiation you lose just a few cells. This in itself is not dangerous. After all, our bodies contain lots of cells. The danger comes from the fact that the radiation can *change* some of the chemicals we have in cells. These are the complex molecules of DNA and RNA that control how the cell works (see *The Living World*, topic F6).

As a result, the cell may go out of control. It grows and divides just as if it was an independent organism that feeds on our bodies. It becomes a group of **cancer cells**.

Also, if sperm or ova cells in the reproductive system are changed, the result could be children with birth defects. They could be born with badly formed or missing limbs, or brain damage.

Table 1 Exposure to radiation in everyday situations

Type of exposure	microsievert
1 radiation due to nuclear power stations for a year	10
2 watching television for a year	10
3 wearing a radioactive luminous watch for a year now (not very common)	30
4 having a chest X-ray	200
5 exposure to fall-out in Britain from nuclear bomb testing in 1959	350
6 radiation from a brick house, per year	750
7 working from a month in a uranium mine	1000
8 typical dose received by a member of the general public in a year	1500
9 maximum dose allowed to general public per year	5000
10 maximum dose allowed to workers exposed to radiation per year	50 000

Table 1 shows the radiation doses you might get in different situations, together with the 'allowed doses' for ordinary people and workers in the nuclear industries. They are measured in **microsieverts**, which is a measure of how much energy the radiation delivers to the body.

We cannot escape from ionising radiation. Some comes from outer space, as fast moving particles. Most rocks contain a tiny amount of some radioactive elements. In many parts of the country radioactive decay produces the radioactive gas **radon**. This seeps into buildings and can be breathed in. The effect of all these sources is to produce the natural **background radiation**.

Granite rocks, which have come from deep inside the Earth, contain more than the average amount of radiocative elements (picture 1). People living in granite areas should make sure that their houses are well ventilated. They should get their homes checked now and again for radon, and if too much is found they should seal their floors.

Biologists believe that background radiation is one of the causes of changes to genes in the reproductive cells of all living things. These changes help to produce **variation** in living things, allowing evolution to take place.

Half life and nuclear waste

As explained in topic D7, nuclear reactors produce **nuclear wastes**. Picture 2 shows waste being transported. Radioactive materials are also used in industry, and produce waste. Some of these waste materials have very short half-lives (see topic D5). This means that they give out their radiation very quickly. They are very dangerous — for a short while. But most of the radiation is gone after a few months.

For any radioactive material, after 10 half-lives the material is only a thousandth as active as it was at the start. It has been divided in half ten times. But some radioactive waste materials have half lives running into hundreds of years. These materials are called low-level waste. Storing these safely is a major problem. There is no way we can speed up the rate at which they decay. Half-lives are fixed and do not change.

Storing radioactive waste

Beause it is dangerous to humans, low level radioactive waste has to be stored safely, so that the radiations don't get out. It is stored inside containers made of metal, glass or concrete which absorb the radiations. If low-level waste is to be stored underground, great care will have to be taken to make sure that the containers stay unbroken, perhaps for thousands of years.

Picture 3 shows how radioactive waste might be stored underground.

These problems of waste disposal have caused many countries, including the UK, to draw back from building a lot of new nuclear power stations. The cost of dismantling the power stations safely and storing the waste seems to make nuclear energy less economic than people once thought.

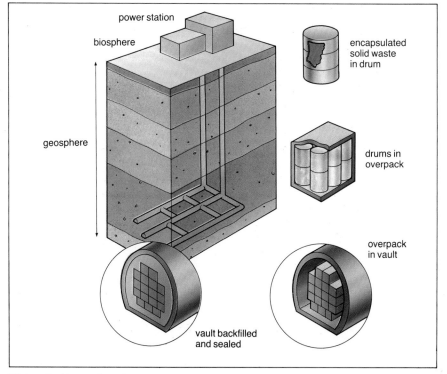

Picture 3 How radioactive waste might be stored

Activities

A Radiations all around us
Use a library to find out about one of the following uses of radiation. Make notes or a poster so that you can report back to the rest of the class.

1 Smoke detectors.
2 Radioactive tracers in hospitals.
3 Using radioactive materials to check for faults in metal objects.
4 Finding the age of old objects using carbon 14.
5 Using radiation to treat cancer
6 Cosmic rays.

B Are you at risk?
Radon gas is radioactive. It can be dangerous if it is allowed to build up in houses or other buildings. Find out what kind of radiation it emits.

Trace a map of the British Isles. Use an atlas to find out which areas are over granite or other igneous rocks. Mark these areas on your map. Find out whether your local council has any plans for checking or dealing with radon gas.

Questions

1 You can't escape from ionising radiation on Earth. Name two natural sources of this radiation.
2 Describe briefly two effects of ionising radiation on living things.

3a What are the three kinds of ionising radiations that are emitted by a radioactive substance?
 b Describe how the human body can be protected from each type.
 c From which of the sources of radiation given in table 1 are you personally most likely to be at risk?
4 Why are radioactive substances with a short half life more active than those with a long half life?
5 Describe two ways in which radioactive waste (e.g. from hospitals) could be stored safely.
6 Ultra violet light is also an 'ionising radiation'. What dangers does this present to human beings on Earth? How can these dangers be reduced?

Good tidings?

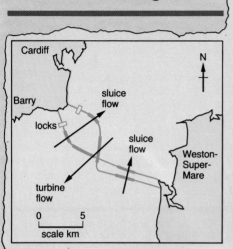

Picture 1 Map of the proposed barrage

Map labels: Cardiff, Barry, locks, sluice flow, sluice flow, turbine flow, Weston-Super-Mare, N, 0 5 scale km

New Energy Source Planned for South Wales!

Picture 2 Artist's impression of the Severn Barrage

Picture 3 A turbine blade

Plans have been published for a major project that might revolutionise Britain's energy resources. This will use the fashionable idea of *renewable energy*. The idea is to build a huge dam (called a **barrage**) across the estuary of the River Severn, between Cardiff and Weston-super-Mare (see map, picture 1).

The estuary acts as a funnel for the tides as they flow in. At the site of the proposed dam the sea rises and falls a remarkable 11 metres twice a day. This is more than the height of a 3-storey house. The barrage will be 16 km long. Millions of tonnes of sea water will be channelled to flow through turbines as the tide comes in and out.

The result will be a supply of 8000 MW of electricity. This would make it the largest tidal power station in the world. Its power output would be equivalent to four nuclear power stations, or a 'wind farm' of 2500 wind generators, each 90 metres tall!

Picture 2 shows what it would be like if it was built. The turbine blades would be huge, as our artist's impression shows (picture 3). This is the most exciting project for pollution-free energy ever!

Letters to the Editor

Threatened wildlife

I was horrified to read the article about the dam across the Severn Estuary. Don't these people know that this is the one of the best wildlife sites in Britain? Every year millions of wading birds and wildfowl feed and nest in the estuary. The mudbanks are one of the few remaining wetland areas that these birds can use. If this barrage is built these mudbanks will be lost for ever. Is Man's greed for ever more energy to waste on unnecessary luxuries like colour TV and central heating yet again going to kill millions of other living things on this planet?

Think again, planners. Save, don't waste!

Yours sincerely,

Amanda Williams, President
Severn Society for the Protection of Birds

Cheap at the price!

This proposal for a Severn Barrage is to be greatly wecomed. We have already wasted billions of pounds of taxpayers' money on dangerous and unsightly nuclear power stations. It wouldn't be so bad if they could produce cheaper electricity!

Over the years thousands of Welsh coalminers have lost their lives producing coal which is burned and lost for ever. Think of how much harm this is doing to the Earth through the Greenhouse Effect! This barrage will cost just 8 billion pounds! Worth every penny, I'd say.

Yours sincerely,

D. Evans, (Ex-miner, Monmouth).

Power for the future

Don't these moaners and whingers realise that if this country of ours is going to play its rightful part in the new Europe we need plentiful supplies of energy. Otherwise we are going to be left behind even places like Spain and Italy!

And don't they realise that coal has had its day? Haven't they heard of the Greenhouse Effect? Would they rather have four nuclear power stations on the estuary? Which is more important, human beings or birds?

Come on! Let us move boldly into the future!

Yours very sincerely,

Jane Smithers
Managing Director, New Age Electronics Ltd

Nuclear benefits

So what is wrong with having four nuclear power stations on the Severn Estuary? The one already there at Hinckley Point has done no harm whatsoever. Billions of pounds have been spent already in developing this highly efficient and safe (yes safe!) source of energy. How many people have been killed in or around any nuclear power station in the UK? None! Think of how many miners have been killed in the mines — thousands!

The very tides that will be tamed in this ridiculous, anti-wildlife and unsightly project are carrying away the waste heat from the nuclear power station. The warmth encourages fish and will improve fishing in the area. The dam will in fact stop any of the young salmon swimming up the estuary to the head of this world-famous fishing river.

A well-regulated nuclear power station is pollution-free, takes up little space and will produce electricity at a third of the fuel cost of coal.

This is the real future for Britain's energy. After all, there is only one Severn Estuary — we can build as many nuclear power stations as we like.

Yours sincerely,

L. Marshall
Department of Engineering, Avon University

Barry beach

What about the beach at barry island? All the sand will be under water. Its not fair.

J. Lewis (age 7)

Role play — the public inquiry

By law, a Public Inquiry has to be held before the Barrage can be built. Anyone with an interest can ask to give evidence or make a case at this inquiry. It will be chaired by an eminent lawyer, who should not take sides and is willing to help inexperienced people.

You will be given (or may be allowed to choose) the role of one of the letter writers above, or a representative of the Department of Energy (Mr Nigel Smythe or Ms Louisa Davenport).

You may need a small team of researchers to help you prepare your case. You may be able to think of other groups of people with an interest in the development, such as local anglers, councillors and hotel keepers from one of the many holiday resorts in the area.

At the end of the activity, the class can vote for or against the proposal.

A sun-warmed house

Picture 1 Using the Sun's heat to keep warm

Picture 1 shows some houses in Chorley, Lancashire. They have been designed to make best use of the Sun's heat. The houses face south, and they have a special energy-trap, called a **sunspace**, at the front.

This is made mostly of glass. In winter the energy that enters the room makes it comfortably warm as long as the sun shines. The floor is made of special tiles that can 'hold the heat'. Picture 2 shows the principle of two kinds of sunspace.

The rest of the house still has to be heated in winter, but even so the sunspace is useful when the sun isn't shining. Energy that would normally escape through an outside wall is kept inside, because the sunspace acts as a good insulator.

In these specially designed houses there are only small windows on the north side of the house. This is to help keep down energy loss. Also, the walls are built to be good insulators. Once the house is warm, its walls and furnishings act as a large 'heat store'. This means it stays warm for a long time. Heating costs are cut to about a half of what they are in a normal house. Try answering the questions.

1 Most energy leaves a house by means of *thermal conduction*.

a Name (i) three materials which are often used in buildings and are good insulators, (ii) three materials which are bad insulators.

b Explain briefly how 'thermal conduction' actually happens.

2a How does 'solar energy' actually get from the sun into the house?

b What problem might the people who live in these 'sunspace' houses have in the summer? How could they solve this problem?

3 What physical property decides whether or not a material is 'good at holding heat'?

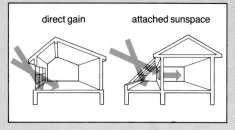

Picture 2 Two kinds of sunspace

4 The walls of an energy-efficient house might be 'cavity walls', which means that they are double walls with an air space in between.

a Describe how such walls help to reduce energy losses.

b What other part of the sunspace (other than its brick walls) is likely to use the same principle as a cavity wall?

5 A normal house might cost its owners £400 in heating costs during the winter. An energy-efficient house will reduce this by at least 50%. But it might cost £6000 more to build.

a How long will it take this extra building cost to be paid back in energy saving? Do you think it is worth building such houses?

b One of the main causes of the greenhouse effect (see *The Material World*, page 14) is burning fossil fuels. How could building energy-efficient houses help counteract this effect? What could a responsible government do to help people make their houses more efficient?

E1 Electric signals

The brain, computers, traffic lights, telephones — all these deal with messages. They all use electricity to make them work.

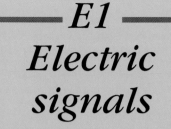

Picture 1   Electric currents can carry messages

Picture 2   Electricity is used for controlling light, as well as producing it.

Picture 3   A simple circuit

Messages and moving charges

When you pick up a telephone an electric current starts to flow. When you speak into it the microphone in the handset changes the strength of the current to match the sounds you make (see page 72).

When a car comes up to some traffic lights it affects a magnetic sensor in the road. This sends an electric message to a control box to let it know that the car is there. The control box has to decide how and when the lights need to change (picture 2).

When you tread on a drawing pin in bare feet your pain sensors send a message to both brain and muscles so that you react pretty quickly. The message is carried by a moving electric pulse through very long nerve cells (see *The Living World*, topic D1).

What is needed for electricity to be a messenger?

These devices use electricity to carry a message, and for this to work three things are needed:

- electric charge which is free to move,
- energy to move the charge,
- a material that charges can move through — a conductor.

Circuits

The examples above are quite complicated devices, using many conductors connected together in complicated ways. But however complicated it may look the basic plan of any electrical device is quite simple: it is a collection of **circuits**. A simple circuit (picture 3) has the energy source (a battery), a switch, some wires and, say, an electric buzzer. The wires are made of a metal which is a good conductor — usually copper.

Luckily, the electric charges are already in the conductor! A metal contains many millions of charged particles that are free to move. These particles are called **electrons**, and all atoms contain them.

For a signal to get to the buzzer, the buzzer it has to be connected to the battery by conductors. All the battery does is to provide a *force* to make the electrons move. How it does this is explained later (page 165).

The switch is there to make a gap in the circuit. Air is not a conductor, so electrons cannot move across the gap. It doesn't matter where the gap is. It is like a blockage on a single road system (picture 4). If cars pile up at the gap, they cause a traffic jam that will tail right back to stop cars getting any further. For traffic to flow it has to be able to get into and *out* of the system.

Exactly the same principle applies to electric charges. They need a complete path which runs from one side of the battery to the other. A gap anywhere in the circuit will stop the charge moving. When you close the gap by pressing the switch, the charge starts moving and carries its message to the buzzer, which makes a sound.

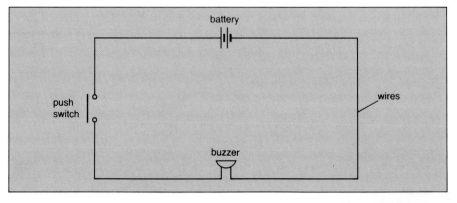

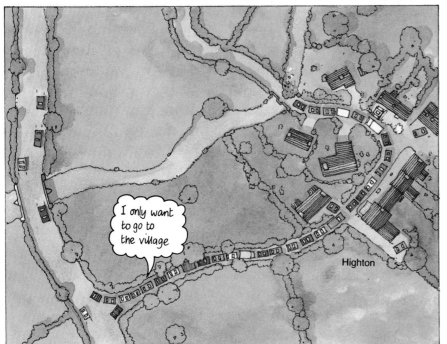

Picture 4 Electricity is like traffic; a gap in a circuit is like a blocked road

What can electricity do?

We recognise electricity by its effects. The moving charges in a circuit can:

1 Make conductors hot (in lamps, heaters).
2 Produce magnetic forces (in electromagnets, motors, loudspeakers).
3 Make special chemicals give out light (in LEDs, fluorescent lamps, TV screens).
4 Cause chemical changes to take place (in electroplating, charging batteries, etc).

The first three effects are especially useful in everyday life. Most electrical devices in the home use one or more of these effects (see picture 5). The rest of the topics in this section describe and explain the uses of these effects more fully.

Electric charge and electric current

An electric current is a flow of charged particles (see also topic E5). Current is measured in amperes, by an instrument called an ammeter. A small torch bulb may carry a current of 0.2 amperes (0.2 A), a car headlamp bulb may carry 20 times as much (4 A).

The current in a lamp is a flow of electrons. The charge on a single electron is very very small. It takes many millions of moving electrons to carry the charge that flows through an ordinary torch bulb in just one second. This large number is hard to think about and work with, so we need a more sensible unit of charge. Instead, we use **the quantity that moves when a current of 1 ampere flows for one second. This amount of charge is called a** *coulomb* **(or C for short).**

Obviously, if 1 ampere flows for 2 seconds then 2 coulombs of charge will have moved; if 2 amperes flows for 2 seconds then 4 coulombs will move (picture 6 overleaf).

Charge, current and time are linked as follows:

$$\text{charge moved} = \text{current} \times \text{time}$$

$$(\text{coulombs}) = (\text{amperes} \times \text{seconds})$$

(a) Lighting

(b) Keeping cool

(c) Listening to music

(d) Ironing

(e) Watching television

Picture 5 Electricity can be used for many different jobs

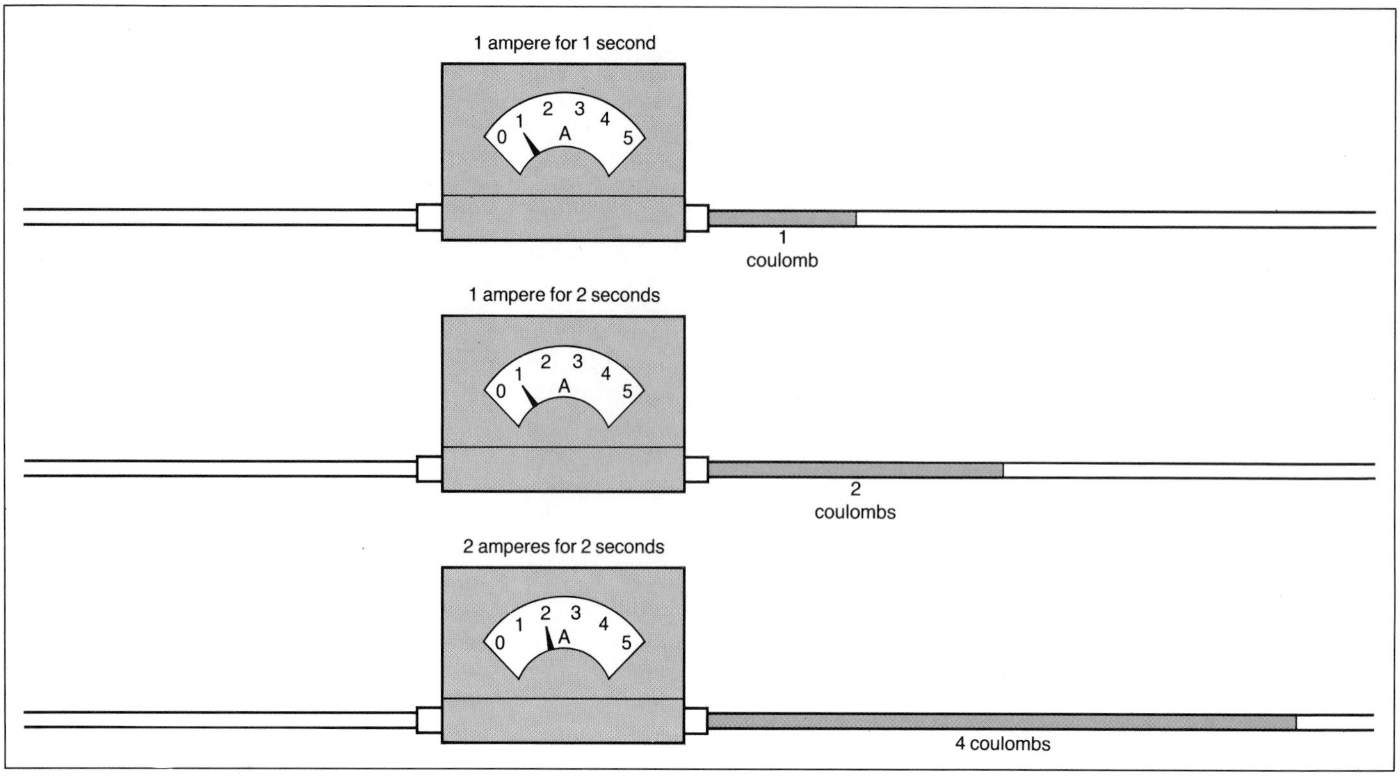

Picture 6 Charge = current x time

Symbols, circuits and formulate

The best way to learn about electricity is by doing experiments and investigations, solving problems, making things work — and talking about them. The activities at the end of the topics will help you to do this. You will meet a lot of new ideas and have to work things out for yourself — with help from your teacher.

Electricity is so useful that it has a language of its own, mostly using circuit diagrams and special symbols to stand for the devices that it uses.

Formulae are used to work out what is happening, or going to happen, in a circuit. All the information that you will need in this section is summarised in table 1 below. Use it when you need to look things up.

Table 1 Symbols and formulae in electricity
The rest of the topics in this section explain what these words and formulae mean, and how to use them. They are all collected here for you to use when you need to.

Circuit symbols			
cell	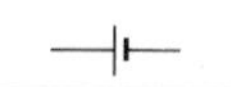	capacitor	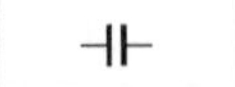
battery	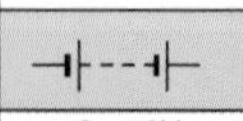	microphone	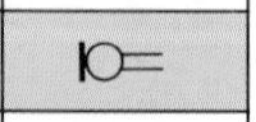
variable power supply	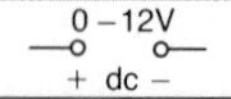	motor	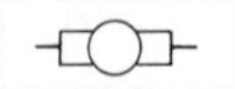
lamps	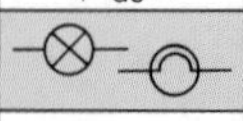	loudspeaker	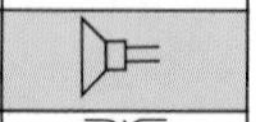
resistor	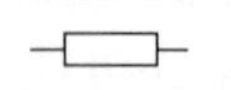	transformer	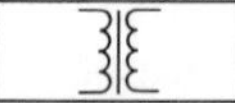
switches	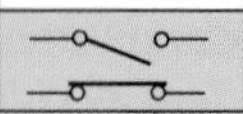	bell	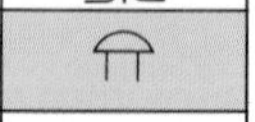
LED	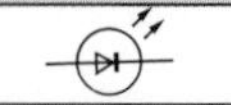	buzzer	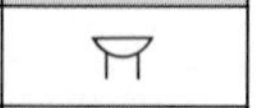
variable resistor	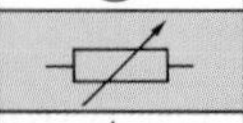	ammeter	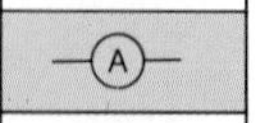
voltage divider	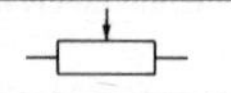	voltmeter	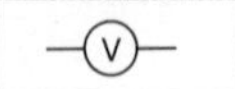

Symbols for quantities

current I potential difference V power P
resistance R (voltage)
charge Q energy E
              time in seconds t capacitance C

Symbols for units (measurements)

amperes (current) A volts V (ohms) resistance Ω

Useful formulae in electricity

Charge, current, time $\qquad Q = It$
Voltage, current, resistance $\qquad V = IR$

Power $\qquad P = VI \quad P = I^2R \quad P = \dfrac{V^2}{R}$

Resistors in series $\qquad R_t = R_1 + R_2 + R_3 + $ etc

Resistors in parallel $\qquad \dfrac{1}{R_t} = \dfrac{1}{R_1} + \dfrac{1}{R_2} + \dfrac{1}{R_3}$ etc

Activities

A   What uses electricity?
Make a list of as many things as you can see in the room that use electricity.

B   What do people think about electricity?
1 (Home activity.) Find ten people and get them to tell you in one sentence what they think electricity is. Write down what they say and tick the statements that you agree with.
2 (Group activity.) Get together with three or four other pupils in your class and look at the thirty or forty statements you have obtained between you.

Choose the ten most important — or interesting — and write them out on a poster-sized piece of paper for the rest of the class to see.
3 (Class activity.) Decide, as a class, the final ten most important, accurate or interesting statements about electricity that have been produced.

The result could seriously worry your teacher!

Questions

1 Make lists of:
a five materials that don't conduct electricity (insulators),
b materials that do conduct electricity (conductors).
2 What units and instruments are used to measure:
a current,
b voltage,
c resistance?
3 What carries the electric charge through a lamp filament?
4 Electric current in wires is a flow of electrons. Why then don't scientists and electricians measure electric current in 'electrons per second'?
5 Why don't electric charges flow through a conductor unless it is part of a 'complete circuit'?
6 Why are metals usually good conductors of electricity?

7 Use table 1 to help you decide what will happen when the switch is pressed in each of the following circuits. (picture 7)

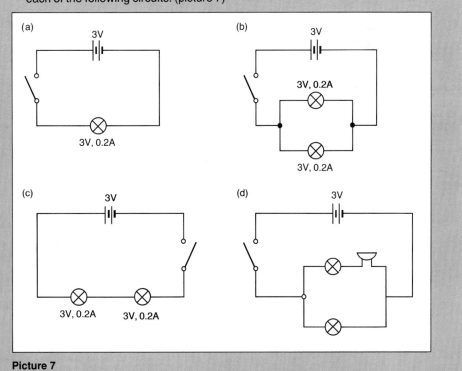

Picture 7

E2 Magnets

Magnetism is a mysterious kind of force. It comes from two sorts of magnets, permanent magnets and electromagnets.

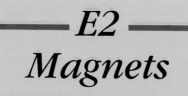

Magnets and magnetic forces

Most people play with magnets when they are children. They are useful for finding lost pins — but aren't any good for finding lost coins. These magnets are the kind that keep their 'magnetic power' for a long time, and so are called **permanent magnets**.

Permanent magnets are usually made from a special kind of steel. They can attract other things made out of iron or steel. But when you have two magnets together another strange effect can be seen. They can repel each other.

The magnetic compass

A magnet tied to a piece of thread so that is free to swing will line itself up in a roughly north-south direction. This effect was discovered by the Chinese over a thousand years ago; they soon used it to help them find their way at sea and in unknown country. This effect is used in the magnetic compass. (See picture 1.)

The mysteries of magnetism began to be solved about 400 years ago, in the time of the first Queen Elizabeth of England. It was a time of great sea voyages, when Western Europeans began to sail the world in search of trade, plunder and conquest.

Queen Elizabeth had a doctor called William Gilbert. He experimented with magnets and compasses and produced a theory to explain how compasses worked. He claimed that it must be because the whole Earth is a magnet. The huge Earth-magnet attracts and repels the small compass magnets so that they always line up in the same way (picture 2). On the whole, he had the right idea.

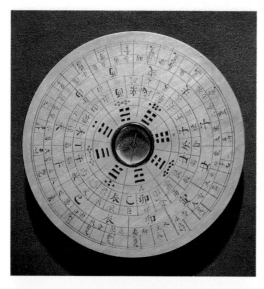

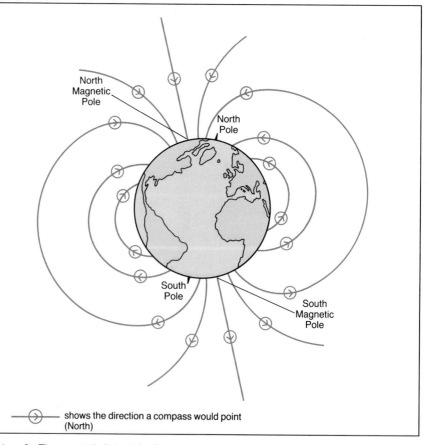

Picture 1 Compasses were invented by the Chinese

Picture 2 The magnetic field of the Earth

Magnetic fields

You can make magnets any shape you like. The simplest shape is a bar magnet. Cover a bar magnet with a sheet of white paper, carefully sprinkle iron filings over the paper and tap it gently with a pencil. You will see the filings forming into a pattern. What has happened is that the small bits of iron have been turned into little magnets and lined up by the forces produced by the big bar magnet.

The pattern shows us the direction of these forces, and give an idea of how strong they are (picture 3). It shows what is called the **field** of the magnet. Activity B asks you to set up some fields, plot them and draw them.

The iron filings just give a rough idea of the field. You need to use your imagination to draw the field lines sensibly, so that you go from the 'real' iron filings to the imaginary field, as shown in picture 3.

The Earth's magnetic field

The Earth has a magnetic field of its own, but it is quite weak. It is too weak to line iron filings up. But if you set up a thousand compasses all around your school, they would line up to show what the Earth's field is like in your area. It wouldn't look very interesting (picture 4).

The direction of the field lines is the same as the way a compass would point (to the north). The arrows in picture 3 are in the direction that a small compass needle would point if you put it in the field.

A suspended bar magnet, or a compass needle, lines up so that one end points north, the other south. The end (or 'pole') that points north is called the north-seeking pole, or N-pole. The other end is the south-seeking or S-pole. They show the direction of the field lines (or lines of force) of the Earth's magnetic field.

Attraction and repulsion

The rule about magnets is quite simple:

like poles repel; unlike poles attract

This means that N-poles attract S-poles, and vice versa. N-poles repel N-poles and S-poles repel S-poles. This seems to be a basic law of nature; much the same applies to electric charges (see topic E5). The field-patterns of like or unlike poles near each other seem to show this (picture 5). The lines go as directly as they can from N-pole to S-pole, but veer away from each other when like poles are placed close to each other.

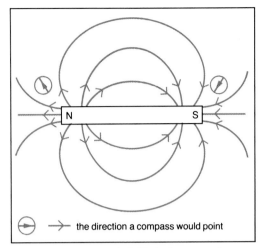

Picture 3 The magnetic field of a bar magnet

the direction a compass would point

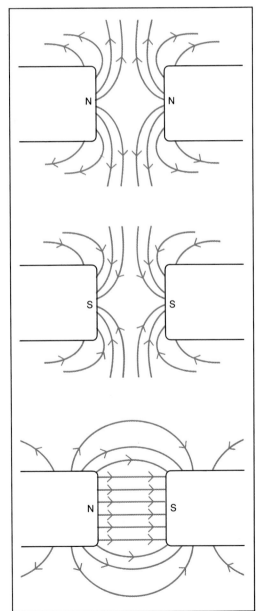

Picture 5 The fields between magnetic poles

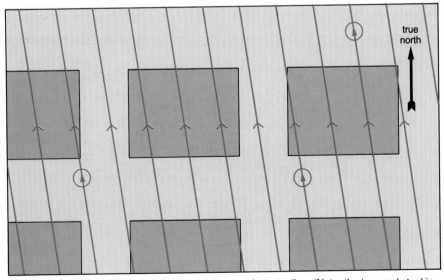

Picture 4 The Earth's magnetic field in the streets where we live. (Note: the iron and steel in buildings might distort the field)

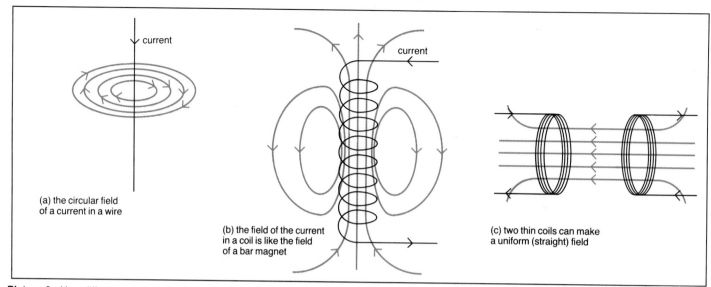

(a) the circular field
of a current in a wire

(b) the field of the current
in a coil is like the field
of a bar magnet

(c) two thin coils can make
a uniform (straight) field

Picture 6 How different magnetic fields can be made

Electromagnets

Electromagnets are easy to make, and can be very strong. There is a magnetic field around every conductor carrying a current. Picture 6 shows how different kinds of effect can be made by winding the wire in different ways.

An electromagnet can be made to have a field just like the field of a bar magnet. All you have to do is wrap the wire around a pencil to make a coil — see picture 6(b).

The field can be made a lot stronger if the coil is wrapped around a piece of iron. This is because the iron is turned into a magnet, and adds its strength to the field of the coil itself (picture 7).

You can use a small compass or iron filings to investigate the direction of the field near electromagnets.

Why electromagnets are useful

Electromagnets can be switched on and off, so that you can have a magnet only when you want it. Also, by making the current larger or smaller you can make the force field stronger or weaker. Electromagnets are more controllable than permanent magnets. Permanent magnets and electromagnets are used in many everyday devices, and you will meet them again (topics E6, E7, E9 and E10).

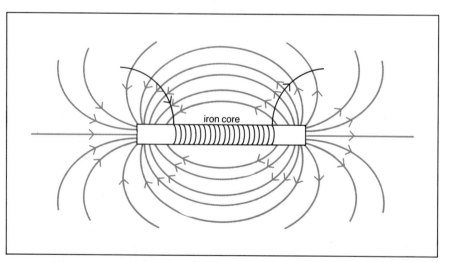

iron core

Picture 7 The field of an electromagnet

Activities

A   Where do we use magnets?

Look for many places or things in your home where magnets are being used. List all the ones you find. Are they permanent or electromagnets? Draw or describe *one* use in as much detail as you can.

B   Looking at magnetic fields

To show magnetic fields with iron filings you have to work carefully. The following rules might help.

1   Wear goggles in case iron filings get blown into your eye.

2   Use fairly stiff paper and make sure that it is level and well supported above, but close to, the magnet(s). Use books or pieces of wood.

3   Sprinkle the filings gently, from at least 20 cm above the paper. Don't use too many — a few iron filings go a long way, and make a clearer pattern. Tap the paper gently with a pencil to make the pattern as clear as you can.

4   Keep the filings away from direct contact with the magnets — they are hard to get off again!

5   Use your imagination to draw the field pattern as a set of lines — don't try to draw every filing!

6   When you've finished, pour the filings on to another sheet of paper and fold it gently to pour the filings back into the sprinkler.

Fields to look at and draw:

i)   a bar magnet,

ii) a horseshoe magnet,

iii) the magnets in a door catch or refrigerator seal,

iv) two N-poles near each other,

v)   two S-poles near each other,

vi) an N-pole and an S-pole near each other,

vii) the field of an electromagnet or coil (your teacher will probably have a special mount for these, to make it easier to set up).

C   Maps and compasses

One of the snags about using a compass to find your way is that it doesn't point exactly north-south.

1   Find out why this is.

2   Use a good compass and a local Ordnance Survey map (1 to 25 000 or 1 to 50 000). Line up the map so that north on the map is aligned with true north on the ground.

3   Identify some local landmarks from the map.

D   What decides how strong an electromagnet is?

You will need: some insulated single strand wire, iron or alloy to use as a core, an electricity supply that can produce a large current at a low voltage, a way of controlling and measuring the current, plus other equipment that you will have to work out for yourself.

1   Think of two or three things that might decide how strong an electromagnet is. Then think of how you might measure how strong it is.

2   Plan out an investigation for testing your ideas, then check with your teacher to see if what you want to do is practicable.

3   Carry out you experiments and write a report about them and what you find out.

Questions

1   Which of the following objects or materials would be attracted by a magnet?

a   a coin,

b   glass,

c   copper wire,

d   wood,

e   iron,

f   carbon,

g   a knife blade.

2   Copy the following arrangements (picture 8) and sketch the field patterns you would expect to find. Mark in the field directions (the way the N-pole of a compass would point) of:

a   a bar magnet,

b   a long coil of wire carrying a current,

c   two N-poles near each other,

d   a large N-pole opposite a similar S-pole (e.g. a pair of Magnadur magnets).

3   Explain the following as clearly as you can, using words or diagrams or both.

a   What is the 'law of poles' for magnets.

b   What is a magnetic field?

c   Why does a compass needle line up north-south?

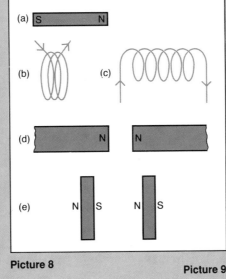

Picture 8

4a   Give two differences between permanent magnets and electromagnets.

b   Why aren't electromagnets used in compasses?

c   Powerful electromagnets are used in scrapyards and steel works to lift up iron and steel scrap and put it down somewhere else. Why aren't permanent magnets used for this job?

Picture 9

E3
Controlling electricity: current and resistance

To make the best use of electricity, we need to be able to control it. We can also use electricity to control other things.

Picture 1   Resistors — the main way we control electricity. Each knob alters a resistance

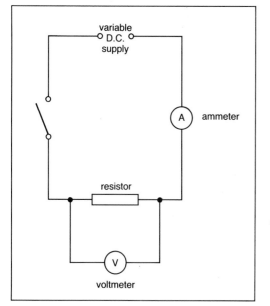

Picture 2   A circuit for measuring resistance

Resistance

With a given energy source, such as a battery or a generator, the size of current that flows is decided by the **resistance** of the circuit. All conductors resist the flow of electric charge to some extent, but some are better at it than others. The bigger the resistance of a conductor the harder it is for electric charge to flow through it. For a given voltage applied to it, the current would be less.

It is like water flowing downhill in a river. If the bed of the river is smooth the water can flow easily, and more can get through in a given time. But if the bed of the river is rocky the water can't flow so easily. It will move downhill more slowly, and a lot of energy is wasted — you can hear the noise and see the water being thrown up in the air. We find the same kind of effect in a conductor with a high resistance. It cuts down the flow of charge — and energy is released. The conductor gets hot.

Resistors

A **resistor** is a special type of conductor made from a high resistance metal or alloy, or perhaps carbon.

A resistor is designed to have a fixed resistance, so that for a fixed voltage exactly the right size of current goes through the circuit. Some different types of resistor are shown in picture 1.

Measuring resistance

Resistance is measured in units called **ohms** (Ω). A 10 ohm resistor would have twice the resistance of a 5 ohm resistor. For a given applied voltage the current in the 10 ohm resistor would be half of that in the 5 ohm resistor.

The size of current is worked out using the formula:

$$\text{current} = \frac{\text{voltage}}{\text{resistance}} \quad \text{or} \quad I = \frac{V}{R}$$

We can switch this formula around to calculate resistance, knowing the current for a given voltage: $R = V/I$

A simple test circuit (picture 2) allows you to measure current and voltage and so calculate the resistance of a resistor.

Variable resistors

Picture 3 shows examples of a very useful type of resistor. By moving a sliding contact, more or less resistance material is put in the way of the current, so making the current smaller or larger. Picture 4 shows how it does this.

These devices can be used in simple circuits to change the size of a current, but they are most often used to control the size of the applied **voltage**.

Voltage

Voltage is not an easy thing to imagine. It measures what the battery or supply actually does, which is to give energy to make charges flow through conductors. We measure the voltage of a supply in terms of how much energy it gives to a **coulomb of charge.** In a lamp circuit the energy would be mostly given to the lamp, heating up its filament.

If a coulomb of charge delivers 6 joules of energy as it goes around a circuit then the supply voltage is **6 joules per coulomb**. We call this **6 volts**.

Another way to look at it is in terms of current and power. This is more practical because the instruments we use measure amps, not coulombs.

The voltage of an energy source such as a battery decides how much **power** a given current can deliver. Power is the rate at which energy is provided, in joules per second or **watts**. See topic D3 for more about power and energy.

A 1 amp current from a 6V battery can deliver 6 watts (6 joules of energy per second). In comparison, a current of 1 amp from a mains supply at 240V will deliver 240 watts.

So the voltage of a source tells us how many watts it could provide per ampere of current, according to the formula

$$V = \frac{P}{I} \text{ or volts } = \text{ watts/amperes}$$

You choose the energy source to suit the power you want to use. The mains supply to a house in the UK is 240 V. This means that small currents can be used to run most of the devices we have in our homes. We can see this from the formula, if we rearrange it as $P = VI$. The bigger V is, the smaller I can be to get the same result.

In the USA the mains voltage is 110 V. Their devices need a bigger current if they want the same power. This means they have to be made with lower resistances

Controlling voltage

When you turn the volume control of a radio or cassette player you are using a variable resistor to control the voltage somewhere in the amplifier circuit. This decides how loud the sound that comes from the loudspeakers will be.

The volume control is a variable resistor. The voltage of the signal being amplified is first of all fed across the whole resistor as shown in picture 5. By moving the sliding contact you can get all, or just a part, of the signal voltage fed to the amplifier. The bigger the signal that gets to the amplifier, the louder the sound will be in the loudspeaker. When used like this the variable resistor is called a **voltage divider** (potential divider). It divides up the total voltage into smaller amounts.

Picture 3 Resistors can be 'variable': you can control both current and voltage by varying the resistance

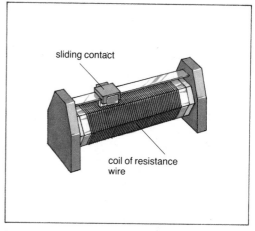

Picture 4 A rheostat — 'flow controller'

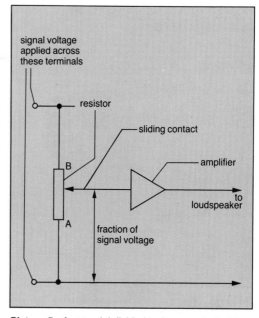

Picture 5 A potential divided 'voltage controller'

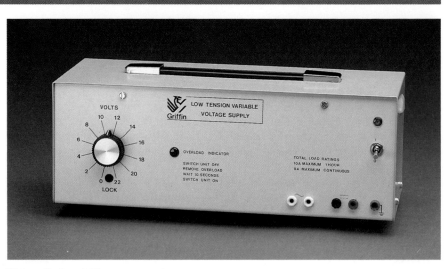

Picture 6 A variable power supply

Variable power supplies

Picture 6 shows a typical low-voltage power supply. Its output can be changed, using a control knob, from 0 to 12V. The main part of the power supply gives a fixed voltage. The control knob is connected to a variable resistor which changes the output to anything between 0 and the maximum, using a voltage divider.

Measuring current and voltage

Current is measured with an **ammeter**, and voltage with a **voltmeter**. Both of these instruments look much the same, and may in fact work on the exactly the same principles. This can be confusing, because what they measure is very different. Also, they are placed differently in circuits.

Ammeters tell us how much electric charge is passing though a circuit per second. Remember that 1 ampere is a flow of 1 coulomb of charge per second. Ammeters have to be put directly in the path of the current, so that they can check everything that goes through the circuit (see picture 7).

Voltmeters are trickier. They take a sample of the current in a device or circuit and then **calculate** the voltage that must be across it. They have to keep the 'sampling current' small if they are not to change the circuit too much. This means that they have high resistances and are connected **across** the device being tested. (See picture 8).

Picture 7 An ammeter will read the current going through it

Ohm's Law

For many useful conductors there is a simple rule which connects current, voltage and resistance. If we double the applied voltage, the current is doubled. If we halve the voltage, the current is halved. This effect doesn't work with all conductors, but is true for metals and for carbon, if they don't get too hot.

For most circuits we can use the rule to calculate in advance what will happen when things change.

The rule is: **for a given conductor at a constant temperature the current in it is proportional to the applied voltage.** This is known as **Ohm's Law**.

We can write this as a formula $I = V/R$. We can use this formula to make calculations with all values of V for a given conductor. But we have to be careful, because the resistance might change if the conductor gets hot.

Activity C is about this useful 'law' of electricity.

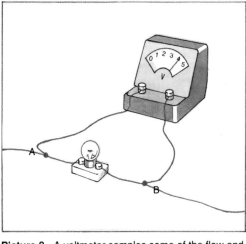

Picture 8 A voltmeter samples some of the flow and 'works out' the voltage between two points (A and B)

What is a law?

Some laws in physics are unbreakable. Gravity always behaves in the same way (see topic B4). Mass-energy cannot be created or destroyed. In every energy change, some always gets less usable. Of course, the universe is large and can come up with some surprises. By physicists are pretty sure that if any of these laws were broken it would only be because there was a better or stronger law to take its place.

Ohm's Law is not like these. It doesn't say what *must* happen. It just describes how some materials behave. It applies only to metals, ionic solutions and perhaps carbon. There are lots of conductors that don't 'obey' Ohm's Law. This doesn't worry anybody! If the Law of Conservation of Mass-energy was broken, the Universe would be a different place!

Activities

A Measuring resistance

Use the circuit shown in picture 2 (or a similar one given to you by your teacher) to measure the resistance of as many of the following as you can. Make sure that you check the rated voltage of the device so that you do not exceed it and possibly damage the device you are testing.

1 A 12 V lamp,
2 a standard ('electronic') resistor,
3 a small electric motor,
4 a voltmeter (**think!**),
5 a small electric heater.

B Looking at the label

All electrical devices used in the home must by law have certain information printed on them. Look at some devices and find this information.

1 Copy out what electrical information is given on any one device (e.g., a food mixer, battery radio, electric razor).

2 Use the information to make the following calculations for any three devices:
 i) the current it takes in normal use,
 ii) its resistance.

You may need to use the formulae given in this topic:

$$R = \frac{V}{I} \quad P = VI \quad (\text{or } V = \frac{P}{I})$$

$$\text{or } I = \frac{P}{V}$$

C Investigating Ohm's Law

Ohm's Law says that, for a given conductor at a constant temperature the current in it is proportional to the applied voltage.

1 What does 'proportional' mean ? How could you test two sets of measurements to see if they are proportional to each other?

2 You will need the equipment as used for activity A (see picture 2) and a metal (e.g. wire) resistor. Set up a circuit to measure the current in the wire for six different voltage values. Does the wire obey Ohm's Law?

3 Replace the wire with a 12 V lamp and repeat the experiment. Does the lamp obey Ohm's Law?

Questions

1a Name the instruments used to measure current and voltage and draw their circuit symbols.

b Copy the circuit in picture 9 and label the empty circles with A or V to show which would be ammeters and which voltmeters.

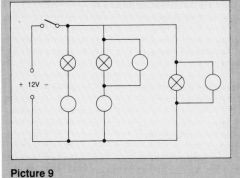

Picture 9

2 You find two wires made of the same metal. One wire is long and thin, the other wire is short and thick. Which do you think would have the bigger resistance?
Give a reason for your answer.

3 People often try to understand about electricity by comparing it with water flowing in a river.

a In what ways is water in a river like electricity in a wire?

b In what ways is it different?

c Can you think of anything else that is like electricity flowing in a wire?

4 Batteries with the same voltage marked on them are sometimes very different in size. Suggest a reason for this.

5 Use the formula $R = V/I$ to work out the resistance of the following devices:

a a lamp that takes a current of 2 A from a 12 V supply,

b a lamp that takes a current of 0.5 A from a 12 V supply,

c an electric toaster that takes a current of 3 A from a 240 V supply,

d a cake mixer that takes a current of 2.2 A from a 240 V supply.

6 Car headlamps are powered by a 12 V battery or alternator and have a typical resistance of 2.5 ohms. How much current must each lamp take from the supply?

7 Why must ammeters be connected in line (in series) with a device, but voltmeters have be connected across it (in parallel)?

8 Draw the circuit you would use to measure the resistance of a torch bulb to be used in a 3 V torch.

9 A resistor used in an electronic circuit must obey Ohm's Law.

a One such resistor is labelled '20 ohms'. What current would be in it when the following voltages are applied across it: (i) 20 V, (ii) 10 V, (iii) 2 V.

b Another resistor is labelled '2 kΩ'. What voltage would need to be applied to produce a current of 10 mA in it?

E4
Using circuits

The key to using electricity is the **circuit**.

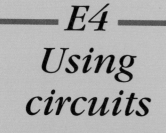

Picture 1 A Christmas tree. How are the lights controlled?

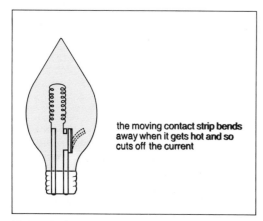

the moving contact strip bends away when it gets hot and so cuts off the current

Picture 2 How a flashing lamp works

How do flashing lights work?

Picture 1 shows a Christmas tree, decorated with coloured lights that can flash on and off. The flashing is controlled by just one bulb that is different from the others. When the filament in it gets hot enough it switches itself off. Then it cools down and switches itself on again (see picture 2). When this control lamp is out all the other lamps go out as well. They only work when the control lamp is on.

Sometimes a lamp in your home stops working and goes out. But this doesn't make all the other lamps go out. Why is this?

Series and parallel connections

The Christmas tree lights are connected in a line, one after the other, as shown in picture 3. When a connection is broken in one lamp the charge cannot flow, so all the lamps go out. This way of connecting things in a circuit is called **series** connection.

In a house the lamps in a room are connected in a different way. It would be very annoying if all the lights went out just because one lamp wasn't working. In the home, each lamp is connected separately to the mains supply, as shown in picture 4. Each lamp can have its own switch, and we can have any of the three lamps on, or none of them. This is called **parallel** connection.

Why do we use parallel circuits in the home?

Electrical devices in the home are connected in parallel because we want to control them independently of each other. Each device is connected directly to the mains, so that it gets its proper voltage. Also, we can easily connect other devices into the circuit, without having to 'break it'.

Of course, the more appliances we connect, the more energy we will use, and the more it will cost. This is because when we connect more devices in the circuit in parallel we must take more current from the supply in order to make these extra devices work. This means that the total resistance must get less!

Resistance in parallel: more means less!

We can understand this idea by seeing what happens to the current drawn from the supply when we add more things to a parallel circuit. Take a simple circuit that you could try for yourself in the lab (activity A, picture 7). If we start with just one 12-volt lamp, say, we can calculate the current through it by using the formula $I = V/R$.

For one lamp this might be just 1 A. If we add another lamp it too will need 1 A to make it work — and the supply will have to deliver 2 A.

In the first case the circuit resistance is 12 ohms:

$$R = \frac{V}{I} = \frac{12 \text{ volts}}{1 \text{ amp}} = 12 \text{ ohms}$$

In the second case, with two lamps, the supply delivers 2 A, so the circuit resistance is now 6 ohms:

$$R = \frac{V}{I} = \frac{12 \text{ volts}}{2 \text{ amps}} = 6 \text{ ohms}$$

What would be the circuit resistance if another similar lamp were to be added?

You can work out the answer (4 ohms) in this simple circuit by common sense, but in more complicated circuits you will need to use the formula for resistors in parallel: $1/R_t = 1/R_1 + 1/R_2 + 1/R_3$ etc. This formula is needed when the devices added in parallel have different values of resistance.

Resistors in series

Picture 5 shows resistors connected in series. If you try this (activity B) you will find that the more lamps you add the dimmer they get! This is another reason why lamps are usually connected in parallel. The ammeter shows that the current gets less as more lamps are added. This means that the resistance of the circuit has **increased**. When we add resistors in series the total resistance goes up:

$$R_{total} = R_1 + R_2 + R_3 \text{ etc} \quad \text{(see page 135)}$$

Picture 6 shows how complicated circuits can get! But all the components you can see in this computer are connected in series and/or in parallel, and the engineers who designed and made it know what current each component has to carry.

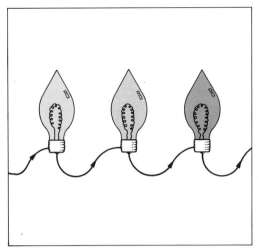

Picture 3 Christmas tree lamps are connected in series. The same current goes through each in turn

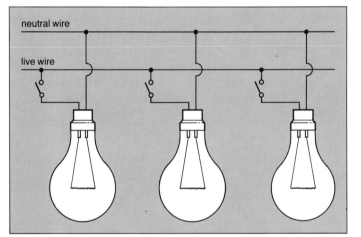

Picture 4 House lights have the lamps connected separately to the mains (i.e. in parallel)

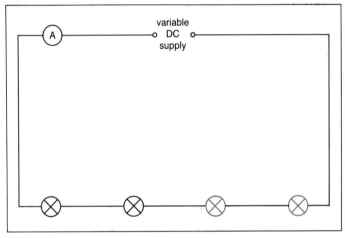

Picture 5 Adding lamps in series

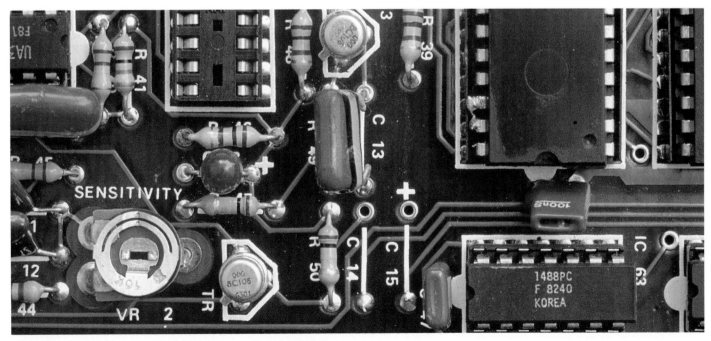

Picture 6 Inside a computer — some circuits are quite complicated!

Activities

A   Lamps in parallel

Picture 7 shows a simple circuit for testing what happens when lamps are connected in parallel. Unless your lab is very well equipped you may have to make do with just one ammeter and move it around the circuit to wherever you need it.

1   Set up the circuit with just one lamp, set the supply to 12 V and **check it with a voltmeter**.

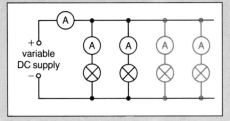

Picture 7   Adding lamps in parallel

2   Measure the current that goes through both the lamp and the power supply and note it in a copy of the table below.

3   Add another, similar lamp in parallel with the first. Check that the voltage of the supply is still 12 V — it may need adjusting — and record the current in each lamp and the current delivered by the power supply.

4   Repeat with as many lamps as you and your teacher consider is safe.

Analysis of results

i)   What is the connection between the current delivered by the supply and the sum of the currents in the lamps?

ii) Calculate the resistance of one lamp

iii) Calculate the circuit resistance for each arrangemnent.

iv) Check that the circuit resistance R can be calculated by the formula
$$1/R = 1/L + 1/L + 1/L \text{ etc,}$$

where **L** is the resistance of one lamp.

B   Investigating a series circuit

Picture 5 shows how conductors can be connected in series.

1   Use up to four similar resistors and investigate what happens as you connect them to a power supply in series. Each time you add a new one, measure:
   i) the current in the circuit,
   ii) the voltage drop across each component.
       Set out your results in a table:

2   Describe clearly what happens to the current and the voltage.

3   Calculate the resistance of each component and the total circuit resistance in each arrangement.

4   Check that the total circuit resistance **R** is given by the formula
$$R = R_1 + R_2 + R_3 + R_4 \text{ etc}$$

Number of lamps	Current in each lamp				Current delivered by supply
	1st	2nd	3rd	4th	
one		–			
two			–	–	
three				–	
four					

Number of resistors	Current in circuit	Voltage across each component
one only		
two		
three		
four		

Questions

1   Draw circuit diagrams showing:

a   a battery driving two lamps in series with a switch,

b   a battery driving two lamps in parallel, with a switch for each one.

2   Give *two* reasons why the lamps at home are connected in parallel rather than in series.

3   Draw circuits showing how you could connect:

a   a lamp in parallel with a small electric motor, run off a 12 volt battery, with a switch that controls both lamp and motor.

b   a lamp and motor as above, but with each having its own separate switch.

4   Some cars have heated rear windows, with the heater switched on and off by the driver. The makers of the car guard against the driver leaving the window heater on whilst the car is parked overnight.

a   Why would this be a stupid thing to do? (Try to think of two reasons.)

b   The simplest way for the car manufacturers to stop this is by making sure that the window heater can't be on unless the ignition switch is also on (the ignition switch has to be on for the engine to work). Draw a simple circuit showing how the two switches are connected.

5   For this question you may need to use the formulae given in the topic for calculating resistances in series and in parallel.

a   A 6-ohm lamp and a 2-ohm motor are both run from a 6 battery. They are connected in parallel. How much current is drawn from the battery?

b   Ten Christmas tree lights are connected in series to the 240 V mains supply. A current of 0.5 A is drawn from the mains. What is the resistance of each lamp?

c   A 12 V car battery has to supply a heater and a starter motor. They are connected in parallel and have resistances of 12 ohms and 0.5 ohms respectively. What is the effective resistance when both are switched on? What current does the battery have to supply?

Looking for old ruins

One of the ways archaeologists find the buried remains of buildings — or people — is to make a resistance survey. Two electrodes are placed a fixed distance apart in the ground and the resistance between them is measured. See picture 1. The current flows between the electrodes through the ground, but not in a straight line. Any buried material will usually have a different resistance than plain soil. It may hold more or less water, or have carbon particles in it.

Table 1 shows the results of such a survey. In this example, the field being investigated is divided into a grid pattern at intervals of 5 metres. The ground resistance is measured between metal electrodes 10 cm apart at each point of intersection of the grid lines. The units are relative.

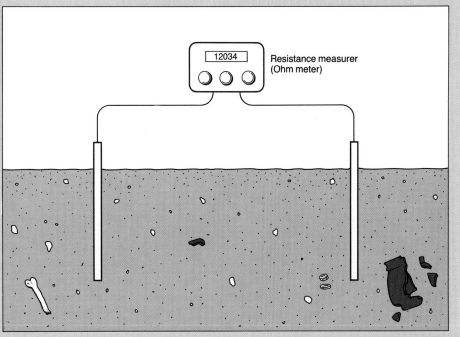

Picture 1 Carrying out a resistance survey

Table 1 Results of resistance survey

Grid distances West to East from point 0 North to South	0	5	10	15	20	25	30	35
0	10	10	10	10	10	10	10	10
5	10	13	8	8	8	8	8	10
10	10	13	8	8	8	8	8	10
15	10	13	8	8	13	13	13	10
20	10	13	8	8	13	13	10.5	10
25	10	13	8	7.5	7.5	8	8.5	10
30	10	13	8	7.5	7.5	8	8.5	10
35	10	9	13	10	9.5	10	10.5	13
40	13	13	13	10.5	13	10	10	7.5
45	13	13	13	13	13	8	7	6
50	13		13	13	8	7	6	9

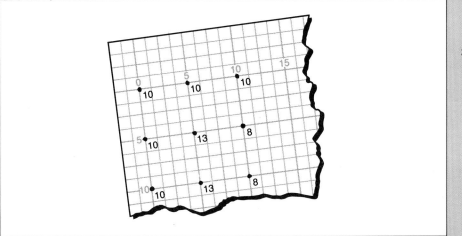

Picture 2 How to map your results

1 Mark out a sheet of graph paper in a grid as in the table above. At each intersection write in the resistance value. Picture 2 shows you how to begin.

looks! You will have to make some guesses here and there, because the measured values fall between whole numbers.

2 Draw in 'resistance contours' by joining up lines of equal resistance. it is best to keep this simple, and just draw them in for whole number values (e.g. integers, 7, 9, 10, 13). This is not as easy as it

3 You should get some clues about what might lie buried under the field from your contour map. Another technique to make things clearer is draw a *resistance profile.* This is rather like a transect in a biology field study. Take a line running across the field from east to west — say the one at 20 metres south of the base point 0.

Plot a graph of the resistance values at each grid intersection of this line as you go west to east. Picture 2 shows you how to start.

The result should give you some ideas about the buried structure. If you have the time you can draw profiles at other lines.

4 Use the evidence from your graphs to write a brief report to your field workers suggesting where they might start digging and what they might find.

E5
What is electricity?

We use 'electricity' every day, but it is more than just a good way of making things work . . . . .

Picture 1 Amber is fossilised tree-sap

Picture 2 Static electricity is easy to produce

The amber mystery

People have known about electricity for thousands of years. The word electricity comes from the Greek word for amber — *elektron*. Amber is the fossilised sap of pine trees, and is used to make beads and jewellery. It is a shiny, clear, golden material, which may even contain the fossils of insects trapped in it (picture 1).

One snag with amber jewellery is that it seems to get dusty very easily. The Ancient Greeks noticed this, and worked out that amber had the mysterious ability to attract small objects to it. This happens when a piece of amber is rubbed by cloth. If you don't happen to own any amber, you can get the same effect by rubbing a plastic pen or comb (picture 2).

The rubbed amber produces a force of attraction which came to be called 'amber force' — or **electricity**. The force was very small, and the effect was completely useless, unlike the forces, say, of magnetism. At least magnets could tell you which way north was. So electricity was forgotten about.

The rediscovery of electricity

When electricity did become important it wasn't for good scientific reasons. This renewed interest began just over 200 years ago. At that time 'science' as we know it didn't exist. 'Scientists' called themselves 'natural philosophers'. Some of them were people who were rich and had money and spare time to spend on their hobby, like Robert Boyle who discovered 'Boyle's Law' for gases.

Many earned a living as doctors, like William Gilbert who did experiments on magnets. Galileo (page 44) started off as a doctor of medicine.

Some were clergymen, like Copernicus, who put forward the strange new theory that the earth went round the sun, and not vice-versa. Some were astronomers or mathematicians, like Isaac Newton.

Some discoveries were made by 'artisans', people who were good with their hands and earned a living making clocks and other instruments. Michael Faraday (see page 160) began life as a bookbinder.

Most of the ones we know about lived in Europe or the British colonies in North America. There were very important discoveries made in China and India, but few of these were heard about in the West.

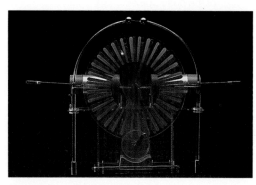

Picture 3 Being charged with electricity is like falling — it doesn't hurt until you touch the Earth!

Picture 4 A Wimshurst Machine

The shocking history of electricity

In the early days the best fun was to be had from 'electrifying' people. Small boys were easily persuaded to be hung up by silken ropes and 'charged' up (picture 3). They were stroked with dry cloths until their hair stood on end. In the dark, you could see sparks leaping from them to anybody standing close by.

Picture 5 Shocking people has always been fun!

Later, machines were invented that used amber-like materials to produce electricity just by turning a handle (picture 4). It was quicker than stroking people with a dry cloth.

The whole thing became so popular that you could earn a living by electrifying people in fairgrounds. People thought that getting charged up was good for the health, and paid money to be strung up and electrified.

In Holland in 1746, Peter van Musschenbroek tried to electrify water in a bottle, and by accident found a new effect. The 'charge' could be stored in a jar. This was the first **capacitor**, and was called a Leiden Jar. This was because he did the experiment in his home town of Leiden.

Now things got more serious — enough charge could be stored to give people a severe shock. The shock could kill small animals and birds, and melt thin wire.

In Paris, 180 of the King's guards were lined up, holding hands, and connected to a charged jar. They all leaped high in the air, at the same instant. The experiment was repeated with 300 Carthusian monks, formed into a line 100 metres long. The results were equally spectacular and even more crowd-pleasing (picture 5). History does not record what the monks thought of it.

Shocks became popular as a cure for gout and rheumatism and electricity was suddenly all the rage.

A theory of electricity

What was going on? What *was* electricity? Where did it come from? Many interested people tried to answer these questions, and it wasn't until the early years of this century that they could be fully answered. But a very good start was made by an American called Benjamin Franklin, who lived from 1706 to 1790 (picture 6).

He started life as a printer's apprentice and then became a succesful printer. It was only at the age of forty that he saw some experiments in electricity, and was so fascinated by them that he sold his printing press and spent his whole time experimenting. He was the first man to realise that lightning was a huge electric spark. He flew a kite into thunderclouds, at great risk to his life, to collect charge from them. He learned enough to invent the first lightning conductor, for which he became famous page 237.

Later on he became even more famous as a revolutionary politician when the American colonies rebelled from Britain and became the United States of America.

Positive and negative

Franklin's theory of electricity was simple. He said that every object, even animals and small children, contains electricity. If it has more than its normal

Picture 6 Benjamin Franklin — rebel, scientist, ambassador and inventor of lightning conductors

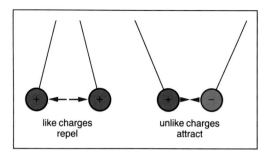

Picture 7 Like charges repel!

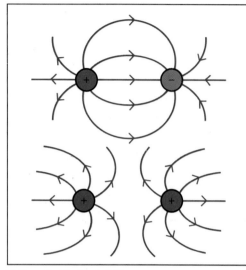

Picture 8 Electric field lines

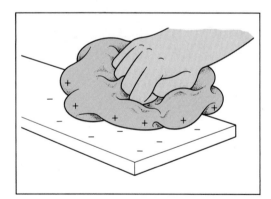

Picture 9 Charging by rubbing. The cloth leaves electrons on the plastic strip. The strip becomes negatively charged and the cloth becomes positive.

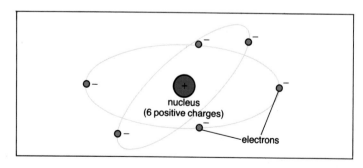

Picture 10 Atoms contain equal numbers of positive and negative charges

share of charge he called it 'plus', if less it was called 'minus'. Since then we have learned that in fact there are two kinds of electric charge, but we still call them 'plus' and 'minus' or **positive** and **negative**.

Electric forces are caused by the attraction or repulsion between these two kinds of charge (picture 7):

like charges repel, unlike charges attract.

Just as with magnetism and gravity, we imagine there to be an **electric field** near charged objects. We draw field lines (lines of force) with the direction positive to negative (picture 8(a)). This is the direction a positive charge would move under the forces of attraction or repulsion.

In magnetism, we never seem to get N-poles or S-poles on their own. They come together, in opposite pairs. But we can get free electric charges, both positive and negative ones. This is because all matter is made of charged particles which can be separated.

Electricity and the atom

All matter is made up of atoms. In turn, atoms are made of a central massive nucleus surrounded by a cloud of very light particles called electrons. The nucleus is positively charged and attracts the cloud of moving electrons because they are negatively charged. The charge on the nucleus is exactly balanced by the charge on the electron cloud. This means that the atoms, and the material they are made of are normally uncharged. They are **neutral**. See *The Material World*, topic J3 for more about the structure of the atom.

How objects get charged

Electrons are very small and are easy to move. When a piece of almost any material is rubbed with a cloth electrons are pulled off. Sometimes they go from the cloth onto the material, sometimes they go the other way, depending on the materials. (Picture 9)

This movement of electrons makes both materials electrically unbalanced. One will gain electrons and it becomes negatively charged. The other loses electrons and becomes positively charged (picture 10).

Moving air can produce the same effect as moving cloth. Cars travelling at speed can have electrons rubbed off them as they pass through the air, and become charged up. Hot air rising on sunny days can become charged, and the charge can be given to water drops in clouds. Sooner or later the clouds have enough charge to cause lightning and thunder (picture 11).

Conductors and insulators

Any substance can be electrically charged, but if the material is a conductor the charge might flow away. The most likely place for charge to flow to is the

Picture 11 Lightning

Earth, simply because the Earth is so big that it can hold a huge amount of unbalanced charge without becoming noticeably charged up.

Human beings are quite good conductors, and for a body to be 'electrified' it has to be kept off the ground by good insulators. Otherwise the charge would flow away to the ground. This is why the children who were charged up in fairgrounds were strung up on silk ropes. Silk was then one of the best insulators known.

Picture 12 shows a girl who has been charged up to about 100 000 volts. The force of repulsion between the like charges on her head and on her hair causes it to flare out.

Picture 12 A hair-raising experience

Static electricity

When electric charges do not move the effects they produce are called **electrostatic**, and the objects are said to be charged with **static electricity**.

Chemistry

Chemistry only happens because atoms and molecules combine together or break apart. This involves making or breaking chemical bonds. The bonds that hold atoms to each other are electric forces.If these forces didn't exist there would be no molecules, no chemical changes, no chemistry and no life on earth.

Activities

A Getting things charged

It is easy to charge things up with electricity. The main thing to remember is that the following experiments will only work well if the materials are *dry*. The best way to dry them is to use a hair dryer. Good plastics to use are polythene, nylon or acetate.

1 Take a strip of any plastic material and rub it with a dry cloth. Test that it has become charged by trying to pick up small pieces of paper.

2 Now try some of the following:
   i) Hold the charged end close to your ear. Do you feel or hear anything?

   ii) Recharge the strip and hold it close to a thin stream of water falling from from a tap. Describe or draw what happens.

B The laws of charge

To investigate these you can use some light metallised balls and two different kinds of plastic (e.g. polythene and acetate). The balls are held hanging from a ruler by thin nylon thread (picture 13).

Try the following experiments and draw or describe what happens. Do they agree with the rule that like charges repel, unlike charges attract?

1 Charge up one of the strips by rubbing it with a dry cloth and touch both of the metallised balls with it. They should become charged with the same sign of charge, and you will see the effects of the electrostatic force.

2 Charge up the other strip and bring it gently up to the balls. What happens?

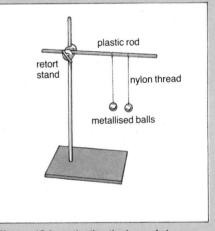

Picture 13 Investigating the laws of charge

Questions

1 What are the two kinds of electric charge?

2a What do people mean when they talk about 'static'?

b What happens when an object become electrically charged by 'friction'?

c Give two examples of things that become 'accidentally' charged in this way in everyday life.

d Sparks from charged objects can give a shock, or possibly start a fire. For *one* of the examples you gave in part (b), state
   i) how you could stop the charges building up
   ii) how you could let the charge get away safely.

3 Explain what the following mean:
a conductor,
b insulator,
c static electricity,
d earthing.

4 Why do you think 'static electricity' wasn't much use in the 17th century? Describe any practical use you know for it nowadays.

5 Why is electricity so important in chemistry? Write a short essay explaining how chemical changes in atoms and molecules involve electric charges.

E6
Using electricity: heating and lighting

Electricity is useful because it can carry energy from one place to another and use it to do so many different jobs.

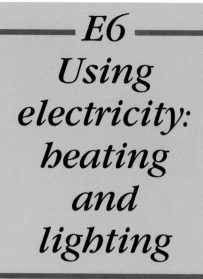

Picture 1 Electrons give energy to atoms in a conductor by colliding with them

Electric heating

When electrons flow through a conductor they collide with the atoms in the conductor. As they do this they give energy to the atoms. The energy is simply movement energy, transferred from the moving electrons to the atoms. The atoms are fixed, and just vibrate a little more (picture 1). We feel this extra vibration as a rise in temperature — the conductor warms up.

The electrons soon speed up again after the collisions, pushed on by electric forces. These forces are provided by a battery, or a cycle dynamo or the generators in a large power station. Most of the work done by electrical devices in home and factory uses the energy from fuels burnt in power stations — see page 168.

Resistance wires

Electric heaters need special kinds of wire as conductors. The wire in an electric fire element is usually an alloy of different metals. The alloy needs to have the right properties:

- a high resistance,
- a high melting point (to stand the high temperatures produced),
- chemical stability (so that they don't burn or corrode at high temperatures in air).

Types of heater

Picture 2 shows four types of electric heating device. Only one of them gets to be red hot, with a temperature of 600 to 700°C. This is the ordinary electric fire, which radiates energy as electromagnetic waves from the hot coil of wire.

An electric kettle has an element which doesn't need to get hotter than about 100°C, the boiling point of water. If it does, perhaps when the kettle boils dry, an automatic switch cuts off the current.

The night storage heater is very heavy because it is full of special bricks which can store a lot of energy when they get hot. They take a long time to

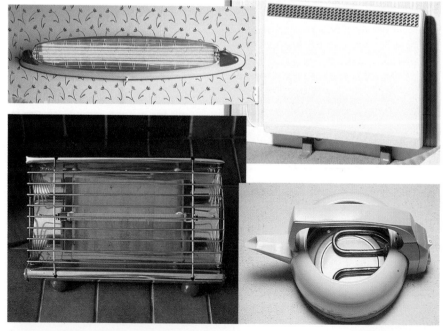

Picture 2 Different ways of using electrical heating

Picture 3 The parts of an ordinary filament lamp

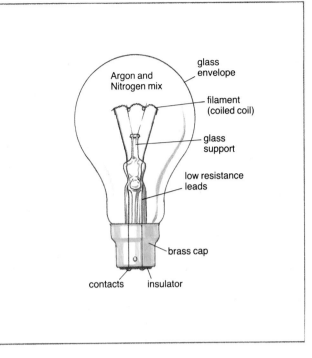

heat up, but take an equally long time to cool down. This is why they are so useful. They release their energy to heat up the room slowly, so that the room is heated over a long period of time. Inside the bricks there is an electric heater, which heats up the bricks at night. Electricity is cheaper at night — see page 173.

Infra red heaters are another kind of low temperature heater. Unlike ordinary electric fires the hot wire is embedded in a special glass. This means that it can't be touched, and so these heaters are safer for use in bathrooms where the danger of electric shock is greatest.

The glass gets hot, but the heater relies on the fact that radiation energy can get through the glass. It is this **infra red** radiation that warms up the room and the people in it.

Electric lighting

There are two main kinds of lighting used in the home. The oldest is the filament lamp (picture 3). This contains a very thin and long piece of wire (the 'filament'). It is made of a metal, tungsten, that can be heated to such a high temperature that it becomes white hot but doesn't melt. At this temperature it would burn in air, so it is kept inside a glass bulb filled with gases that don't react with it, such as argon and nitrogen.

Although it is glowing white hot, and sending out a lot of energy as radiation, most of the energy comes out as invisible (infra red) radiation. You can feel this if you put your hand near the bulb. In fact, only 2 or 3% of the energy supplied to the lamp is turned into visible radiation (light).

Fluorescent lamps

Fluorescent lamps are more efficient. A 40 watt fluorescent lamp produces as much light as a 150 watt filament lamp — and far less heat (picture 4). They work on a completely different principle.

The lamps are filled with a gas (mercury vapour) at low pressure. Electrons flow through the gas and collide with the gas atoms. When collisions take place the mercury gives out invisible (and dangerous) ultra-violet radiation. But don't worry, this radiation doesn't escape from the

Picture 4 Fluorescent tubes being checked

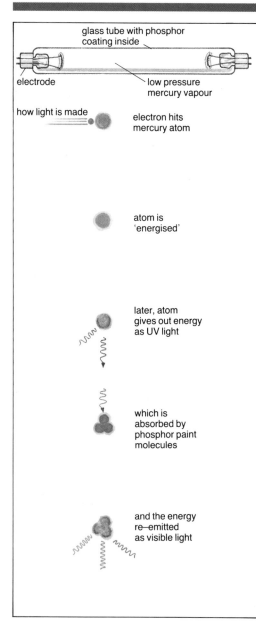

glass tube with phosphor coating inside

electrode

low pressure mercury vapour

how light is made

electron hits mercury atom

atom is 'energised'

later, atom gives out energy as UV light

which is absorbed by phosphor paint molecules

and the energy re–emitted as visible light

Picture 5   How a fluorescent tube works

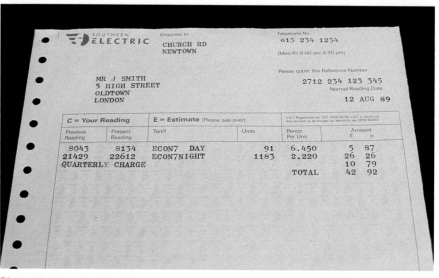

Picture 6   Your electricity bill

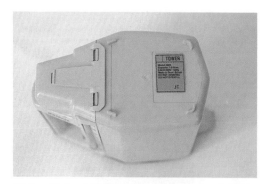

Picture 7   Look for the power rating on your kettle or hairdryer

lamp. It hits a special **phosphor** paint on the inside of the lamp and makes it glow white (picture 5). It is much the same as what happens in a TV tube (see page 174), where high-speed electrons are used to make the phosphors glow diferent colours.

How much does electricity cost?

The electricity boards don't make us pay for 'electricity'! We get that free. What we pay for is the **energy** that the electricity carries to us.

In science we measure energy in joules, but this is too small a unit to be easily used in everyday life. Of course we could use kilojoules or megajoules, but the electricity boards use a unit of energy called a kilowatt-hour. An electricity bill is shown in picture 6.

The 'units' are in fact kilowatt-hours, and as you can see some cost more than others. The cheap units are the ones used overnight, for putting energy into storage or water heaters. The kilowatt-hour is quite a sensible unit, in practice. It is equivalent to 3.6 megajoules (3 600 000 joules). It is the energy transferred when, for example, you use a 1 kilowatt electric fire for an hour.

Power ratings

The power rating of a device is given in watts or kilowatts. It tells us how much energy the device uses each second. Thus 1 watt is 1 joule of energy per second, a kilowatt is a thousand joules per second. A piece of electrical equipment, like a TV set or a lamp, will usually have its power rating printed on it (picture 7). The power of a device is decided by how much current it takes at its correct operating voltage.

In the home the voltage is kept at 240 V, and the current is ther decided by the resistance of the equipment. We can use the formula $I = V/R$ to calculate the current, if we know the value of the resistance.

We can use the formula:

$$\text{power} = \text{voltage} \times \text{current} \ (\boldsymbol{P = VI})$$

to calculate the power rating if we know current and voltage. We can also use it to calculate the current needed to provide a desired power. Then we can use the first formula to find out what resistance the device needs to have. This is what the manufacturers have to do when they produce the electrical equipment we buy.

The chart in table 1 gives some typical power and current ratings for mains-operated equipment in the home.

Table 1 How much power do appliances use

Appliance	Power rating /kW	Voltage /V	Current /A	Resistance /ohms
High power				
storage heater	2	240	8.3	30
cooker (total)	14	240	58	0.03
microwave oven	0.65	240	2.7	89
3-bar electric fire	3	240	12.5	19
Medium power				
1-bar electric fire	1	240	4.2	58
electric kettle	2	240	8.7	26
hairdryer	1	240	4.2	58
vaccuum cleaner	0.8	240	3.3	73
toaster	0.9	240	3.8	64
iron	1	240	4.2	58
drill	0.3	240	1.25	192
Low power				
refrigerator	0.12	240	0.5	480
lamp	0.06	240	0.3	800
hairstyling brush	0.020	240	0.08	2880
radio cassette player	0.012	240	0.05	4880
calculator	0.0005	6	0.08	75

Activities

A How much power do you use?

1 Do this activity at home at 7 p.m. one evening. Find the power of each of the electrical devices that you are using at that time. You can either look at the labels or use the table above. Work out how much total power you are using.

2 Find out how much you have to pay for 1 'unit' (kilowatt-hour) of energy. (If you can't, assume it to be 6p). Now calculate how much it would cost to run all all these devices for 2 hours.

B Measuring the power rating of a device

Use the following circuit (picture 9) to measure the power required to run some or all of the devices listed below.

Circuit	Device
	6 V lamp
	12 V lamp
	low-voltage heater
	electric motor

Picture 8

Questions

1 Suggest a reason why electrical energy is cheaper at night.

2 Use the formula:
power = voltage × current to work out the power needed to run the following devices:

a a 12 V car headlamp that takes 4 A,

b a vacuum cleaner motor that takes 3 A from the mains supply at 240 V.

c a washing machine heater that takes a current of 11 A from the mains supply at 240V.

3 Use the formula resistance = voltage/current to work out the effective resistance of each of the devices in Question 2.

4 Use the data given in table 1 to work out the cost of using some electrical appliances as described below.

Assume the cost per unit (kilowatt-hour) of electrical energy to be 6p.

a Using a 3-bar electric fire for 4 hours,

b Using a hair dryer for half an hour,

c Using a computer for 10 hours,

d Leaving two 100 W electric lights on for 10 hours a day for a week,

e Using a microwave oven to cook something for 10 minutes at full power.

5 Copy out the following passage, filling in and underlining the missing words. Choose the words you need from the following list:

current, voltage, thousand, power, energy, resistance, second, hour, cost.

Your electricity bill asks you to pay for the _____ you have used, not the voltage or the _____ supplied. The _____ rating of an electrical appliance tells you how much _____ it uses per _____. The bigger the power rating, the more it will _____ to use. A label on the device that says '1 kW' means that it will use _____ at the rate of _____ joules per second.

Electricity can be very dangerous; in fact it is probably the most dangerous thing you let into your home.

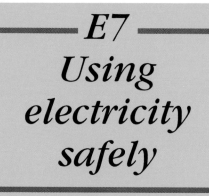

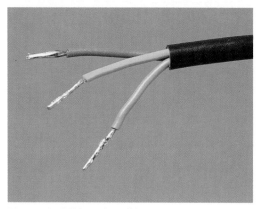

Picture 1 A three-wire cable

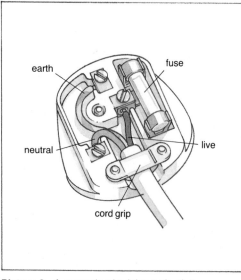

Picture 2 A correctly wired three-pin plug

Stick to the rules!

There are very strict rules about how houses should be wired and how electric appliances should be made. This is because electricity can give a strong electric shock which may kill. It can also cause fires. Most of the fires started in homes and buildings are caused by electrical faults. How can we guard against these dangers?

The three-wire system

The mains supply uses three wires (picture 1). One pair carries the current 'out and back'. These are the **live** and **neutral** wires. The household supply is **alternating**. The '240 volts' of the mains supply is actually a kind of average of the alternating voltage. Actually, mains voltage swings from +340 volts to -340 volts fifty times a second.

All household devices would work perfectly well with just these two wires, but there is a third wire, the **earth** wire. This is for safety. The three wires are colour coded, as shown in picture 1.

If an electric appliance is properly made and connected using a three-pin plug (picture 2) it should be impossible for you to get a shock, or for a fire to start, even if something goes wrong.

The earth lead connects the metal case of an appliance to the ground inside or just outside the house. If a fault occurs which would make the case live, electric charge will flow harmlessly into the earth. This stops the charge going through you to earth, or through another part of the appliance which might get too hot (picture 3).

Fuses and circuit breakers

The path to earth has a very low resistance, so that as soon as a fault occurs the current that flows is very large. This could be dangerous, because it could make the conductor too hot and start a fire. To guard against this a **fuse** or **circuit breaker** is built into the circuit. Modern wiring systems have both.

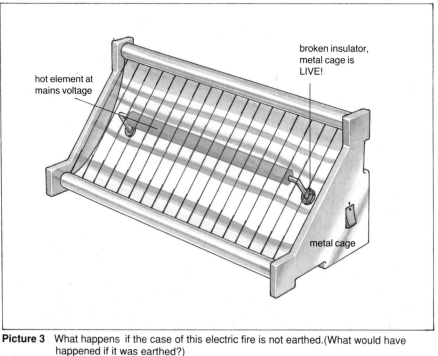

Picture 3 What happens if the case of this electric fire is not earthed.(What would have happened if it was earthed?)

There is a fuse in each three-pin plug, and many devices have fuses built into them as well. A fuse is simply a short length of wire inside a protective case (picture 4). The wire is made of an alloy with a low melting point. A current larger than its 'rated' value will make the wire hot, and it melts. This breaks the circuit and current stops flowing.

A circuit breaker is usually built into the central distribution board where the mains supply enters the house (picture 5). It is usually near the fuse box. There are quite small ones that fit into mains sockets so that you can use them to protect you whenever you use especially dangerous equipment. You should use them for electrical devices used out of doors, such as lawn mowers.

Circuit breakers use an electromagnetic or electronic device to cut off the current if it gets too large because of a fault, or if current is leaking to earth. They are very sensitive and can detect quite small faults. They work much more quickly than a fuse. They give extra protection because even a small current going through the wrong part of a device can cause overheating, with the risk of fire. An ordinary fuse might not 'blow' under these conditions.

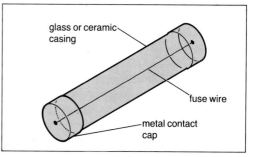

Picture 4   A fuse

Double insulation

Some equipment is 'doubly insulated', so that the live wire cannot reach any outside metal part at all. This means that the equipment does not have to be earthed. Even so, it still needs to be protected by a fuse or circuit breaker. In case the lead (cord) is damaged.

Picture 5   Modern house circuits use a 'leakage' switch for extra safety

Electricity and the human body

The human body is quite a good conductor of electricity — once the electricity gets inside it. This is because the human body is largely water, with all kinds of salts dissolved in it.

Luckily, when the skin is dry it is quite resistant to electricity. Also, if you are wearing rubber or plastic soled shoes the charges can't get out easily. So people may touch bare wires at mains voltage and still survive, although they will certainly feel the shock and it's not a lot of fun.

The most dangerous condition to be in is to have a wet skin and bare feet. (picture 6). This makes bathrooms very dangerous, electrically, and special care is needed. To start with, all switches in bathrooms have to be operated by 'remote control', using lengths of insulating cord. Heaters with bare wiring are banned. Bathrooms are not the places to watch TV or listen to mains radios! The makers of electric showers have to make sure that the water is kept totally insulated from the mains electricity.

What happens when you get an electric shock?

The nervous system of the body works by means of electricity (see *The Living World*, topic D1). Muscles are controlled by electrical messages from the brain or from nerve sensor cells. When a current enters the body it can override the nerve signals. You lose control of muscles. This means that you might be unable to let go of a live wire, your body might quiver uncontrollably, and you might be unable to speak. This can happen with a current as small as 15 *thousandths* of an ampere.

If the current is larger it could cause burning, due to its heating effect. But the real danger is that the heart stops beating. The heart, like any other muscle, is controlled by electrical nerve pulses. If these are overridden by an electric shock the heart might stop beating, and you will also stop breathing. Death follows in a very short time.

Picture 6   Water conducts electricity well: wet bodies are at risk. Why are bathroom light switches safer than an ordinary switch?

Picture 7 Danger! This is a serious fire risk

Thus the first aid treatment for electric shock is similar to that for drowning — heart massage and mouth-to-mouth resuscitation. But first-aiders must be careful to switch off the electricity first, otherwise there would be two patients to deal with.

Good advice

The local electricity board produces pamphlets which give very good advice on the safe use of electricity. They can be got free from any showroom. The most common dangers are due to old, frayed wiring, or cuts in new wiring. Everyone should know how to fit a mains plug correctly, and to use the right fuse in it. Many appliances come with the plug already wired and with the correct fuse fitted.

It is dangerous to 'overload' a circuit. This might be done by connecting high-current appliances (like heaters or even TV sets) to low-current lighting circuits. Instead, they should be connected to the correct 'ring-main' circuit. This is the circuit that has three-pin sockets and the wiring is thick enough to carry the current without overheating. But even this circuit can be overloaded if you use adaptors which allow you to connect too many appliances. Picture 7 shows a selection of dangerous fittings.

Activities

A Danger warning

Make a poster warning either:

1 cooks or,

2 children about **one** electrical danger they have to guard against.

B Make a safety check

Make a survey of the electrical appliances, wires or fittings in your home that might be a source of danger and need attention. List them and say what needs doing to each.

C Learn to get it right

Practise wiring a 3-pin plug. Do this under supervision at school, so that your work can be checked.

D Are you prepared for shocks?

You go into your kitchen. To your horror a member of your family is lying on the floor, still holding an electric iron. You see that the iron has a frayed cable. It is clearly a case of severe electric shock.

What would you do?

If you don't know, find out!

Questions

1 Why are the 3 wires in a household electrical cable colour coded? What are the colours for:

a live,

b neutral,

c earth?

2 Cartridge fuses (as in picture 4) are normally available as 3 A, 5 A or 13 A.

a What would probably happen if you used a 3 A fuse in the plug for a 3 kW electric heater?

b Why is it bad practice to use a 13 A fuse in the plug for a 60 watt desk lamp?

c What happens when a fuse 'blows'?

d You buy a second-hand hair dryer, in good condition, but without a plug fitted. The dryer is labelled '240 V, 800 W'.

What fuse would you choose to put into the plug? Explain how you worked out your answer.

3 You can get a deadly shock from the 240 V mains, but can be charged up to over 100 000 V by a Van de Graaf Machine (page 151) without danger. Explain these facts.

4 Give four precautions used to cut down the risk of electrical accidents in the home. Write a sentence about each of them, so that a younger person could understand why they are used.

5a What is a 'short circuit'?

b What does 'earthing' mean?

c Why are many electrical appliances 'earthed'?

d Why is a short circuit especially dangerous when it happens in a device which isn't earthed?

6 Circuit breakers (page 157) are often used nowadays instead of fuses – especially with appliances that take a large current to make them work.

Why are circuit breakers better than fuses in these cases?

7 Design and sketch a way you could use electromagnetism to switch a current off if it got too big.

8 Most fires in homes and offices are caused by electrical faults. Explain how an electrical fault can start a fire.

A steam iron

Picture 1
A modern iron

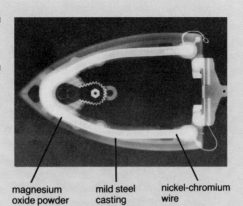

magnesium oxide powder — mild steel casting — nickel-chromium wire

Picture 2 Inside a steam iron

Picture 1 shows a modern steam iron. It is made in three separate parts, or *sub-assemblies*. Each sub-assembly is made of lots of smaller parts, totalling more than a hundred altogether. The sub-assemblies are:

■ the handle,
■ the soleplate,
■ the thermostat.

The handle sub-assembly

This is made mostly of plastic. It has to be light and be a good insulator for both heat and for electricity. It also houses the terminal block for the electrical connections. It has to be well-designed, both to look good and to make it comfortable to use.

The soleplate sub-assembly

This is made mostly of metal. It has a smooth outer case, particularly underneath, where the metal is in contact with the clothes being ironed. The soleplate has to heat up quite quickly and also allow the energy to transfer from the heating element to where it is useful.

The soleplate contains the heating element, which is a coil of resistance wire that gets hot when an electric current flows through it. It has to be insulated from the main soleplate to avoid electric shocks to the user. This is done by packing the coil in a compressed powder of magnesium oxide (picture 2)

The coil also heats the steam chamber, in which water is boiled to provide steam for making clothes easier to iron. Water is a good conductor of electricity, which is another reason for having the heating coil so well insulated.

The thermostat sub-assembly

This controls the temperature of the ironing surface of the soleplate. It does this by switching the current on and off as needed. This means that it has to be made out of very reliable components. In the ordinary life-span of the iron the current will be switched on and off many thousands of times.

Picture 3 shows how the thermostat switch works. It uses a bimetallic strip. This consists of two different metals welded together. One metal expands more than the other when it is heated. Because of this the strip bends when is heated. Thus when the iron gets hotter than it should be, the contact is broken and the heating coil is switched off. The thermostat control works by moving the contacts nearer together or further apart, so that the contact is broken at the required temperature. Answer the following questions.

1a Give three reasons why plastic is used in the handle, rather than metal.

b Suggest another material that could be used instead of plastic, and suggest a reason why it *isn't* used.

2a Why is the heating coil surrounded by an insulator?

b Plastic could be used instead of magnesium oxide as an insulator around the heating coil. Suggest two reasons why the magnesium oxide might be a better choice.

c What kind of chemical bonding is present in magnesium oxide? Why does this type of bonding make it suitable for this use?

3 Use the diagram (picture 3) to decide which of the metals used in the bimetallic strip expands the most when heated.

4a In use the bottom of the iron is surrounded by steam, and it is used over a wide range of temperatures. Suggest what properties the material used to make the base of the plate must have if the steam iron is to be usable for a number of years.

b The material used for the outside of the soleplate is a *mild steel* plate covered with a smooth coating of *zinc*. The inside of the soleplate is made of cast *aluminium*. Give reason one reason in each case for the use of these materials (printed in *italics*).

c Which part of the steam iron is most likely to break down after many years of use? Give a reason for your answer.

5 How did 'irons' get their name?

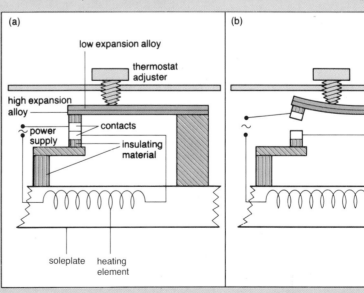

(a) low expansion alloy / thermostat adjuster / high expansion alloy / power supply / contacts / insulating material / soleplate / heating element

(b)

Picture 3 How the thermoset switch works

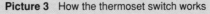

E8
Using electricity: motors and dynamos

Electric currents and magnetism can work together to make things move.

Picture 1 Michael Faraday lecturing to his students

Poor boy makes good!

Two hundred years ago it wasn't easy for a poor boy to get an education, even if he was a genius. Michael Faraday (born in 1791) was the son of a blacksmith. He learned to read and write and do arithmetic but left school at 13 to work as an errand boy for a bookbinder.

At 14 he was promoted to apprentice bookbinder. Keen to learn, he started to read the books that he bound. He was fascinated by the science books, especially the ones about physics and chemistry. He spent some of his small pay on materials and started experimenting at home. He made lots of smells — and an electrical machine.

Later on, he went to lectures put on for the public at a great research laboratory in London called the Royal Institution. He was very interested by four lectures given by a famous scientist called Sir Humphrey Davy, who was the director of the Institution (see *The Material World*, topic J1). He made careful notes of what he heard and wrote them out neatly. He used his skills to bind these notes and sent them off to the great man, with a letter asking for a job as a lab assistant.

He was given the job, and at the age of 21 he was able to give up bookbinding for ever. Twelve years later Michael was made Director of the Institution, and had become one of the most famous scientists in the world. Nearly everything you will learn about in this chapter was first discovered by him. The equipment he used can still be seen at the Royal Institution.

Electricity and magnetism working together

An electric current produces a magnetic field (see page 138). Put a wire in the field of a magnet and pass a current through it. Both wire and magnet will try to move. There is a force between them caused by the interaction of their magnetic fields (picture 2). This is the **motor effect** which is used in all electric motors, from toy cars to large electric locomotives.

Strangely, the force is exerted at right angles to both the current direction and the magnetic field lines. When you change the direction of the current in the wire, the force on it changes to the opposite direction too. Changing the poles around so that the field is now in the opposite direction also changes the direction of the force on the wire.

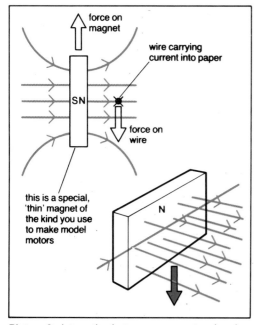

Picture 2 Interaction between a magnet and a wire carrying a current

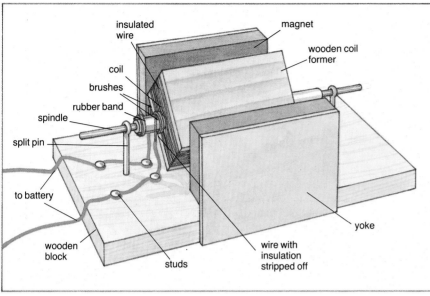

Picture 3 A model electric motor you could make

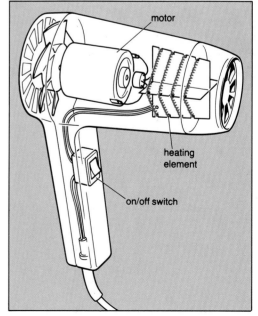

Picture 4 The forces on a coil in a magnetic field make it spin

Electric motors

Simple electric motors are scaled up versions of ones you can build yourself from a length of wire and two magnets (picture 3, and activity 2). A model motor uses a special pair of slab magnets that give a straight field between them. The force makes a turning effect that is greatest when the spinning coil is at points A and B (picture 4). The design lets current into the coil only when it is at this position.

Small working motors

Picture 5 shows the inside of a small electric motor. A motor like this is used in electric drills or vacuum cleaners. It has a much longer coil than your 'home made' one, and its magnetic field is made larger by wrapping the coil around some soft iron. This turns the coil into an electromagnet (topic E2). It is called an **armature** when it is used like this.

The problem with a spinning electromagnet is to get the electric current into it and out again. This is done by using sliding contacts called **brushes** which just touch the ends of the spinning coil. A real motor has more than one coil with more than one pair of brush contacts. As the motor spins round, one coil after another is brought into action so that the force is almost continuous.

The field is not straight as in the motor that you might make in the school laboratory. It is radial, like the spokes of a wheel. This means that the coils are always at right angles to the field lines and so they always have a strong force on them. In a larger motor the magnets producing the field are electromagnets, not permanent ones.

Electricity from magnetism?

Michael Faraday didn't stop at only making magnetism and electricity work together to produce movement. He thought that there ought to be a way of using movement and magnetism to produce electricity. It took him seven years to think of a way of doing it, but what he discovered is the scientific basis of the electricity industry — **electromagnetic induction**.

Picture 6 overleaf shows a modern version of his experiment. When the current in the first coil is switched on the meter shows a pulse of current in the second coil. Nothing happens when the current in coil 1 is flowing steadily, but another pulse occurs when it is switched off.

If you alter the current in coil 1 steadily, by using a rheostat in the circuit, you get a steady current in coil 2.

Picture 5 An electric hair dryer showing the electric motor inside

Picture 6 A modern version of Faraday's discovery of induced electricity (the transformer effect)

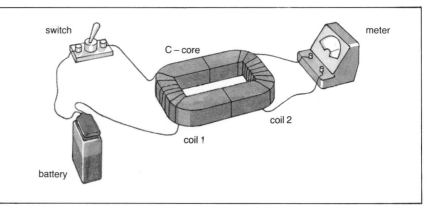

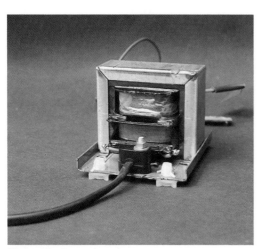

Picture 7 A transformer

Picture 9 A bicycle dynamo

All this happens because a voltage is **induced** in coil 2 whenever a current is changed in coil 1. The modern application of this is the **transformer** (picture 7). There is more about transformers in topic E10.

Faraday tried to imagine what must be happening. He had already thought up the idea that magnetic fields could be explained in terms of lines of force. He explained this new effect could by imagining that as the current was being switched on and off so were the lines of force of its magnetic field. The voltage in the second coil was made only when the lines grew and died away.

A generator

So what would happen if he made the lines of force come and by a different method, by moving a magnet into and out of a coil? He tried this (picture 8) and it worked. Once more a current was produced. The same thing happened when he kept the magnet still and moved the coil instead.

He had invented the first **electric generator**, or **dynamo**. A simple dynamo (picture 9) has the same parts as an electric motor. Both have spinning coils surrounded by magnets. In the motor, putting a current through the coil makes it spin. In a dynamo, making the coil spin produces an electric current. Large dynamos, called generators, are used in power stations to produce the mains supply that we use to run so many things in home and in industry.

Using electricity to carry energy long distances

Most railways now use locomotives driven by electric motors. These are powered by generators which can be many hundreds of kilometres away (picture 10).

The main advantage of electricity is that it is so easy to move it from one place to another. A disadvantage is that it is hard to store. Topic E10, *The electricity industry,* deals with the production and supply of electricity.

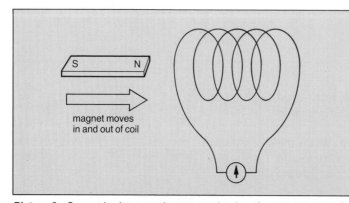

Picture 8 Current is shown on the meter only when the magnet moves in or out of the coil (the dynamo effect)

Picture 10 Electric locomotive

Activities

A Investigating the motor effect

You will need a pair of slab magnets fitted to a U-shaped piece of soft iron, a low voltage, high current power supply, some wire and some wire cutters.

1 Cut out three pieces of wire as in picture (i), and fix the two longer ones to the red (+) and black (-) terminals of the power supply (picture 11(i)).

2 Hang the small piece of wire over the other two, as in picture 11 (i).

3 Slide the soft iron with the magnets over this cross-wire so that it is in the field of the magnets. Switch on the current and see what happens.

4 How can you change the direction in which the sliding wire moves?

B Making an electric motor

To do this you will need a special kit. It may look like the model motor shown in picture 3. Follow the instructions given in the kit carefully. When you have got the motor to work answer the following questions.

1 Find two ways to make the motor spin in the opposite direction.

2 How can you make the motor spin faster?

3 Give two differences between this model motor and the ones used to power an electric drill.

C Making a dynamo

Relax. You have already made one. Disconnect the power supply from the motor you made in activity B and hand it back to your teacher. Collect a sensitive meter and connect it to where the power supply was connected. When you spin the motor the meter should show a small current.

D Investigating transformers

Try to repeat Michael Faraday's experiments with a pair of coils wound around an iron bar or ring. You may be given special apparatus for this.

Things to explore:

What happens if you have many more turns in the secondary (output) coil than in the primary? Can you make the lamp connected to the secondary coil light up?. How do you know that magnetism is involved? What happens if you use alternating current (a.c.) instead of direct current (d.c.) in the input?

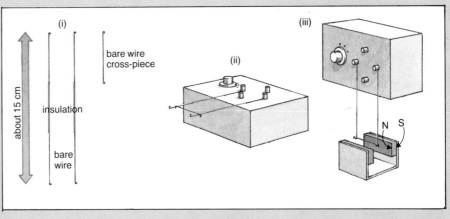

Picture 11 Investigating the motor effect

Questions

1 Picture 11 (iii) shows how you might investigate the force on a wire carrying a current in a magnetic field. What would happen if:

a the magnets were swapped round, so that the S-pole was on top and the N-pole was on the bottom?

b the battery connections were then changed so that the current flows in the opposite direction?

c the current was made bigger?

2 Name (a) two devices that use an electric motor to move something from one place to another (b) two devices that use electric motors to do something else.

3a Give one place where you would expect to find a transformer being used.

b What is it being used for?

4 Look at picture 12, which shows a coil connected to a meter, and a magnet.

When the magnet is moved slowly towards the coil the meter shows a current as shown.

a What would happen to the current if the magnet were moved more quickly?

b What would happen to the current if the magnet was kept still and the coil was moved towards the magnet?

c What would happen to the current if the magnet was moved, but this time away from the coil?

5 You have made a model electric motor and it works! What could you do to:

a make it turn faster,

b make it spin in the opposite direction,

c make it work as a dynamo?

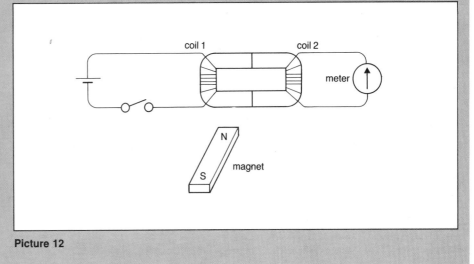

Picture 12

E9
Electricity from frogs?

*To the ancient scientists all electricity was 'static'. Then by accident, an electric **current** was discovered . . .*

Picture 1 'You cannot be serious!'

Picture 2 An electric eel

The first electric current

When electricity was 'rediscovered' 300 years ago (topic E5) it was 'static' electricity. Apart from entertaining people in a shocking way it wasn't much use. But in 1780 an Italian doctor, Luigi Galvani, discovered a way to produce a continuous flow of charge — in fact the first electric **current**. Like many discoveries in science it was a kind of accident.

Galvani was interested in electricity and had all kinds of equipment in his home. The story goes that Galvani was preparing some food for his sick wife, who happened to like to eat frog's legs. When they were laid out ready for cooking his wife noticed that whenever a spark was produced by a nearby 'electric machine' (see page 149) the frog's legs twitched.

A missed meal

She never got to eat the frog's legs. Galvani started experimenting with them to find out why they twitched. He found that it happened when the leg nerves were stimulated by electric sparks. Then, to his great surprise he found that the legs twitched even when the machine was not working. But this only happened if the ends of the nerves were touched by metals.

He then drew the wrong conclusion. He thought that the electricity came from the frog's leg. It just needed the metals for it to be conducted away. After all, if you can get electric eels (picture 2) why not electric frogs?

The first battery

Galvani did his experiment in 1780, and it aroused great interest amongst the electrical hobbyists of the day. But it took another 14 years for the true cause of 'frog electricity' to be explained. This was done by another Italian, Alessandro Volta. He showed that the source of the electric current was not the frog's leg at all. It was caused by the fact that Dr Galvani had used *two different metals, in the presence of salt water*. The frog's legs had been preserved in salty water. And Dr Galvani had indeed used a zinc dish and steel scalpels, and he had also touched the legs with copper wires.

Picture 3 Dr Galvani experimented with frog's legs

Volta used this idea to make the first battery. He made it from alternating pairs of zinc and copper discs separated by cloth soaked in salt water (picture 4). This produced a continuous flow of what he called 'artificial electricity'.

Modern batteries work on the same principle (picture 5). There are many different kinds. But each uses a pair of metals with a solution in between. The most common batteries use carbon and zinc separated by a solution (**electrolyte**) of ammonium chloride. Carbon often acts like a metal, electrically. Ordinary torch batteries are of this type.

Other cells use such metals as nickel, iron, cadmium or mercury.

Cells

A pair of electrodes separated by an ionic solution is called an **electric cell**. The solution has to contain ions so that charge can flow between the metal electrodes. A battery is really a collection of more than one cell, although we often use the word for any kind of cell.

Galvanometers and **volts** are named after these two Italian scientists who did the early work in making 'artificial' electricity.

Electricity and chemistry

The news of this battery soon spread across Europe and it was realised that electricity had a lot to do with chemistry. Michael Faraday's boss Sir Humphrey Davy became famous by using the new 'current' of electricity to break down compounds by what is called **electrolysis** — see *The Material World*, topic J1. He discovered new elements like sodium, potassium, calcium and magnesium.

The use of various metals in electric cells led to the idea that metals can be arranged in an **electrochemical series** as shown in table 1. Compare this series with the reactivity series in *The Material World*, topic E2. The voltage of a cell is decided by how far apart its metal electrodes are in this series.

Picture 6 shows a variety of batteries in use today.

Table 1 Part of the electrochemical series

element	voltage compared with hydrogen
calcium	−2.76
sodium	−2.71
magnesium	−2.37
zinc	−0.78
cadmium	−0.40
hydrogen	0.00
copper	+0.34
mercury	+0.79

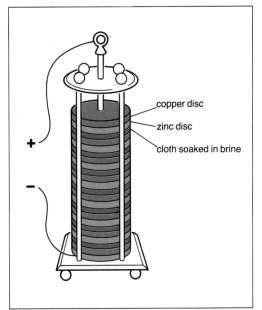

Picture 4 The first battery used zinc, copper and salt water

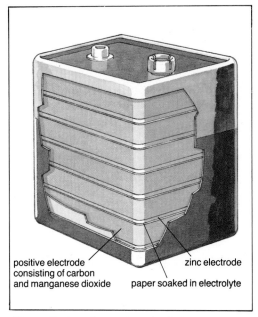

Picture 5 A modern carbon-zinc battery

Picture 6 All these batteries rely on the electrochemical series

Activities

A Fruit electricity

It is possible to get electricity out of a lemon. All you have to do is to stick in two electrodes made of different metals. Design an investigation to see how effective this source of electricity is. Check your plan with your teacher before you carry it out.

Will it work with other fruits? What metals work best? Is this a practical, economic source of electricity?

B Using batteries 1

1 Make a list of everyday devices that use batteries to make them work.

2 For each one, say why batteries are used instead of mains electricity.

3 Find out the cost of a battery and how long it will run one of the devices you have listed.

4 Look at the labelling on the device. Try find out its power in watts or kilowatts. Calculate the cost of using batteries to run it, per kilowatt-hour. Is it cheaper or dearer than using mains electricity?

   (Mains electrical energy costs about 6p per kilowatt-hour).

C Using batteries 2

Find out the answers to the following questions. You can ask people, read pamphlets, use a library.

1 List as many different types of battery as you can (i.e. based on different combinations of chemicals).

2 In what way are car batteries different from ordinary 'torch type' batteries?

3 Many pocket calculators use solar cells. How do solar cells work?

4 What is 'animal electricity'?

D The electrochemical series

Design an experiment to check the order in which elements appear in the electrochemical series. You can use a voltmeter (preferably a digital one) and a solution of dilute acid. Suitable elements include: copper, zinc, nickel, magnesium, aluminium, iron and tin.

Do not carry out your experiment until you have checked it for safety with your teacher.

(*Hint*: to get reliable results make sure that the samples are clean Use some emery cloth to scrape off any dirt, grease or layers of oxide).

Questions

1 Name four metals that are used in making batteries.

2 How could you use four 1.5 V cells to make a 6 V battery? Draw a simple diagram of the arrangement.

3 What do you think happens when a battery 'runs down'?

4 A one-cell battery is labelled 1.5 V. It is used in a device that takes a current of 0.2 A from it. It runs out after 10 hours continuous use.

a What power does the battery supply?

b How much energy does it deliver?

c How many coulombs of charge does it supply?

   (You will need to use some of the formulae given in table 1, topic E1)

5a What are ions?

b Why does the solution in a battery have to contain ions?

How does it work?
An electric bell

The diagram below shows the main parts of an electric bell. Your task is to describe how it works. You will need some clues, which are given below the diagram.

Facts

1 When you press the bell-push (a switch, **B**), current flows through the circuit that includes the electromagnet.

2 The core of the electromagnet (**M**) is made of an alloy (e.g., 'soft iron') which is easily magnetised in a magnetic field, but loses its magnetism very quickly when the field disappears.

3 The bell hammer is connected to a piece of springy steel (**S**), which has another piece of soft iron attached to it (**A**).

Clues

What happens to the circuit when the electromagnet pulls **A** towards it?

Then what happens to the electromagnet?

Tasks

1 Now describe as clearly and logically as you can how the bell keeps on ringing as long you keep your finger on the bell-push.

2 The circuit also contains a capacitor (**C**). This helps to stop damage that might be caused by high-voltage sparking. Where does this voltage come from?

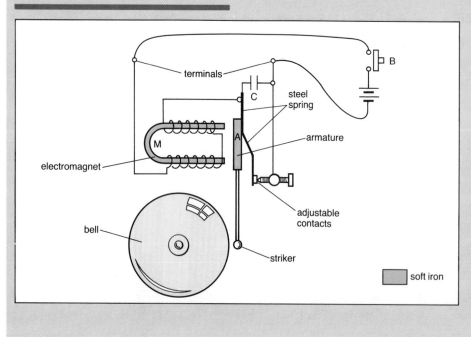

Seeing further and clearer

The Hubble Space Telescope was launched in April 1990. It had been designed over a period of 20 years, and will cost $8 billion. It will be effective in space for just 15 years. Picture 1 shows what it looks like in space. The main tube is a Newtonian telescope, with a very large main mirror.

The telescope is designed to see further and more clearly into the depths of space than ever before. The telescope can be used to send its images to one of several detecting instruments in turn. The most spectacular results will probably come from the 'wide field' camera. This can take large, clear pictures of large objects like nearby galaxies and planets.

It will get its advantages by being outside the Earth's atmosphere. Earthbound telescopes have to look through many kilometres of air. Stars twinkle — which means that their images dance about and can't give a clear image. Using bigger telescopes just means getting a bigger blur. Also, some radiations are absorbed by the atmosphere, and so never reach the Earth's surface.

Astronomers rarely look through telescopes. The images of stars and galaxies are usually recorded on film. But this method is no good for the Hubble Telescope! Instead, it uses silicon chips.

These chips are very thin layers of semiconductor material. They contain

Picture 1   The Hubble Space telescope

thousands of tiny photon detectors, the *pixels*. There are a quarter of a million of these to every square centimetre of chip. The photons of light trigger off each pixel when they hit it. They are over a hundred times more sensitive than the best film, so they can detect much fainter images. They are also small enough to make accurate pictures.

They are called *charge-coupled devices* (CCDs), and picture 2 shows a part of one of them. Four of them are used together in the wide-field camera in the Space Telescope. When photons of light hit a pixel, it releases an electron. This charged particle then moves to form a current. The currents from all the pixels are very carefully aligned to form a signal which can

be sent back to Earth. The number of electrons from each pixel tells us how much light hit it. A computer is used to convert the signal current into a picture.

This is just like the way TV works. Indeed, modern TV cameras use CCDs, which means that they are sensitive enough to take pictures in the dark — you can see this in nearly every TV news bulletin.

The image is made on the array of CCDs by the main telescope. They can detect all kinds of radiation — light, UV and even X-rays. The telescope can't focus X-rays — they go straight through it! But light and UV can be focused very accurately. The mirror is curved, and 2 metres in diameter. Its aluminium surface is smoothed to a tenth of a wavelength of light. If it was scaled up to be the size of the USA the biggest bump on it would be just 2 centimetres high. This means it can make best use of the accuracy given by the CCDs.

Now try the following questions.

1   What are the advantages of having a telescope in space, compared with one on Earth?

2   Give a reason why cameras with photographic film would not be much use in the Hubble Telescope.

3   Explain what you understand by the word *photon*.

4   Why does the mirror surface have to be so smooth?

5   What is the advantage of having the main mirror as large as possible?

6   Give two advantages of using CCDs in this telescope.

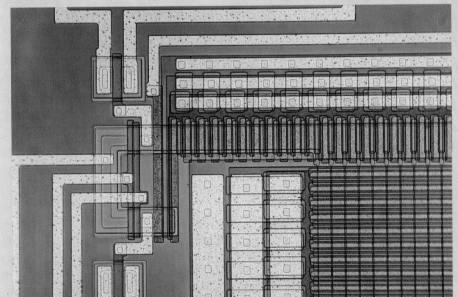

Picture 2   A CCD screen

E10 The electricity industry

This topic deals with how electrical energy is generated and then transported to where it is needed.

Picture 1 Drax Power Station near Selby, North Yorkshire

The power station

Picture 1 shows a large power station. You can see the huge stock of coal that will be taken in at one end. At the other end are the pylons holding the wires through which the electricity will be carried away. The diagram (picture 2) shows the main parts of the power station where the energy conversions take place.

The energy changes in a power station

Energy is released when coal burns with the oxygen of the air. This energy is used to boil water and then heat the steam to a high temperature. Burning coal also produces large amounts of carbon dioxide which goes into the atmosphere, together with other waste gases such as sulphur dioxide (see *The Material World*, topic I2).

The steam is made very hot so that it is at a very high pressure. This means that it can provide very large forces to turn the huge steam turbines. This takes energy from the steam, which cools down, but doesn't become so cool that it condenses back into water.

The spinning turbines (picture 3) are connected to the coils of large generators. These coils carry current and act as large electromagnets. As they spin they induce a high voltage in the fixed coils surrounding them (see page 161 and picture 4). This causes a current which is fed into the **National Grid** system that carries the electricity to wherever it is needed (picture 5).

Picture 3 The turbine - generator room of a power station

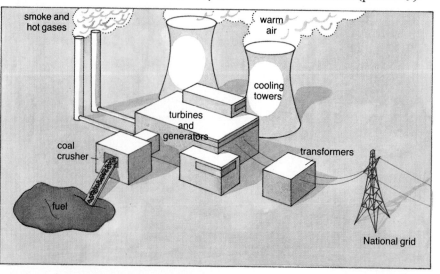

Picture 2 The main parts of a power station

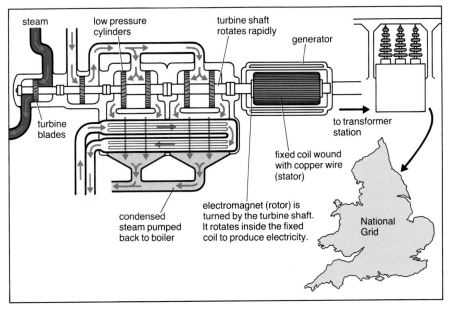

Picture 4 The structure of a large generator

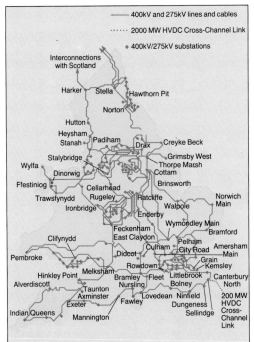

Picture 5 The main power lines of the National Grid

Energy flow in a power station

Picture 6 shows the energy flow through a typical power station. A large power station might be rated at 1000 megawatts. This means that every second it delivers 1000 million joules of energy. This is about the same as the total power output that could be produced by every human being in the United Kingdom working flat out. No modern industrial country could survive on slave labour!

But to produce this energy, fuel equivalent to an energy of 3000 million joules per second has to be supplied. *Two-thirds of the energy input is wasted.*

How is the energy wasted?

Some of the waste is 'accidental', because energy leaks out to warm up the air. For example, it moves as hot air from the boiler chimneys, and wiring gets hot.

But most of the waste is *necessary* waste. This is because the turbines can't take all of the energy out of the steam. This leaves lots of steam which still warm, but cooler than it was when it went in, and too cool to make any turbines work.

This is the reason for the cooling towers. They are used to take energy from the 'used' steam, so condensing it back to water. This energy usually ends up warming a river or the sea.

This waste seems a great pity, but it is a consequence of one of the most ruthless laws of physics. This is the Second Law of Thermodynamics, which says that whenever you try to do something useful with energy some of it always ends up in the wrong place (see topic D4).

In this example, the wasted energy goes into the cooling system and eventually into the surroundings.

Can we make better use of this waste energy?

The waste energy can be put to good use. It is stored in warm water. The water is not hot enough to be useful to make electricity — it could not make the steam turbines work. But is quite warm enough to heat homes and other buildings.

In Germany most towns have their own small power stations, and they often pipe the 'waste' hot water to people's homes to keep them centrally heated. These are called **combined heat and power schemes**. They

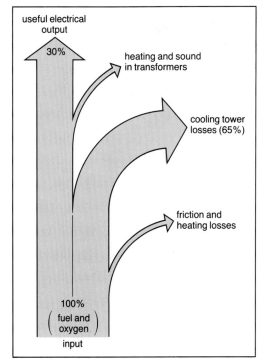

Picture 6 Energy flow in a power station

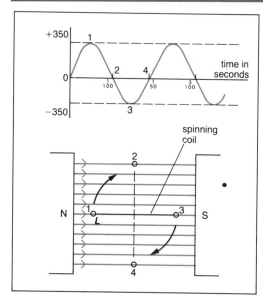

Picture 7 The voltage of the mains supply swings between +340 V and −340 V, 50 times a second
Numbers 1, 2, 3, 4 show where one side (L) of the spinning coil is to produce the outputs marked 1, 2, 3, and 4 on the graph

reduce waste and make electricity cheaper.

In the UK most power stations are larger and have been built a long way from towns and cities. This means that using them to produce 'combined heat and power' is not economic.

Alternating current

The turbines spin the field coils of the generators at high speeds. The wiring is arranged to give **alternating current**. This is done by making the voltage produced by each coil change direction every half-turn. This means that the current changes direction as well. It flows one way for half the time, then in the opposite direction for the next half. This change-over occurs every 1/100 of a second. How this is done is shown in picture 7. Only one coil is drawn, to make the idea clearer.

The rate of spinning and the number of coils is designed to make each coil change (**alternate**) its voltage and current completely 50 times a second.

The speed of movement, the large magnetic field and the large number of turns used means that the voltage produced is quite high: 25 000V.

There is a good reason for producing alternating current (a.c.) rather than direct current (d.c.), like a battery produces. It is because it makes it easy to change the voltage of the supply, using transformers. This is explained next.

How do we get our electricity?

The generator produces current at an output voltage of 25 000V, which is extremely dangerous. It could be arranged for this to be 240V, as used in the home, but this would be very uneconomic. In fact the voltage is made even higher as it leaves the power station. It is raised to 275 000V or even 400 000V. The reason for this is the resistance in the cables which take the electricity from the power station to a home or factory.

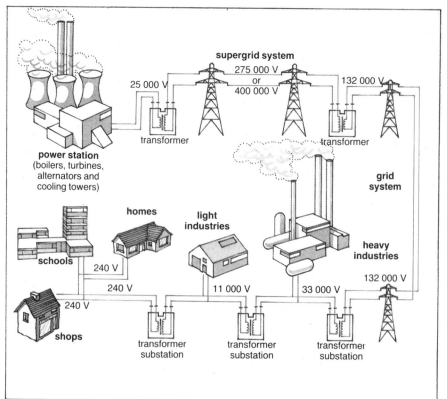

Picture 8 Voltages in the National Grid system

Picture 9 A power cable

Why is electricity transmitted at very high voltages?

Most people live and work many hundreds of miles from the power stations that produce our electricity. This is because the stations are built near good supplies of coal and cooling water.

Whenever a current flows in a conductor some power is lost in heating the conductor. Engineers can cut down this energy loss by supplying the current at very high voltages.

The power loss in a conductor is given by $P = VI$, where V is the voltage drop across the conductor. We can change this to give power loss in terms of the current I and the resistance R of the cable. This gives $P = I^2R$ (because $V = IR$). The engineers want the loss in the cable to be as little as possible. They do this by making the resistance (R) of the cable as small as possible. The resistance can be cut down by using very thick cables (see picture 9).

Even so, there will always be some cable resistance, and there comes a point when the cost of making the cable is greater than the value of the energy we are trying to save. The only other thing they can change is the current in the cable. This is done by using very high voltages.

High voltage, low current

The delivered power is also given by $P = VI$. This time V is the voltage drop at the end where the customer wants to use it. The power is the same for a low current at a high voltage as for a high current at a low voltage. Suppose the customer wants 100 000 W of power to be delivered. This could be done, for example, either by sending 200A at 500V or by sending 1A at 100 000V.

Now suppose the cable resistance is 2 ohms. The cable loss in the first case (voltage 500V, current 200A) is:

$$\text{power loss} = I^2R = 200 \times 200 \times 2 = 80\,000 \text{ watts}$$

This would leave only 20 000 watts for the customer!

In the second case (voltage 100 000V, current 1A) we get power loss $= I^2R = 1 \times 1 \times 2 = 2$ watts!

It makes very sound economic sense to send the electric power down the cable at high voltages. But you would not be too happy at having a mains supply at 100 000 volts. This is a very dangerous voltage. This is where transformers come in.

What transformers do

Faraday's experiment with two coils (page 162) gives the basic idea of how transformers work. We can switch a current on and off in the primary coil. This produces and then takes away a magnetic field in the iron core (picture 10). The changing field induces a voltage in the secondary coil.

Now, when an alternating current is supplied to the primary coil it also produces a changing field in the iron core. This field changes as the current changes. Just as with the switching experiment this changing field also induces a changing voltage in the secondary.

But, if you have investigated transformers (activity D, topic E8), you will know that the voltage in the secondary coil depends also on **how many turns** there are in the coils. If there are more turns in the secondary than in the primary coil, the output voltage is bigger than the input voltage. If there are less turns in the secondary than in the primary, the voltage is reduced. Thus we can have **step-up transformers**, which increase the voltage, and **step-down transformers**, which do the opposite (picture 11).

This leads to the transformer rule:

$$\frac{\textbf{Output voltage}}{\textbf{Input voltage}} = \frac{\textbf{secondary turns}}{\textbf{primary turns}} \quad \text{or} \quad \frac{V_{out}}{V_{in}} = \frac{N_{sec}}{N_{pri}}$$

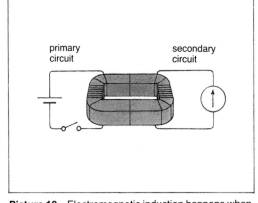

Picture 10 Electromagnetic induction happens when a magnetic field is changing or a conductor cuts through the field

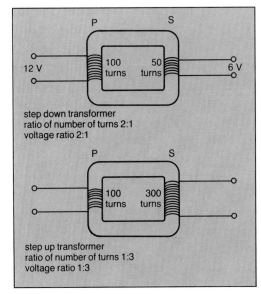

Picture 11 Step down transformer: ratio of number of turns 2:1, : voltage ratio
Step up transformer: ratio of number of turns 1:3 : voltage ratio 1:3

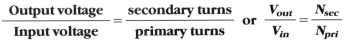

Transformers and the Grid System

Step-up transformers are used at power stations (pictures 1 and 2) to raise the 25 000V produced by the generators to the higher voltages which are so efficient for transmitting electricity over long distances. The network of cables carrying current at these very high voltages is called the **National Grid**.

The cables are carried on tall pylons which are easy to spot when you travel about the country. Very many people live in Southern England, and they need a lot of electricity. Some of it comes from as far away as Scotland or France (by undersea cables). The normal Grid supply voltage is 275 000V, but for moving electricity very long distances the **Supergrid** is used, using a voltage of 400 000V.

At the customer end, step down transformers are used. They lower the Grid voltage to the 550V used in some factories and the 240V used in the home, in shops, etc.

Clean and efficient energy?

Electricity is the most efficient of all the energy types we use in the home or in industry. Electric motors are about 90% efficient, compared with a petrol engine which may be only 20% efficient. Similarly, electric trains are three times as efficient as diesel engines. An electric fire uses 100% of the energy it receives to heat the room. Coal fires and oil central heating systems lose some energy in hot air and fumes going out of the chimney.

These figures are impressive, but a little misleading. Back at the power station the laws of thermal energy take their inevitable toll (see above). Producing the electricity in the first place is only 30% efficient, which is about the same efficiency as a diesel engine. This loss is there even if we use nuclear energy to power the turbines. The loss is due to the nature of all **thermal engines**, like turbines, that rely on heating to make them work.

Thermal power stations also produce much of the polluting gases, like suphur and nitrogen oxides, that cause acid rain. This is because coal and oil contain impurities. But what is more, all fuels produce carbon dioxide when they are burnt. It has only recently been realised that this 'harmless' gas may be seriously affecting the environment on a global scale. It is the major gas involved in the 'greenhouse effect'. This is dealt with more fully in *The Material World*, topic I4.

Another problem is that sooner or later, the fossil fuels that we use to run power stations will be used up.

Nuclear power stations

Nuclear power stations are just about as efficient (or inefficient) as oil or coal-fired power stations. They normally produce little pollution — unless there is an accident, as happened at Chernobyl in 1986. But they do generate radioactive waste, which needs to be stored for perhaps many thousands of years (see topic D7).

Are there other ways of getting electrical energy?

These facts have led to an interest in 'alternative' or renewable sources of energy, which will not run out, and may also be safer and less polluting.

This is dealt with more fully in topic D5.

Electric power from water and wind

One of the most common renewable sources of energy for producing electricity is falling water. It is renewable as long as it keeps raining in the hills where the water is stored in dams.

Picture 12   The Genissiat hydroelectric power station, on the River Rhone, France

In the UK only 1.2% of the electrical energy we use is produced in such **hydroelectric** power stations. But countries with more mountains, and plenty of water, make more use of this energy source (picture 12). All of Norway's electric power is produced in this way. In these stations the power of falling water is used to turn the turbines, instead of steam.

Research is being done to make more or better use of the energy of moving air (wind) and of waves and tides. Topic D5 deals with these issues.

Cheap electricity?

Coal-fired power stations can't be switched off easily. The furnaces have to be kept going all the time, because if they cool down they get badly damaged. Thus many power stations are running all the time, both day and night.

At night most people are asleep. Factories are closed down, few trains are running. The demand for electricity is much less than in the daytime.

But the power stations are still burning fuel, even if the turbines are not working at full capacity. It is more economical for the electricity supply industry to sell electrical energy cheaply at night than not to sell it at all. Thus many homes and factories have heaters which are time-switched so that they use only night-time electricity (see page 153).

(see page 153).

Activities

A Somewhere near your house or school there will be a transformer which reduces the voltage of the electricity supply to 240 V. Find the transformer, sketch it and describe it. Write down any official notices that might be on it and explain what they say.

B Ask questions, read pamphlets, look up books in the library to help you answer one or more of the following questions.

1 Where is your nearest power station. What fuel(s) does it use?

2 Why is electricity cheaper at night?

3 In the mountains of North Wales, at Ffestiniog and Dinorwic, there are special kinds of power station called 'pumped-storage' stations. What are these and how do they work?

4 Where in the British Isles would you expect to find:

   i) hydroelectric power stations,

   ii) nuclear power stations?

   What factors decide where such stations are built?

5 How does the electricity industry cope with the problem that much less electricity is used in summer than in winter?

Questions

1 Explain what jobs the following do in a coal-fired power station:

a generators.

b cooling towers,

c step-up transformers.

2a What is the National Grid?

b Why are the cables that carry electricity held so far above the ground?

c Electricity is sent along the National Grid at very high voltages. Why must they be so high?

3 The graph (picture 13) shows the cost per metre of cable of different thickness. It also shows the cost per metre of the power loss in the cable due to resistance heating. Both of these quantities are plotted against cables of different thickness.

a Why does the cost of lost power go down when the cable gets thicker?

b Why does the cost per metre of cable rise when the cable gets thicker?

c What is the combined cost of cable-plus-power-loss for cables of thickness: (i) 10cm, (ii) 15 cm?

d From the graph, what is the most economical cable thickness to choose?

e The graph for the cost of the cable rises much more quickly than the power saving cost falls. Suggest a reason for this.

4a Calculate the current that needs to be taken from a 25 mW power station at a generating voltage of 2500 V.

b What would this current be reduced to if the voltage was stepped up to 400 000 V for the National Supergrid?

c What would be the effect of this reduction in current on power loss in the cables?

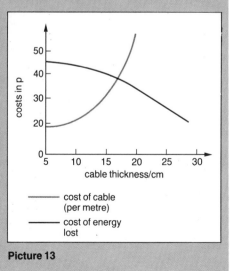

Picture 13

E11
Electrons in space

So far we have looked at electric charges flowing in wires, liquids or gases. Interesting things happen when we get the charges out on their own, in empty space . . .

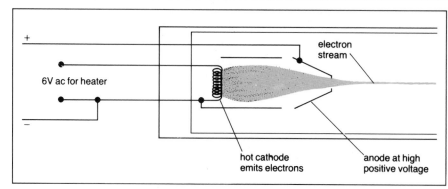

Picture 2 The electron gun in a TV tube

A TV tube is mostly empty space. The picture screen is covered on the inside by a **phosphor**. This is a chemical which glows when it is hit by electrons (picture 1). The electrons have to be travelling fast. In a colour tube they need to hit the phosphor at over 15 million metres a second.

The electron gun

The electrons are fired fom the narrow end of the tube by the device called an **electron gun** (picture 2). The electrons are produced by heating a metal oxide so that it just glows red hot. This is called **thermionic emission**.

This part of the gun is kept negative and is called the **cathode**. The electrons that 'boil off' the cathode are attracted by a metal cylinder which is kept at a very high positive voltage — the **anode**. This produces a force which accelerates them to a very high speed.

The space inside the tube must be completely empty of any gas, otherwise the electrons would collide with the gas molecules. The stream of electrons is focused into a narrow beam which has to be very accurately aimed so that the phosphor screen glows in exactly the right places at exactly the right times to make a picture.

Making the picture

The picture is made by **scanning** the electron beam across the screen in a series of lines which are then moved from top to bottom of the screen. There are 625 lines in a screenful, and the screen is completely scanned by the beam 25 times a second (picture 3). If you wave your fingers to and fro in front of your eyes whilst watching TV you can see odd 'gaps'. You don't normally see these gaps because it takes about a twentieth of a second for your eye to wipe off a picture from its retina. The pictures come quicker than

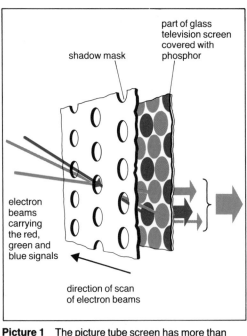

Picture 1 The picture tube screen has more than 300 000 coloured phosphor dots arranged in groups of three on its surface. A metal mask behind the screen has holes which keep each of the three electron beams in line with its own colour dots and away from dots of other colours

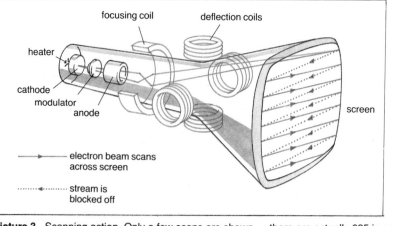

Picture 3 Scanning action. Only a few scans are shown — there are actually 625 in each 'screenful'

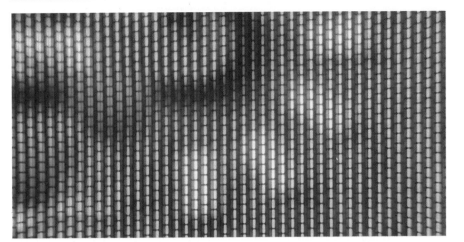

Picture 4 The 'pixels' in a TV screen. A TV picture is made of thousands of these

that, so one picture merges neatly with the next. Better pictures can be made with more lines in a screen, and the new 'high definition' TV sets use many more lines.

Controlling the electrons

The electron beam has to move across the screen very quickly and very accurately. In a colour tube the beam is split into three, so that it can hit three separate picture cells (pixels). The pixels glow red, blue or green (picture 4). These are the primary colours (see page 96). These can be combined to give the illusion of all the colours of the rainbow.

The electrons are directed to the right place by a strong magnetic field. In picture 1 you can see the coils of wire that produce this field. Its strength is controlled by the signal sent out by the TV station and picked up by the TV aerial.

The electron stream is of course just a flow of charge — an electric current. Just like the current in the coils of an electric motor the electron stream can be moved by a magnetic field. The field can direct them to where they have to go to make the picture.

Electrons are the lightest objects in the universe, so they change direction very quickly. The magnetic fields have to change very quickly to produce a new picture 25 times a second, but the electrons are light enough to follow the changes.

Picture 5 shows the direction of the forces on a stream of electrons in a magnetic field. Just as in a motor, the field has to be at right angles to the electron stream, and the direction of movement is at right angles to both field and current direction.

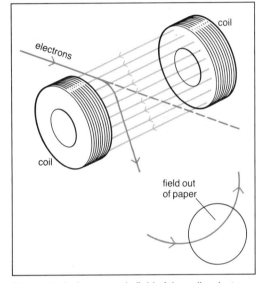

Picture 5 In the magnetic field of the coils, electrons are deflected as shown

The cathode ray oscilloscope

Another way of changing the direction of a stream of electrons is to make use of electric forces. This is the method used in the measuring and display instrument called a **cathode ray oscilloscope** (CRO), (picture 6).

The electron gun in a CRO fires its beam between two pairs of metal plates (picture 7). A positively-charged plate attracts the negatively-charged electrons. The plates are arranged at right angles so that when voltages are applied the electron stream is deflected either up or down or from side to side.

The plates that move the stream vertically are called the **Y-plates**. The **X-plates** move the electrons horizontally. In a CRO the X-plates are used to move the beam steadily across the screen from left to right. This done by a steadily increasing positive voltage applied to the right-hand plate. The

Picture 6 A cathode ray oscilloscope is a very useful instrument

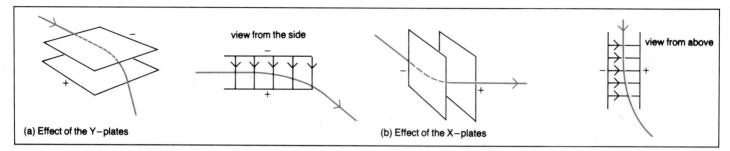

(a) Effect of the Y—plates

view from the side

(b) Effect of the X—plates

view from above

Picture 7 Using electric fields to control beam direction

steady movement of the beam across the screen is called a **time-base**. It is like the 'time axis' you might draw on a graph. It allows us to see the pattern produced by an effect that is changing with time.

The signal to be studied is applied to the Y-plates. For example, if we apply an alternating voltage it makes the beam move up and down. This is because the top plate becomes alternately positive and negative. When this plate is positive the electrons are pulled up. When it is negative they are pushed down.

This up and down movement is very fast, because the mains supply changes direction 50 times a second. If the time-base is switched off we see a straight vertical line. This is because the line is being traced, over and over again, 50 times a second.

But if we use the time-base, set to the correct speed, we see a wavy line which shows us how the voltage is changing with time. This is made visible because the beam is being moved sideways at a steady rate. (See picture 8)

X-rays

X-rays are yet another example of a scientific discovery made by accident. A German physicist, Konrad Roentgen, was investigating electron streams when he noticed that an unused phosphor screen was glowing. The strange thing was that it was on the other side of the lab. It was much too far away for any 'leakage' of electrons to get to it through the air.

Roentgen brought the screen closer and found that glowed even brighter. After testing further he proved that the glow wasn't being caused by electrons but by an unknown kind of radiation. Because he didn't know what it was he called them 'X' rays.

Roentgen worked out that the X-rays are produced when a fast stream of electrons hits glass or metal (picture 9).

Then he discovered the most interesting property of X-rays. When he put his hand in the path of the rays he saw that they cast a shadow on the screen. The shadow of his hand was a very strange one. *It showed the bones as well as the flesh* (picture 10).

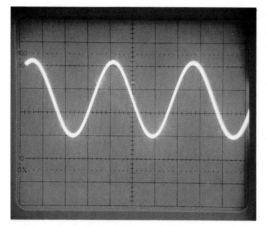

Picture 8 A combination of up and down and sideways movement can draw out a graph

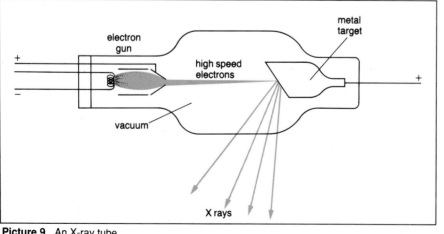

Picture 9 An X-ray tube

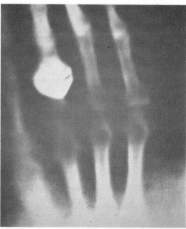

Picture 10 An X-ray shadowgraph of Mrs Roentgen's hand

Within a few weeks of this discovery in 1895, X-rays were being used in hospitals all over the world. Doctors were using them to look for broken bones, swallowed pins, blocked intestines and damaged lungs. But just as with static electricity 200 years earlier, some people went mad about X-rays. They had 'skeleton pictures' taken of themselves and their families. They tried to use them to cure all kinds of diseases.

It took twenty years for doctors to notice that people who used (and misused) X-ray machines tended to get serious cancers. X-rays are invisible, highly energetic electromagnetic waves. These waves interact with matter and ionise atoms. They thus kill the delicate cells of living matter. But they can also alter the cells, making them turn into cancer cells. All ionising radiation, such as that given out by radioactive materials, can do this.

We know now how to manage X-rays so that the 'dose' given to people is as small as possible. Nurses and technicians who operate the machines are well shielded from them (picture 11). X-rays are only used on people when it is medically necessary, after carefully weighing up the balance between harm and benefit.

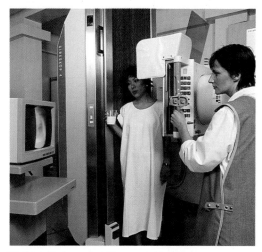

Picture 11 A modern X-ray machine in use

Activities

A Moving electron streams

Electron streams are used in TV sets and in cathode ray oscilloscopes such as the ones used in your lab. Find out what other devices use electron streams in this kind of way. Write a sentence or two about each device that describes or explains what the device is used for.

B Using a cathode ray oscilloscope

1 Look at the control knobs and switches on the front of the CRO. Find the following: on-off switch, focus, brightness, X shift, Y shift, time base.

2 *Find the spot!*
Switch on the oscilloscope. The most difficult part now follows — find the spot! If you are lucky you will see a spot or line already on the screen. If not, switch the **time base** off (or to EXT X), turn the **brightness** to the right to give maximum, and use the X SHIFT and Y SHIFT together to try to get a spot of light in the centre of the screen. Use a combination of FOCUS and BRIGHTNESS control to make the spot small, clear but not too bright. If you can't find it ask for help.

3 *What does a CRO measure?*
Find the Y INPUT. Use the red connection (it may be marked 'dc' or '+') and connect 1 1.5 V cell between it and the **earth connection** (marked '-' or   ). If nothing happens to the spot of light try adjusting the control marked Y GAIN (or 'Y amplitude')

What happens when you (i) reverse the cell connections, (ii) use more than one cell? What does this investigation tell you about what a CRO measures?

4 *Using the time base*
First, remove the batteries you used in part 3 and replace them with an a.c. supply at about 2 or 3 volts. Draw what you see on the screen. Use the X SHIFT to move the spot sideways fairly quickly. Do this once or twice and remember what effect it has. Then . . .. Find the TIME BASE control and switch it on. Try it at slow, medium and fast. use the VARIABLE control. The idea is to get a clear, fixed trace of the a.c. supply on the screen (see picture 8).

5 Try 'looking at' the dc output of the supply. How is it different from the d.c. from a battery that you looked at in 3?

6 If you have time, connect a microphone to the CRO and talk, whistle, play music or sing into it. You may need to adjust the TIME BASE and/or the Y GAIN to get a clear trace.

Questions

1 What is a **vacuum**? Why does a TV tube need to have a good vacuum inside it?

2 Explain briefly how an 'electron gun' works.

3 Draw diagrams to show how electrons would move in the fields given in picture 12 (look at picture 7 to help you)

4 Picture 13 shows a CRO screen with a 50 Hz a.c. mains displayed on it. Draw diagrams to show what the trace on the screen would be like if: (i) the time base were to be switched off (ii) the time base speed were to be halved.

5 Why are X-rays dangerous?

6 X-rays are used in hospitals to look at broken limbs and other parts of the body. Apart from this medical use, what else are X-rays useful for?

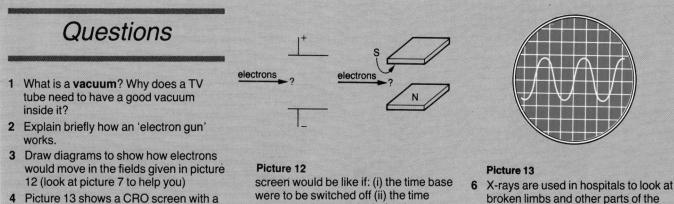

Picture 12

Picture 13

F1

Sensing — human and artificial

We get information from our surroundings by using our senses. More and more, we use artificial sensors to gain information – and to help make decisions.

The human body is covered with sensors. They are special kinds of living cells, which biologists call receptors. They share between them the task of sensing the outside world. They can sense temperature, pain, pressure and touch. Certain parts of the body are even more specialised. Cells are grouped together into sense organs — for example, the eye and the ear, the nose and the tongue. You can read about how they do this in *The Living World*, topic A4.

Sensors as transducers

The sensor cells of an animal are triggered by a **stimulus** of some kind. The stimulus may be a sound, or a change in temperature or light, for example. But before the animal can react to it, the stimulus has to be changed into an electrical signal. This is what the sensor cells do. The electric pulse is then carried along one or more nerve cells into the brain, or some other part of the **nervous system.**

Picture 1 shows an athlete undergoing tests in a sports laboratory. The devices connected to his body take messages to a computer, just as his body sensors are taking messages to his brain.

Devices which change a stimulus, signal or message from one energy type to another are called **transducers**. Transducers are also dealt with in topic C5.

Microelectronics and sensors

The modern world relies more and more on machines of one kind or another. Many of these are 'automatic': they are designed so that they **control themselves**. To do this, they need information about themselves and the surroundings — just like an animal would.

Car engines

A car engine works best at a certain temperature. If it gets too hot it will be inefficient, or seize up altogether. But it is also inefficient if it is too cold — it uses up too much fuel and may also cause extra pollution. The cooling system of the car relies on the flow of water inside the engine (the 'coolant'), and the flow of air over the engine.

When the engine is first switched on it needs to warm up. At this stage the cooling system is not needed. A **temperature sensor** checks the temperature. The circuit decides that it is too cool, and prevents the water from flowing.

When the engine is hot enough the sensor sends a message to allow water to flow. When the engine works harder the water flows more quickly, so that it is kept at the best working temperature. Picture 2 shows the cooling system in a car.

When the car is moving at a normal speed the flow of air over the engine helps to cool it. It also cools the water, by means of the 'radiator' at the front of the engine. In slow moving traffic the cooling effect of the air may not be enough. The engine temperature rises towards danger. The sensor then switches on an air fan to increase the flow of air. If this isn't enough the sensor may switch the engine off altogether, to avoid damage.

A modern car uses **microelectronic** circuits to switch the cooling systems on and off. These circuits are able to measure and compare the signals coming into them. They use the signals to make decisions. And then they can switch parts of the engine system on or off.

The next topic is about how these circuits work.

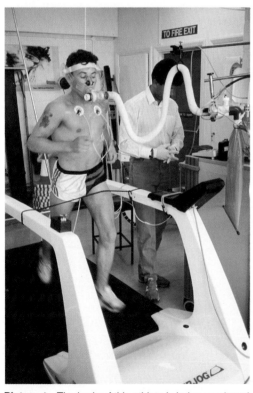

Picture 1 The body of this athlete is being monitored by electric sensors

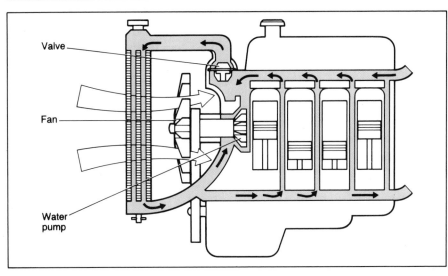

Sensing temperature

A temperature sensor is usually made of a piece of material which is a good conductor of electricity when it is hot, but a poor conductor when it is cold. Heating the material gives energy to electrons which are normally trapped in atoms. The electrons are freed, so they can now move and so more charge can move through the material. In other words, the resistance of the material gets less as it gets hotter. Materials like this are called **semiconductors**.

These temperature sensitive materials are used to make devices called thermistors. See picture 3.

Think about a thermistor connected in a circuit. As the temperature rises its resistance gets less, and the current through it will increase. The current gets less when it gets colder. The changes in current can be used to switch other circuits on or off. Thus thermistors can be used to convert temperature changes into electrical signals.

Activity A is about investigating thermistors. They are used as electric thermometers, central heating controls and fire alarms. They are also used to protect lamps and motors from sudden surges of current when they are first switched on. They start off cold, and provide a protective resistance which stops too much current flowing. After a while the current through them warms them up and they let more current through.

Light sensors

Heating a semiconductor material can free electrons and change its resistance. The same effect can be produced by the energy carried by radiations. **Light** can free electrons from some materials and they can be used to make **light-dependent resistors**, or **LDRs**. Picture 4 shows a typical LDR. In the dark its resistance might be 100 000 ohms (100 kΩ). In bright sunlight so many electrons are freed that its resistance falls to about 100 ohms.

So, by connecting them in the right circuits, we can use LDRs to convert changes in light intensity into electrical signals.

Sensors for sound

We have met these already — they are **microphones**. These convert pressure changes — sound waves — into a changing electric current. See topic C5.

Force and movement

There are many devices that can sense movement. For example, the moving object can be linked to a sliding contact moving over a wire resistor. This is

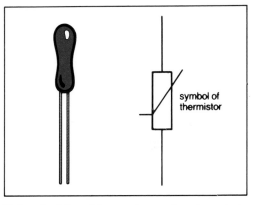

Picture 3 A thermistor — a temperature dependent resistor

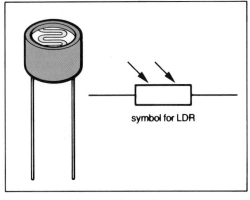

Picture 4 A light dependent resistor

Picture 5 A computer mouse is a movement sensor. Movements are charged to voltages

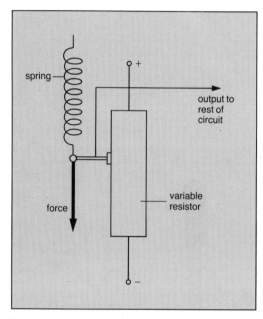

Picture 6 A variable resistor attached to a spring can monitor a force

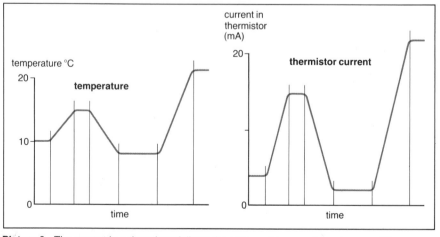

like a rheostat or potentiometer (see topic E3). Many computer games use movement sensors like this as 'joysticks'. As they are moved they alter the resistance and so the voltage applied to the input (sensing) circuit. A computer 'mouse' uses the same principle (see picture 5).

Devices that sense force and movement often use the same idea — after all, a force can produce a movement. The force can be made to stretch or compress something and so change the resistance of an object. This could be a sponge made of conducting material, or grains of carbon (as in the carbon microphone, page 72), or a special **strain gauge**.

For example, a spring could be connected to a variable resistor, as in picture 6. When a force is applied to the spring it stretches and moves the sliding connection of the resistor. This changes the current, or the voltage applied to another circuit.

Strain gauges use this principle. They combine the spring and the resistor together, in one length of thin wire. The wire can stretch like a spring. As it stretches it gets thinner, so its resistance increases. Again, this change can be used to change a current or a voltage. These devices are useful because they can detect very small movements. They may be connected to the foundations of a bridge to monitor what happens as loads of different sizes cross over it, or when strong winds blow (see picture 7).

Other sensors

Automatic sensing is so useful in modern science and industry that many types of sensor have been developed. Temperature can be sensed by bimetallic switches, thermocouples and special diodes. Magnetic fields can be sensed by Hall Effect probes, search coils and resistors that change in magnetic fields. Ultrasound, infra red and radar can detect movement and position. Capacitors and electrets can also detect movements. Light can be sensed by photodiodes and photocells. There are far more than you have time to learn about — but you are bound to have several of them somewhere at home!

Analogue transducers

Most of the sensors described above produces an electric current or voltage which is a **copy** of the change it is sensing. The changes occur gradually, and the sensor output also changes gradually. This is illustrated in picture 8, which shows graphs of changing temperature and the changing current it produces in a thermistor.

The electric current is said to be an **analogue** of the changing temperature. Most transducers, in fact, produce an output which is an analogue of

Picture 7 Strain guages being used to test a girder. The gauges are covered by a layer of plastic glue

Picture 8 The current in a thermistor follows the pattern of temperature changes. It is an analogue of them what they are sensing. But ordinary computers, and the microelectronic devices dealt with in the next topic can't cope with these gradually changing, analogue signals. They respond only to numbers. In other words, they need a digital input. Topic C5 has explained how analogue signals are changed to digital ones.

Activities

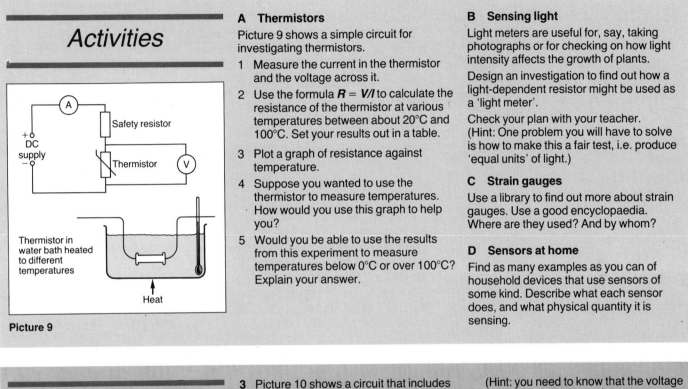

Thermistor in water bath heated to different temperatures

Heat

Picture 9

A Thermistors

Picture 9 shows a simple circuit for investigating thermistors.

1 Measure the current in the thermistor and the voltage across it.

2 Use the formula $R = V/I$ to calculate the resistance of the thermistor at various temperatures between about 20°C and 100°C. Set your results out in a table.

3 Plot a graph of resistance against temperature.

4 Suppose you wanted to use the thermistor to measure temperatures. How would you use this graph to help you?

5 Would you be able to use the results from this experiment to measure temperatures below 0°C or over 100°C? Explain your answer.

B Sensing light

Light meters are useful for, say, taking photographs or for checking on how light intensity affects the growth of plants.

Design an investigation to find out how a light-dependent resistor might be used as a 'light meter'.

Check your plan with your teacher. (Hint: One problem you will have to solve is how to make this a fair test, i.e. produce 'equal units' of light.)

C Strain gauges

Use a library to find out more about strain gauges. Use a good encyclopaedia. Where are they used? And by whom?

D Sensors at home

Find as many examples as you can of household devices that use sensors of some kind. Describe what each sensor does, and what physical quantity it is sensing.

Questions

1a Make a list of the kinds of stimulus that a human body can **sense**. For example, one would be *touch.* We talk about the 'five senses', but you should be able to think of more than five.

b Can you think of any stimulus that can be sensed by machines but not by human beings?

2 What advantages or uses can you think of for being able to sense each of the following *at a greater distance than normal*?

(a) temperature, (b) movement, (c) sound.

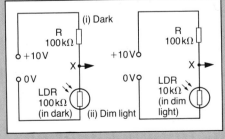

Picture 10

3 Picture 10 shows a circuit that includes an LDR. When light shines on the LDR its resistance becomes a lot less than what it was in the dark.

a The resistor R and the LDR are connected in series. Jamal says, Whatever the resistances of R and the LDR, the same current will flow in both. Do you agree with this? Give a reason for your answer.

b In the dark the LDR has the same resistance as the resistor R. What is the voltage at point X?

c In dim light the resistance of R stays unchanged but the LDR's resistance decreases to 10 000 ohm. What is the voltage at point X now?

d In bright light the resistance of the LDR becomes less than a hundredth of its dark value. Which of the following is a reasonable (nearest) guess at the voltage at point X in bright light?

(i) 10 V (ii) 5 V (iii) 1 V (iv) 0 V

(Hint: you need to know that the voltage across a conductor is proportional to the current in it.)

4 Two students did an experiment in which they measured the resistance of a thermistor at different temperatures. The results they got are shown in the table.

a Plot a graph of resistance against temperature for this thermistor.

b The students wanted to use the thermistor as a 'remote thermometer', to measure the temperature in a bird's nest from a distance. Draw a simple circuit they could use to do this.

c They found that the resistance readings they obtained varied between 85 ohms and 110 ohms. What range of temperatures did this give for the nest?

d Suggest what made the temperature in the nest change like this.

Temperature, T/°C	0	10	20	30	40	50	60	70	80
Resistance, R/ohms	300	200	140	100	70	50	35	25	18

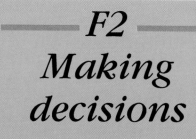

F2
Making decisions

Modern microelectronics can make simple decisions — leaving us to worry about the harder ones . . . .

Picture 1 Life is full of decisions!

Logic — and decisions

'If it's Tuesday today then it must be Wednesday tomorrow', is a **logical** statement. But it is only true as long as we keep the same order for the days of the week! Logic uses words like IF and THEN, and tries to end up with a TRUE statement. If you've ever tried computer programming you will have learned a lot about logic.

Logic is a way of thinking based on clear rules. If we follow the rules, we get the right answer. It is like mathematics, without the numbers. Of course, making decisions isn't always easy — and the ones we humans make aren't always logical (picture 1).

Logic words

Logic words are short and simple, like AND, NOT and OR. Let us think about using logic to make some decisions.

AND decisions

For example:
Decision needed: Do I have to carry an umbrella?
Input facts: A It is raining and **B** I have to go out.
Decision (output): Yes, this means I must carry an umbrella. This decision would only be correct if both of the input facts (A *and* B) were correct. Of course there's no law against carrying an umbrella on a fine day. But you would look a bit silly carrying one to bed with you, even if it was raining outside.

OR decisions

OR decisions are like this:
Decision needed: Shall I go to the school disco?
Input condition: I will go to the disco if: **A** either Jane invites me *or* **B** Tracy invites me.
This means that if neither girl invites me I don't go. (But what if *both* girls invite me? See below!)
Decision: Jane has invited me. Yes, I go.
These are simple decisions to make once you have decided on the 'rules'.

Microelectronic devices use simple logic like this, however. Even so, things can get quite complicated. One way of keeping track of everything is to use a **truth table**.

Truth tables

If a statement, fact or condition is *true* we give it the value one (**1**). If not, we give it a zero (**0**). These are the things we have to put into the electronic device so that it can make a decision, so we call them **inputs**. In the above examples these have been labelled **A** and **B**.

The decision is about *doing something*. It is an **output**. If the decision is *yes* we call the output **1**, if *no* the output is **0**. The truth table for the umbrella decision is this — each row represents a possible situation see table 1:
In the 'output', **0** = no umbrella **1** = carry umbrella

You only carry an umbrella if it is raining *and* you are going out. For the disco decision (table 2):
This says that if neither girl invites me then I don't go to the disco. If any **one** girl does then I do go.

Depending on how we all feel about it, I might decide to go when **both** girls invite me. If I do then we have what is called an **INCLUSIVE OR** kind of logic. The condition is 'A **OR** B **OR** both'. The decision in column three is then 1.

If I feel it would be unwise to go when both girls invite me, then I don't go. I am using an **EXCLUSIVE OR** logic. This would be spelled out as 'A **OR** B but **NOT** both'. The decision is then **0**.

The questions at the end of the topic give you some practice in using truth tables.

Logic gates

A gate is something that lets you through, or keeps you out. It depends on whether it is open or shut. Microelectronics devices use gates which 'open' or 'close' depending on what their inputs and the logic tells them. They are quite complicated circuits containing resistors and transistors. Picture 2 shows part of a logic chip containing many thousands of gates. You would need a course in quite advanced electronics to understand how they are made to work. But we can use them for all sorts of things without needing to know exactly what is going on inside.

Gates can be used to open and close other circuits. In other words they act like switches, and their output is either ON or OFF. For example, an AND gate has two inputs. Its output will be ON only if both of its inputs are ON. If we call its inputs A and B, and its output C, then its truth table is like table 3.

The output is ON only if both A ***and*** B are ON.

If we represent ON by a **1**, and OFF by an **0**, then this is exactly the same as the umbrella decision above.

Using the **1** or **0** system, the **INCLUSIVE OR** gate produces a truth table like table 4:

This is like the disco decision.

In some books you will find that **1** is called **high**, and **0** is called **low**.

Table 1   The umbrella decision

INPUTS		OUTPUT
A	B	
0	0	0
1	0	0
0	1	0
1	1	1

Table 2   The disco decision

INPUT		OUTPUT
A	B	
0	0	0
1	0	1
0	1	1
1	1	1? or 0?

Table 3   The umbrella decision using the AND gate

INPUT		OUTPUT
A	B	C
A		
OFF	OFF	OFF
ON	OFF	OFF
OFF	ON	OFF
ON	ON	ON

Table 4   The disco decision using the INCLUSIVE OR gate

INPUT		OUTPUT
A	B	C
0	0	0
1	0	1
0	1	1
1	1	1

Picture 2   A logic chip

In the diagram (picture 3) input B is from the thermistor. Input A is connected to the ignition switch, so that when the engine is running A is ON.

When the temperature is low, say 20°C, the thermistor has a resistance of 2000 ohms. The standard resistor R has a resistance of 1000 ohms. The same current goes through both the thermistor and the resistor. This means that the voltage across them is proportional to their resistances.

By simple proportion the voltage across R is just one-third of the supply voltage (12V). The voltage is shared, with one-third across the 1000 ohm fixed resistor and two-thirds across the 2000 ohm thermistor. Thus the 'input' voltage at X is 4 volts. This is not high enough to turn input B ON — it needs 10 volts. The output of the AND gate is zero, because only one input is 'ON'

A	B	C
1	0	0

When the cooling water temperature reaches, say, 70°C, the thermistor resistance decreases to 200 ohms. The voltage across R changes to be $\frac{1000}{1200}$ of 12 volts. This is just 10 volts, so input B goes ON and the output is now ON:

A	B	C
1	1	1

Thus the cooling fan is now switched on.

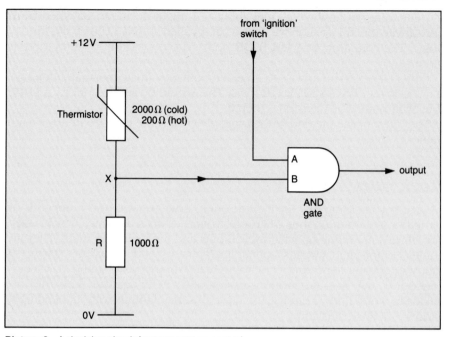

Picture 3 A decision circuit for temperature control

Sensors plus gates = decisions

The fan in some car cooling systems switches on only when the engine gets too hot. But it only works when the engine is running. This means that it must have an AND gate in it. One input to the gate is the engine on/off switch (ignition switch). The other is the temperature sensor.

The output from the AND makes the fan switch on when the engine is running AND the temperature is high. The car manufacturer has to set the temperature sensor to the right value so that it gives its '**1**' at a sensible temperature.

Picture 3 shows a simple circuit for doing this. It uses a thermistor (see topic F1), and one gate input (say **B**) is connected to point X. The resistor R is chosen so that at the 'danger' temperature the voltage input to B is high enough to trigger it. This might be 10 volts.

With the values given this will happen when the temperature is high enough to make the thermistor have a resistance of 200 ohms. The box shows how this is worked out.

Some more gates

The NOT gate. A NOT gate is a very contrary sort of gate. It only has one input connection. When its input is **1 (high)** its output is **0 (low)**. As you might expect, when its input is **0 (low)** its output is **1 (high)**.

It is also called an **inverter**. It is quite a useful device, especially when used with other gates, as described below.

The NAND gate and the NOR gate. These gates are AND gates and OR gates with a NOT gate added to them. That is, they are 'NOT-AND's and NOT-ORs.

This means that while an AND gate gives a 1 output when both inputs are **1**, the **NAND** gate gives a 0 output when both of its inputs are **1**. Similarly, a **NOR** gate gives a 0 output if any input is **1**. This is the opposite (or inverse) of what the OR gate does.

Picture 4 shows the truth tables for the main logic gates that you are likely to meet. It also shows their circuit symbols.

Activity A is about applications of microelectronics which make decisions using simple logic gates with two inputs.

logic gate	symbol	is equivalent to	truth table		the output is high, logic 1 when
			input	output	
NOT	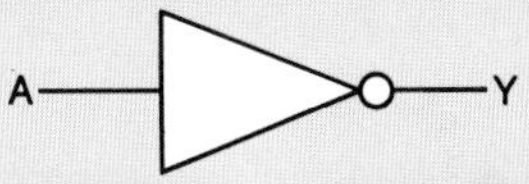 A—[>o—Y	INVERTER	A	Y	input A is NOT high (output is the input inverted)
			0	1	
			1	0	
OR	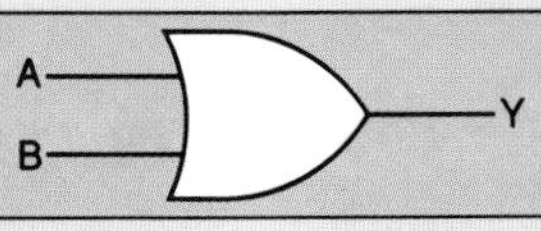 A—B—Y	(inclusive) OR	A	B	Y	input A OR input B is high (or both are high)
			0 0	0	
			0 1	1	
			1 0	1	
			1 1	1	
NOR	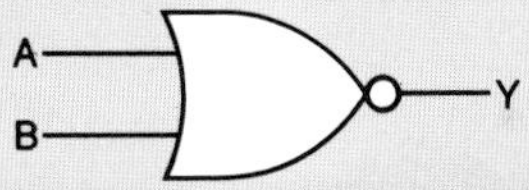 A—B—o—Y	OR-NOT	A	B	Y	neither input A NOR input B is high
			0 0	1	
			0 1	0	
			1 0	0	
			1 1	0	
AND	A—B—Y		A	B	Y	input A AND input B are high
			0 0	0	
			0 1	0	
			1 0	0	
			1 1	1	
NAND	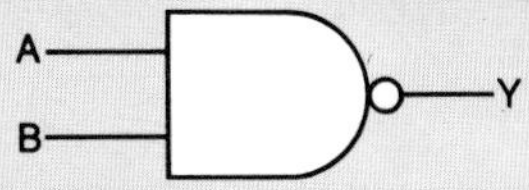 A—B—o—Y	AND-NOT	A	B	Y	input A AND input B are NOT both high
			0 0	1	
			0 1	1	
			1 0	1	
			1 1	0	

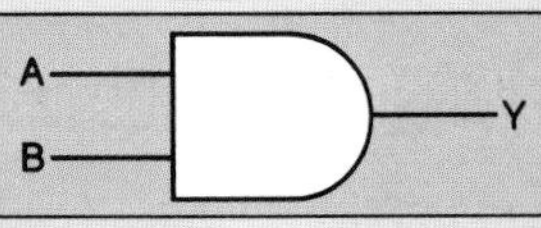

Picture 4 Logic gates

Activities

A Pooling your knowledge

As a group, think of as many household devices or other kinds of everyday 'machines' as you can that **might** use microchips, with sensors and logic gates.

Discuss whether they actually do or not, and list them under three headings:

1 definitely use microelectronics,

2 might use microelectronics,

3 definitely don't use microelectronics.

For each of the items you put in list 1, one of the group should be prepared to give the reasons for putting it there.

B Investigating logic gates

A simple microelectronics kit will have some sensors, an AND gate, an OR gate and a NOT gate. It will also have lamps or even buzzers to show when the output is ON. There are different types of kit and you will need to be given instructions about how exactly you connect them together.

1 Use the three gates above to check their truth tables.

2 Design and then make circuits that:

a use a light sensor so that lamp goes on when it gets dark

b use a temperature sensor that lights a lamp or sounds a buzzer when it gets warm,

c as (b) but gives a warning when it gets cold

d gives a warning when it is cold *and* dark.

Questions

1 You don't need microelectronics to make simple gates. Picture 5 shows two simple circuits using ordinary switches. One of them behaves like an AND gate, the other like an OR gate. Which is which? Explain your answer.

2 Explain the difference between an OR gate and an AND gate.

3 What is the difference between an AND gate and a NAND gate?

4a Describe what a NOR gate does.

b Picture 6 shows a circuit with a NOR gate in it. The indicator lamp goes on when the NOR gate sends it a 1. The light sensor is set to give a 1 when it is light, the temperature sensor gives a 1 when it is warm. What happens to the lamp when:

(i) It is dark and cold? (ii) It is light and cold? (iii) It is light and warm?

c Think up a practical application of this circuit. Describe it, briefly.

d Write out the truth table for this circuit.

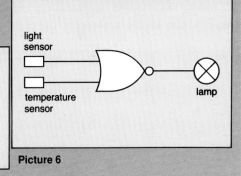

Picture 6

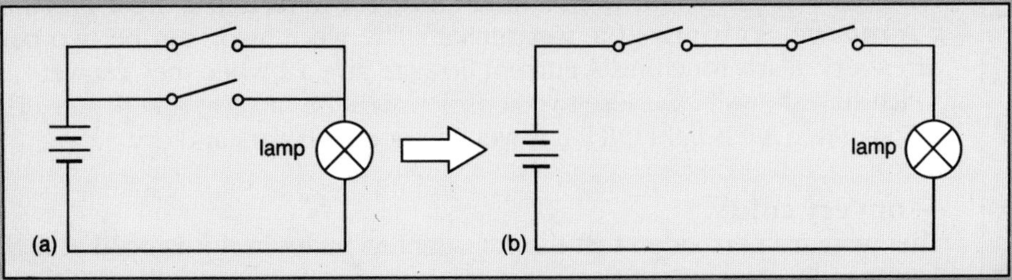

Picture 5

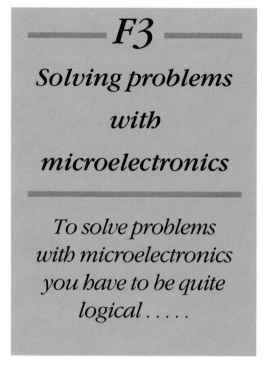

F3

Solving problems

with

microelectronics

To solve problems with microelectronics you have to be quite logical . . . . .

Picture 1 Cats like catflaps

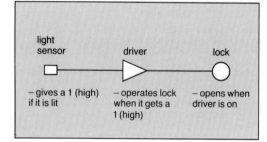

Picture 2 A first attempt

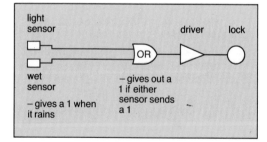

Picture 3 Cats don't like to get wet

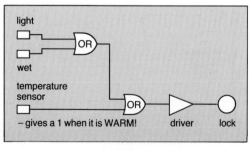

Picture 4 — or to get cold!

The catflap problem

Suppose we want to design an automatic 'cat flap'. This is a small wooden flap cut into the back door. It can be pushed open by the cat (see picture 1). But you can lock it shut, if you want to.

To do this we want to use an electrically operated lock that is opened and closed by a small electric motor. This motor — the **driver** — can be switched on or off.

Keep the cat out at night!

At first you decide that the cat can only come in during the day time, so the lock can be controlled by a simple light sensor. You make a circuit as shown in picture 2. The light sensor can be a light-dependent resistor (LDR). It makes a **1** input when it is light. This lets the door open.

— unless it is raining!

Then you take pity on the poor animal and agree to let it in at night, *provided it is raining.* So we need a 'wet sensor'. The wet sensor can be two bare wires very nearly touching. A current flows to give a **1** when they get wet.

Picture 3 shows how these two sensors might be fitted to an OR gate. The cat can get in if it is light OR if it is wet. But not if it is dark and dry.

— or very cold!

Then you are persuaded that the cat ought to be let in if it is **cold** at night. This needs a temperature sensor. It is connected to another OR gate, connected as shown in picture 4. Now the flap will open if it gets a **1** either from the first OR gate or from the temperature sensor.

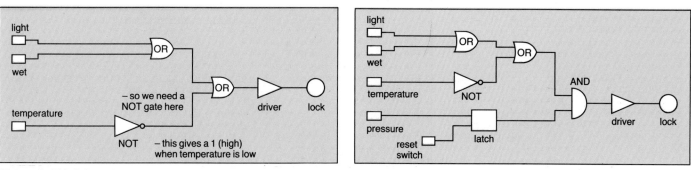

Picture 5 This is better!

Picture 6 The final solution (but see if you can do better)

But there is something wrong with this! When it is cold the sensor you happen to have sends a **0**, and this would not open the flap. Both you and the cat want it to send a **1** when it is cold. This can be done by putting a NOT gate between the temperature sensor and the input (picture 5).

A **0** signal from the temperature sensor is changed by the NOT gate to a **1**. The cat is happy because a **low** temperature sends a **1** to the lock, to open it. Thus the flap will now open on a cold, dry night, as well as on a wet night.

Oh no!

One cold night you wake up to a terrible noise and find that the kitchen is full of cats. What can you do now?

Quite simple — put a **latch** on the cat flap. This means that once the door has closed again it stays closed. But it has to be an electronic latch, which closes the cat flap permanently when **one** cat has got in. It might be the wrong cat, of course, but that's your cat's problem!

You should be able to reset this latch when the cat has been put out each night.

You could fit a latch into the circuit as shown in picture 6. The AND gate unlocks the cat flap only if *both* of its inputs are **1**. The latch is set at **1** and stays there until one cat comes in. Then it switches to a **0** and stays at **0** until you come along and reset it again, using a special reset switch. As long as the latch sends a **0** to the AND gate the catflap lock cannot be opened.

You could set it off like this. The first cat through the catflap steps on a pressure sensitive mat. This makes the latch go to **0**, and it stays at **0**. It is now locked, and will stay locked. This is the key point about a latch — it stays in one state, whatever else happens, until it is reset.

Other names for latch circuits are:

- **flip flops** — because its output flips from one state to another and stays there; then it flops back again when somebody changes it,
- **bistable** — because it has **two** states (on or off) and stays in one (i.e. it is stable) until it is changed.

Latches are one of the most useful microelectronic circuits of all. They are the basis of **memory** devices. Because it stays in one state until it is deliberately changes we say that tt remembers which state it is in. Computers would be useless without their memories, which use millions of bistable circuits in their memory chips.

The rest of this topic explains in more detail how a simple latch can be made from two gates. It is not easy to understand, but it is worth making an effort.

A bistable unit — or electronic latch

A simple latch or bistable can be made from just one gate. Picture 7 shows an OR gate connected to do this. The output (C) is fed back to one of the inputs (R). Both inputs, S and R, are **0** so the output is also **0**. What happens when input S is made a **1**?

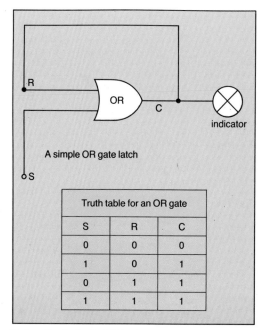

A simple OR gate latch

Truth table for an OR gate		
S	R	C
0	0	0
1	0	1
0	1	1
1	1	1

Picture 7 A simple OR latch gate

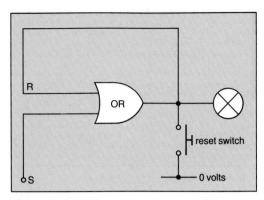

Picture 8   An OR latch gate with reset switch

Table 1

S	R	X	Y
0	0	stable	
1	0	0	1
0	1	1	0
1	1	anything can happen! Avoid doing this!	

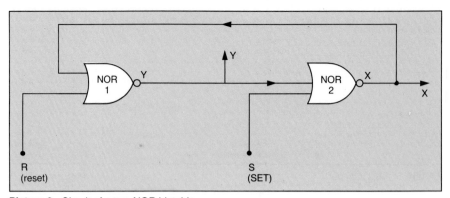

Picture 9   Circuit of a two-NOR bistable

The truth table in picture 7 shows that the output C now goes to **1**. The indicator lamp will light. This also makes the fed-back input R go to **1**. We are now at line 4 of the truth table.

Now we take away the 1 from input S, and the output **stays at 1**.

Check this with the truth table. In fact, whatever we do to S the output stays the same. The system is **stable**, with the output **latched** to **1**. The lamp stays on.

We have used input S to **set** the value of the output. What can we do to **reset** it to **0**, to **clear** it? Remember, we have made S to be **0** (off). The gate is latched to 1 because R is **1**. So to change it we have to make R a **0**.

This can be done by having a switch which is able to connect R, just for a moment, to a low voltage (e.g. earth). Picture 8 shows how this might be done. At this point both inputs are **0**, the output is **0**, and the system will stay like this until input S is given a **1**. This is where we started.

A two gate bistable

Using just one gate means that a mechanical switch has to be used. This might be inconvenient. In most latching or bistable applications two gates are used so that the latching and clearing can be done by electrical signals. NAND or NOR gates are normally used for this.

Picture 9 shows a two-gate bistable. It has two 'free' inputs labelled S and R. S is used to **set** the output X to what is wanted — it will **latch** it. The output stays the same at X until it is changed (or **reset**) by input R.

The key point about the bistable is that the output of one NOR gate is fed to the input of the other. These **feedback** outputs are labelled X and Y.

The relevant truth table for this system is as follows see table 1 (you can check this using the separate truth tables for the two NOR gates, but it's tricky!):

Let's work this out in terms of the catflap problem. To start suppose both S and R are **0**, and output X is **1**. This is fed to the AND gate (see picture 6) so that if any of the other conditions are met (it's light, cold or wet) the motor can work to open the lock of the catflap. The cat can get in.

Then a cat comes in and the flap closes behind it. The cat then steps on a mat just in front of the catflap. This triggers a pressure switch and sends a **1** to input S.

Immediately we are in line 2 of the truth table. Output X goes to **0** and the motor can no longer open the lock. Remember, for the motor to work it needs a **1** at both inputs to the AND gate. So only one cat gets in.

As the pressure switch goes back to OFF when the cat leaves the mat, input S goes back to zero. This puts us in line 1 of the truth table — which is a STABLE line. This means that output X stays at **0**. The catflap is **latched** shut.

Now next evening , when we put the cat out, we want to reset the latch. To do this we push another press switch, R. This is the **reset**. Pushing the switch sends a **1** and puts us into line 3 of the truth table. X is now 1 and the lock can be opened.

When we take our finger off it the press switch R goes back to **0**. S and R are now both **0** so we are back at a STABLE condition. This is shown in line **1** of the truth table. X is now 1, which is where we started.

Using microelectronics

This topic has explained just some of the main ideas about logic gates and how they can be used to solve problems. In your science lessons you should have plenty of chances to use **microelectronics kits** to solve problems. This is the best way to learn what the sensors and gates can do, and how they are used in everyday life. Picture 10 shows some typical microelectronic equipment that you might use in school.

Microcomputers use millions of gates, especially the 'latch' or bistable, which is the basic logic gate in the computer's memory. Most factories now use microelectronic control systems to operate machinery, or to check that it is safe and working properly.

More and more cars use electronic ignition. Cookers and ovens are checked and controlled using logic gates — especially microwave ovens, which even have memories. Cheap calculators use pressure sensors, binary counters and bistables to do arithmetic very quickly. They have made life a lot easier for those of us who need to do quick and accurate calculations. Most of us wear digital watches, which use similar kinds of circuits.

The next topic looks at some of these applications in more detail — especially the way microelectronic devices are used to collect, store and transmit **information**.

Picture 10 Some gates and sensors that you might use

Activity

Using logic gates

You will get a much better understanding of how logic gates and sensors are used by trying to solve some problems with a microelectronics kit. You may already have done this in previous years, but if you find the following problems too easy your teacher can probably think up some harder ones! Design and make systems that can do the following:

1. Let you know if your pot plants need watering on a hot day.
2. Sounds an alarm in your parents' bedroom if someone opens the door of your refrigerator at night.
3. Sounds an alarm if it is windy or raining — but must only work when you have switched it on.
4. A fire alarm which will sound an alarm when it senses smoke or it gets too hot.
5. Switches on a fan when a light is switched on, but only when the window is closed.

Questions

1 Explain what an electronic **latch** is for. Why is it also called a **bistable**?

2 Draw a simple logic circuit that you could use to solve the following problems. Write out the truth table for your solution.

a For a deaf person: who wants a light signal to come on when someone presses a door bell switch during the day only.

b For a blind person: who wants a sound to be made when a teacup is filled to the right level.

c For a gardener: who wants a buzzer to sound when the soil in a plant pot gets too dry — but who doesn't want to be woken up at night!

3 Write out the truth tables for each of the 'catflap' circuits drawn in pictures 3, 4 and 6.

4 What would happen in the circuit of picture 11 when:

a it was cold and light,

b it was cold and dark,

c it was warm and light?

Suggest a use for this circuit.

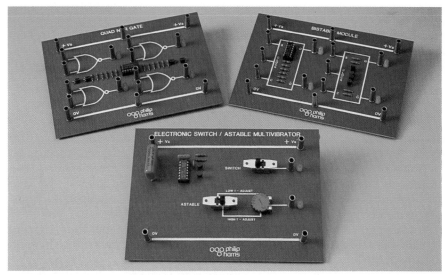

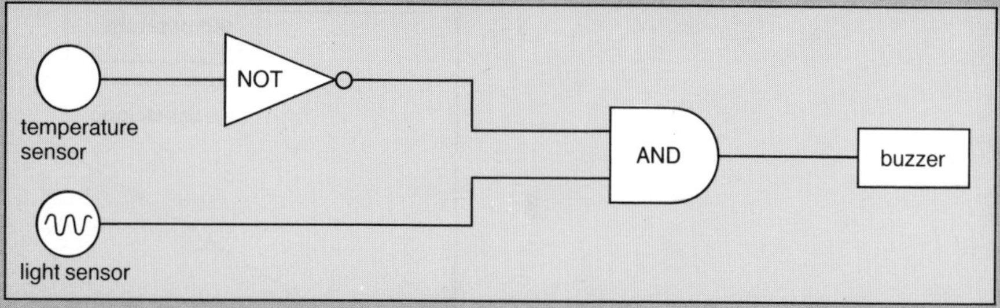

Picture 11

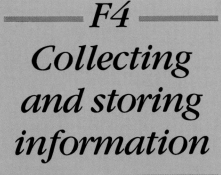

F4
Collecting and storing information

Information is vital to modern life. This topic looks at the information revolution that is happening now.

We all have **memories**. They are **stores** of information that we gather from our senses. Our brains **process** information, getting it from memory and from our sense organs. Human beings have the largest brains (for their body size) of any animal, and can probably store the most information. They also have the greatest capacity for using this information. You can learn more about the human brain in *The Living World*, topic D1.

Humans have invented cameras to do the job of seeing, and microphones and amplifiers to do the job of hearing. The most modern kinds of invention are to do with artificial ways of collecting, storing and processing information. This is called **information technology**.

Simple microelectronic systems

You have learned about **sensors** in topic F1. Topics F2 and F3 are about using **logic devices** to process the information received from sensors. The circuits dealt with in these topics also had one or more **output devices**. These are lamps, buzzers, motors or some other device. They are the devices that actually do something as a result of the information received from the sensors and the decisions made by the logic gates.

Picture 1 shows the block diagrams of these basic control or warning systems. Of course, these systems can be quite complicated, using tens or hundreds of logic gates. If you've tried to make a model traffic lights system with your microelectronics kit you know how complicated they can get.

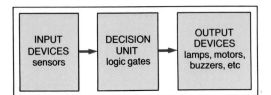

Picture 1   A simple microelectronic system

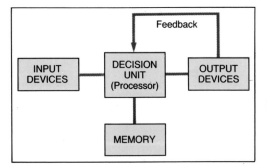

Picture 2   A system with memory and feedback

Systems with memory — microprocessors

To turn a basic system into a **computer** or a **microprocessor** you need an extra device — a **memory**. Picture 2 shows the main parts of such an information processor. Most of them also have the extra feature of a **feedback loop**. Thus the information processor not only 'knows' what is in its memory and what information is being fed to it from its sensors — *it also checks what its output devices are doing.*

As an example, think of filling a bath with hot water. You put your left hand into the bath to check its temperature. It is too hot. You use your right hand to turn off the hot tap and turn on the cold one.

Your **sensor** cells have sent information to your brain. It processes this information and sends a message to its output devices — the muscles in your right arm. The arm moves to work the taps.

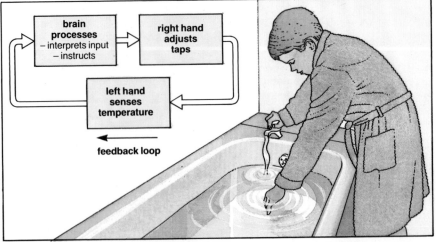

Picture 3   Most living things use feedback

Now your left hand signals to the brain that the water is getting too cold. It responds by getting the right arm to cut off the cold water and put in some hot water. So it goes on, until the bath is full enough and hot enough.

Picture 3 shows the feedback loop involved. It is called **negative feedback**. This is because when the brain gets the signal **hot** it produces the result **cold**. When it gets the message **cold** it arranges for **hot** to appear. Your own body cooling system works like this. You sweat when you are hot and shiver when you get cold.

Negative feedback is seen in many biological situations — see *The Living World*, topics C17 and C18.

Think of what would happen if your brain worked with **positive feedback** as you were filling the bath. If you sensed it was too hot you would keep on making it hotter — result: scalding. If you sensed it was too cold you would keep on adding cold water — result: very invigorating!

There aren't many practical examples of positive feedback. When it happens it is usually unwanted. Picture 4 shows a situation where you will have heard positive feedback. It is the very loud screeching sound you get when a microphone is picking sounds from the loudspeaker system it is feeding.

Every single thing you do, perhaps even when you are asleep, is controlled by a biological information system like the one shown in picture 2.

Microprocessors in industry

Human beings have quite good memories and might even make good decisions. But we are very slow. A typical 'reaction time' in the hot water example given above might be a tenth of a second. This is far slower than a typical microelectronic device will need to sense a change and respond to it. It also makes the same decision every time! This makes them very useful in industry.

Picture 5 shows plate steel being made in a steel factory. A thick piece of hot steel is pressed between rollers which press it until it is exactly the thickness required. To make it thinner, the rollers have to press harder. If they press too hard, or not hard enough, the steel is the wrong thickness and may have to be scrapped. Mistakes are expensive.

In the old days the thickness of the steel was measured mechanically as it came out of the rollers. A system of levers was used to control the pressure exerted by the rollers. It was not very accurate, and the 'reaction time' was quite long. Before any fault was corrected many metres of steel had passed through.

The thickness of the steel is now controlled using a microprocessor. Picture 6 shows how this is done. You should be able to recognise the main parts: input sensor, processor, output device and the feedback loop. Microelectronics makes the reaction time very short and very little steel comes out with the wrong thickness.

Microprocessors in the laboratory

Data collecting instruments are used in many school laboratories. They can be fitted with different types of sensors to measure temperature, light, movement, magnetism, etc. They are **programmed** to collect this information at set times. This could be every thousandth of second or just once a day.

The data is stored in a memory and can be processed in different ways. The device can be connected to a chart recorder and the data used to draw a graph. It may send the data to a computer, which can process it further — by calculating averages, drawing graphs on a VDU screen, comparing it with other data, etc.

Data collecting with these instruments allows us to see what is happening when things change very quickly — such as when chemicals explode, or an

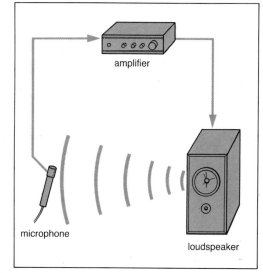

Picture 4   Positive feedback: microphone howl

Picture 5   A steel rolling mill

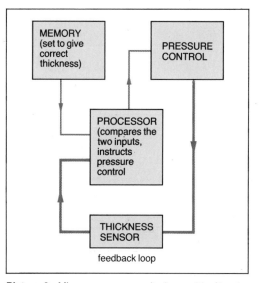

Picture 6   Microprocessor monitoring and feedback in a rolling mill

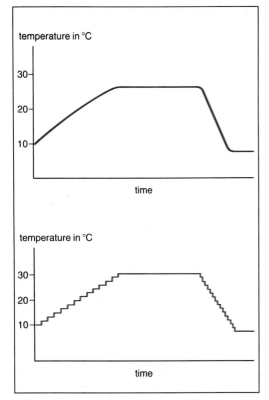

Picture 7 An analogue sensor can change continously and smoothly A digital sensor can only have whole number values, so changes in a jerky way

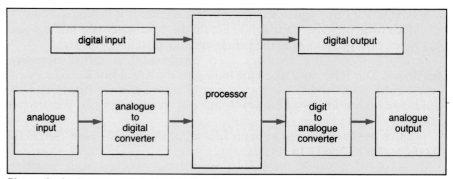

Picture 8 Analogue and digital conversions in a microprocessing system

electric motor is switched on. They can also help to monitor things that take a long time to happen — such as the growth of a plant or the change in acidity of a lake during spring.

Analogue to digital

Microelectronic devices process data in the form of **numbers** — or **digits** as mathematicians call them. These numbers have to be whole numbers, like 1, 2, 3 or 567. They can't cope with numbers like 1.5 or 3.889.

Sensors produce electric voltages which may have *any* value. They are continuous, like a line. Its output is said to be **analogue**. This 'analogue' idea has been mentioned in a number of other topics (C5 and F1).

Digital electronics uses numbers which go up in steps. Picture 7 shows what has to happen to change the continuous output of a temperature sensor into digital numbers. As you can see, the digital version can never be an exact copy of the analogue version. In fact, microelectronic devices use a digital system based on just two numbers, 0 and 1. This is called **binary arithmetic** (or arithmetic to base 2). You may have studied it in maths lessons.

Thus microprocessors and computers need another device fitted in them, an **analogue-to-digital converter** (ADC). Also, because many output devices, like lamps and motors, are analogue devices the output may have to be changed back again, so we need a **digital-to-analogue converter** (DAC). Thus the block diagram of a microprocessor system now looks like picture 8.

Memories

Computer and processor memories use the **bistable** circuit that is described in topic F3. See the explanation and diagram on page 187. One of the inputs to the bistable (see picture 9), say S, is used to make one output of the system, say Y, go to 1. It stays at 1 until it is reset by input R. In other words, the bistable *remembers* the number **1**.

Four bistables in a row can remember a number sequence like 0101 (see picture 10). This is a binary number with four 'places', called a 4-bit number. 0101 is 5 in denary (ordinary) numbers. Sixteen bistables in a row can store a number between 0 and 255. It can be reset to 0 by sending a message to all the reset inputs at the same time. Many computers use 16-bit numbers like this.

Computers are designed to interface with human beings, using a keyboard and often a 'mouse'. They are designed to behave differently with different **programs**. This means that they can do very different jobs — they can be word processiors, spreadsheets, calculators and databases, etc. They can be used to play games.

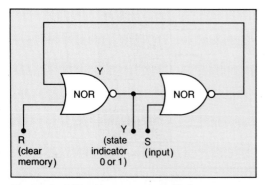

Picture 9 A bistable as a memory device

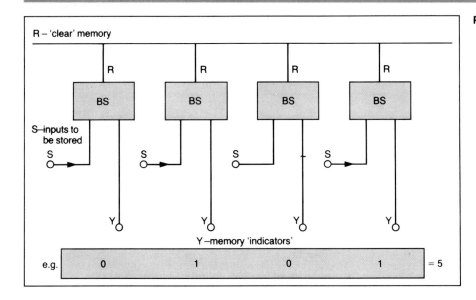

But computers aren't the only devices which have memories and process information. Modern cameras, washing machines, microwave ovens, video recorders, compact disc players and many other kinds of useful equipment also use microprocessors. But they are not **programmable**; they are what is known as **hard-wired**. The buttons do the same job each time they are pressed. Their program is built into them by their circuitry — which is why they are called 'hard-wired'. The user can't change what they do. See picture 11.

Interfacing: how transducers can collect information

Computers can be **interfaced** with outside equipment. This means connecting them to the 'outside world', so that they can both get information and give it out. They have input and output connectors, called **ports**, to do this (picture 12).

The input port gets information as changes in voltage, and this is usually in analogue form (see above, page 192). The information signal then has to be changed into digital form, using a built-in 'A to D converter'.

Similarly, quite often the output information has to be changed into analogue form. The information may be put into a recorder of some kind, or displayed on a monitor screen. But it is very often used to make something work, such as an electric motor. This means that the computer can use information to control a real working device.

The input information can be stored in several different **channels**. This means that information from two or more sources can be kept separate. They can then be compared with each other, or used to plot graphs. For

Picture 11 A CD player is a mix of programmable and 'hard'-wired' devices

Picture 12 Computers are connected to the outside world by ports. So are islands!

Picture 13 Collecting information about light

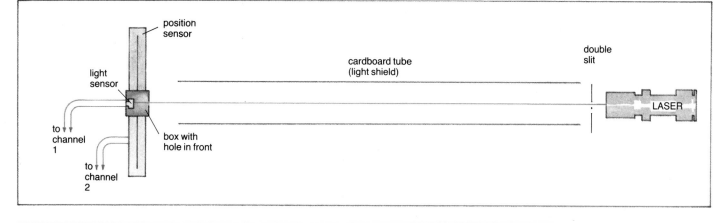

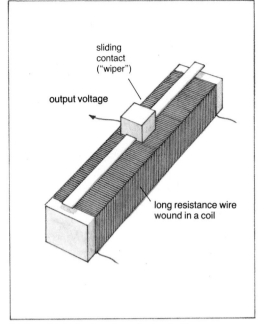

Picture 14 Collecting data electronically: A potentiometer is used as a position sensor, a photocell measures light intensity

example, you could collect information about light coming through a thin slit (as in a diffraction experiment, see pages 90–91) and compare this with the position of a light detector. A graph of light values plotted against detector position would give a graph of the diffraction pattern. Picture 13 shows how this experiment could be set up, using a laser as the source of light.

To collect the information we need to use **transducers**. In this example we need devices which can translate light intensity into voltage, and the detector position into voltage.

The light sensor could be a photocell or a light-dependent resistor (LDR). This is put in a light tight box, with a small hole in the front to let the laser light in.

The light sensor is fixed to a position sensor. This is a variable resistor, used as a **potentiometer**. It is a linear potentiometer (see picture 14). The light sensor is fixed to this.

The computer is programmed to collect and store the information from each sensor, using two different channels. The light sensor is moved across the light pattern, and the information about light intensity is collected and stored as it moves across.

This information can be used to plot a graph of the pattern on the monitor screen.

The position of fast-moving objects can be followed by using a special conducting paper. For example, experiments with dynamics trolleys can be done on this special paper (Teledeltos paper), using sliding contacts. Computers have their own internal clocks, so measurements taken can be plotted against time. Of course, the computers need to be programmed to do this. Examples of these programs can be found in several books, or in the computer manuals.

Activities

A Using electronic data acquisition

Think of an experiment or investigation in which data is hard to get by ordinary means. For example, you might like to:

1 monitor the rate of a chemical reaction
2 find out how much a plant grows over a period of several days
3 investigate how the light from a flash bulb builds up and dies away
4 take weather readings over a 24-hour period

Find out how you could use electronic data collection and storage to solve the problem. Discuss this with your teacher and then carry out your investigation.

B Computer simulations

Many experiments and systems can now be **simulated** using a computer program. Use one of these programs and write a short report about it for use by one of your fellow students who is wondering whether it is worth taking the time to use it. Your report should:

■ explain what simulation means,
■ describe what the program does,
■ say how realistic the simulation is,
■ say what you learned from using it.

C Control

Carry out an investigation or experiment in which you learn how to control a system or a piece of machinery using electronics. You should use low voltage devices (unless you are being directly supervised). You might need to store a simple program in a memory device.

D Can people be replaced by microchips?

Think of a parent, guardian or older brother or sister who is working. Ask them what difference, if any, would it make to their jobs if the electronic storing or sending of information was not invented. Give reasons for your answer.

Questions

1 Make a list of the information you need in order to pack your bag ready to go to school.
2 List the devices you have in your home which: (a) store information, (b) can receive information from a distance, (c) can both send and receive information from a distance. They don't all need to be electronic!
3 Imagine you have pen-pal in a foreign country who has not heard of **electronic data acquisition**. Describe how this is used, by writing about any experiment you have done which used this way of getting and storing information.
4 Explain what **feedback** means. Give an example, one in each case, of how feedback is important in: (a) the control of any body function, (b) an industrial process.
5 What is the difference between **negative feedback** and **positive feedback**? Give an example of each type of feedback.
6 Write an essay entitled *Electronics in Control*.

Fibre Optics

As explained in topic C1, thin glass fibres carrying 'light' signals are much better for carrying information than almost any other medium. The messages are digitally coded (see topic C1, where digital coding is explained).

The 'light' used is normally invisible — it is a form of infra red radiation. This radiation travels at the same high speed as light, but travels better through glass than visible light does. It is produced by a laser, which gives a very pure form of radiation.

The radiation is coded electronically. When you speak into a telephone the sounds you make are first changed into a varying electric current. This is a copy (analogue) of the sounds you make. An analogue-to-digital converter changes the signal to a set of binary pulses. In turn these are used to modulate the laser beam in the same digital pattern.

At the listener's end the infra red pulses are changed to an electric signal again. This can be put back into analogue form electronically and used to make the earpiece work.

Optical fibres are very thin, and are made of very pure and transparent glass. Infra red 'light' is fed in at one end and cannot escape. This is because of **total internal reflection** (see topic C6). A simple fibre works as shown in picture 1. The light bounces off the inside of the fibre. But this produces a distorted signal after the signal has travelled a few kilometres down the fibre.

A better design uses glass in which the speed of the light changes gradually from the inside out. This means that the light path curves gently as shown in picture 2.

Even so, the signal gets distorted sooner or later. **Repeater stations** are built into the line to reshape and amplify the signals, as shown in picture 3. In a modern telecommunications system using fibre optics the repeater stations can be 40 km apart. This compares with having them just 8 km apart when copper wires are used to carry messages electrically.

Another advantage is that optical fibres can carry many conversations at the same time. This needs quite complicated electronics, but a typical optical fibre system can allow 11 000 pairs of people to talk to each other simultaneously. The old copper wire system could only carry about a 1000 conversations at once.

Optical networks can also carry TV signals

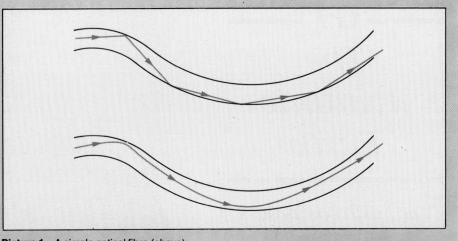

Picture 1   A simple optical fibre (above)
Picture 2   A more sophisticated fibre (below)

and computer data more cheaply and accurately than wire systems can. Thus useful things like home shopping and banking can be much more practicable. But to give everyone an optical fibre connection to their homes is expensive. It isn't cost-effective unless the network is also allowed to carry telephone messages.

Now answer these questions.

1   What other uses are there for total internal reflection?

2   What is the difference between **analogue** and **digital**? Explain briefly why using digital code is more reliable than an analogue system.

3   What does a repeater station do?

4   What kind of light is produced by a laser?

5   Give two advantages of fibre optics compared with metal wires as information carriers.

6   What is **infra red radiation?**

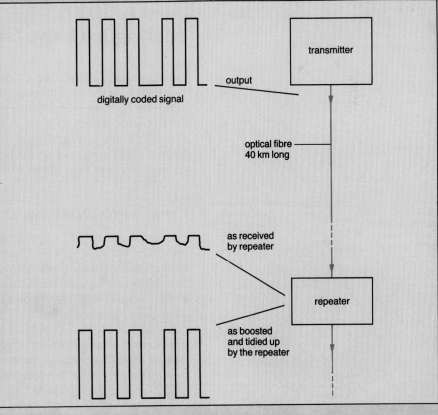

Picture 3   Repeater stations reshape and amplify the signals

G1
The dynamic Earth

The Earth has its own energy source. Deep inside it is active and its crust is continually changing.

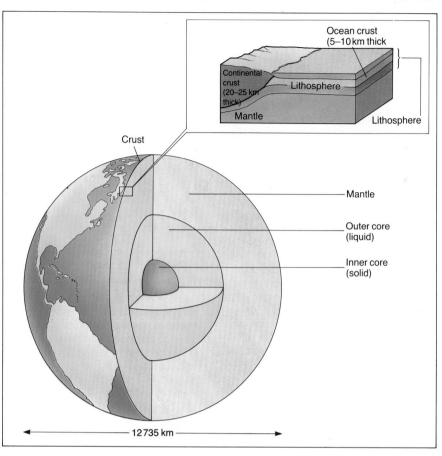

Picture 1 The layered structure of the Earth

Peeling back the layers

The Earth has a layered structure. We live on the topmost layer — the **crust** — which is made of solid rock. Compared to the Earth as a whole, the crust is thinner than the skin on an apple. At most it is 70 km thick. Compare this to the diameter of the Earth which is about 12 735 km. Yet no-one has been able to drill into the crust for more than 14 km — a mere scratch on the surface. So how have scientists so far been able to discover what lies beneath the crust, to discover the structure of the Earth and what it is made of?

Using earthquake waves

The shock waves from earthquakes are recorded at **seismic stations** all over the world. (Seismic stations are places that detect and measure earthquake waves.) Their records have provided us with the bulk of the clues about the structure of the Earth. These waves are rather like sound waves — and just as scientists can 'see' what is inside the body with ultrasound, so geologists can 'see' inside the Earth with earthquake waves (see page 101). For example, scientists noticed that one type of earthquake wave didn't seem to travel through the middle of the Earth (picture 2). It was known that these particular waves cannot travel through liquids, which meant that part of Earth's core must be liquid.

You can see from picture 2 that the earthquake waves change direction as they go through the Earth. Just as light is refracted (bent) if it passes from air to glass, or glass to water, so the earthquake waves are bent as they travel through the different materials that make up the layers of the earth. But unlike glass or water — the density of the Earth changes gradually. This

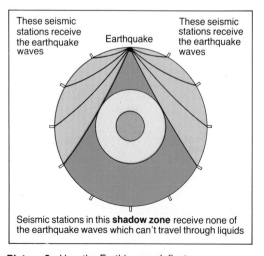

Seismic stations in this **shadow zone** receive none of the earthquake waves which can't travel through liquids

Picture 2 How the Earth's core deflects s–waves

makes the waves follow the curved paths shown in picture 2. Studying earthquake waves has given us more details of the Earth's layered structure. Scientists have used other measurements of the overall properties of the Earth to find out what its interior is like.

Density

For example, density measurements show that the interior of the Earth must be made of very different materials from the crust The average density of crustal rocks is $2800\,kg/m^3$ but the overall density of the Earth is $5500\,kg/m^3$. This suggests that the interior of the Earth is very much more dense than the crust. In fact, the interior has two parts: the **mantle** and the **core** (see picture). It has been found that the average density of the mantle is $4500\,kg/m^3$ and that of the core is $10\,720\,kg/m^3$.

With such a high density, the core can only be made of metal. The measurement of the speeds of earthquake waves which can pass through the core tell us that the most likely metals are nickel and iron.

The mantle is made of a mixture of compounds, containing mainly oxygen, silicon and magnesium. These are all lighter elements than iron and nickel.

Lumps of rock

We do have some more direct evidence about what the inside of the Earth is like. Sometimes lumps of rock from deep in the Earth's mantle are found in the lava from volcanoes. Even rocks from space — **meteorites** — give us clues. It is likely that the Earth was formed at the same time as meteorites such as these. (See topic I3.)

The materials of the Earth seem to become denser the deeper you go. So there are light gases in the atmosphere, water in the oceans, relatively light rocks in the crust, dense rocks in the mantle and very dense metals in the core.

The Earth's crust

The crust itself is made from two layers, a lighter layer which makes up the continents and a heavier layer which goes underneath the continents and makes up the ocean floors. You can see these two layers in picture 1. The **oceanic crust** is made mainly of basalt and the upper **continental crust** has a high proportion of granite rocks. You can find out more about these rock-types in topic G3.

The temperature of the rocks increases with depth. Deep mines are often uncomfortably hot — picture 4 shows two miners working down a mine in Cornwall. They are 1.5km beneath the surface where the temperature is around 36°C. South African gold mines are even deeper and therefore

Picture 4 Cornish tin miners working underground

Picture 3 The extinct volcanoes from which diamonds come are the deepest in the world. Molten rock was spewed up from a depth of 150 km. As well as diamonds, these volcanoes brought up chunks of the material from which the mantle is made.

hotter. They are up to 3 km deep and with temperatures approaching 50°C. Beneath the crust, the mantle is so hot (over 1000°C) that the rocks soften in places and are probably the consistency of stiff Plasticine.

The core is white hot, about 4300°C. The core is believed to be composed mainly of iron and nickel. At the surface of the Earth, both of these metals would boil at this temperature. Iron boils at 3000°C, and nickel at 2900°C, so why isn't the core a gas? The answer is *pressure*. The pressure due to the weight of rocks around the core is so great that the centre of the core is forced to be a solid. The outer section of the core is at a lower pressure and is liquid.

Why is the inside of the Earth so hot?

One of the reasons that the deep interior of the Earth is so hot, is that the Earth formed at a very high temperature (see topic I3). Because the inside is **insulated** by the outer layers some of this original thermal energy is still 'trapped' inside.

Also, some of the 92 elements of the Earth are unstable. They are the **radioactive** elements (see topic D6). The nuclei of such elements (mainly uranium, thorium and potassium) break up, giving out energy as they change into smaller nuclei. These elements are, in fact, quite rare, but the Earth is so large that it contains enough of them to produce huge quantities of energy. Some of these elements are found in crustal rocks.

This energy keeps the Earth hot inside, and every day 2.5 billion billion joules (or 2.5 exajoules) of energy escape from the Earth's surface to help heat the air. This is about four times as much as the energy used by all the people on the Earth.

From heating energy to movement energy

Energy flows from the hottest part of the Earth, the core, outwards to the surface. This energy travels in two ways; by **conduction** and by **convection**.

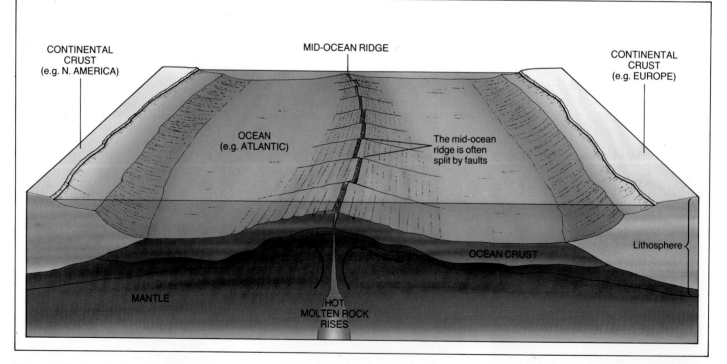

Picture 5   A cross-section of the Atlantic Ocean

When fluids are heated, they can move in warm currents that circulate through the fluid. These are called **convection currents**. The rocks in the mantle are only semi-solid, like Plasticine or very thick custard, so convection currents can circulate through it. These convection currents are very slow and sluggish, but their effect, nonetheless, is very dramatic. They shape the whole structure of the Earth's crust from the depths of the oceans to the high mountain ranges.

Look at picture 5. The convection currents rise through the mantle and reach the curst in the middle of the oceans. Here, hot **magma** (molten rock) bursts through the surface in underwater volcanoes. This magma solidifies to make new ocean crust — a line of towering underwater mountains running along the centre of the oceans. This is called a **mid-ocean ridge**.

The very top of the mantle and the crust form a firm rigid layer around the Earth called the **lithosphere**. The lithosphere 'floats' on top of the semi-solid mantle. So, as the convection currents move horizontally outwards from the mid-ocean ridge, the slabs of lithosphere move along too. New ocean crust is being made all the time from the mid-ocean ridges. Therefore, these slabs, or plates are growing larger all the time. This causes the plates with the continents to move further apart. The Atlantic Ocean is widening in this way by about four centimetres every year.

If the oceans are growing why isn't the Earth getting bigger?

What goes up must come down. This saying applies to convection currents as well as to apples and rockets. At the edges of the oceans, the convection currents dive back down into the mantle. (This is because the material has cooled and is now denser.) Picture 6 shows a section across the Pacific Ocean. You can see that where the convection currents move downwards, next to the continents, they are dragging some of the ocean crust down with them. This is called a subduction zone.

Old ocean crust is being re-cycled back into the mantle at the boundaries with the continents. So while the Atlantic Ocean is widening, the Pacific Ocean is very slowly shrinking. The Earth isn't getting any bigger or any smaller, it's in balance.

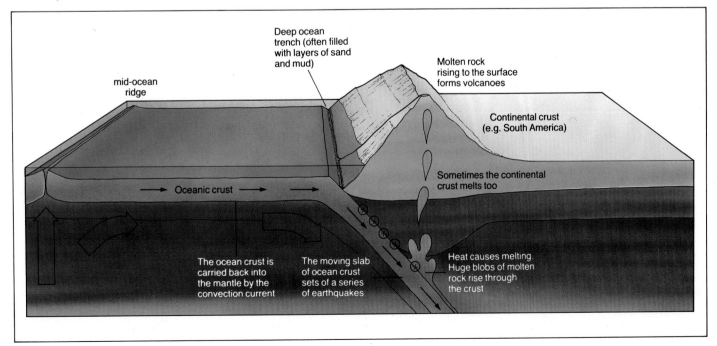

Picture 6   The shrinking Pacific Ocean

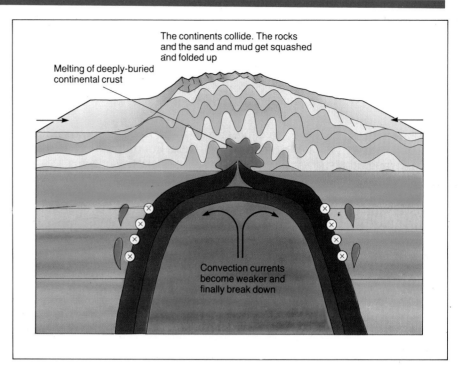

Picture 7 How mountains are formed during continental collision

It is possible, after many millions of years, for an ocean to close up altogether. All the mud and silt on that has formed the bottom of the ocean gets squashed and folded into a thick pile of sedimentary rocks. Finally as the two continents collide, they force these rocks to buckle and fold upwards into mountains. This is how the Himalayas were formed. India's continental crust is moving northwards, and Asia's southwards to create mountains like Everest. (See picture 7.)

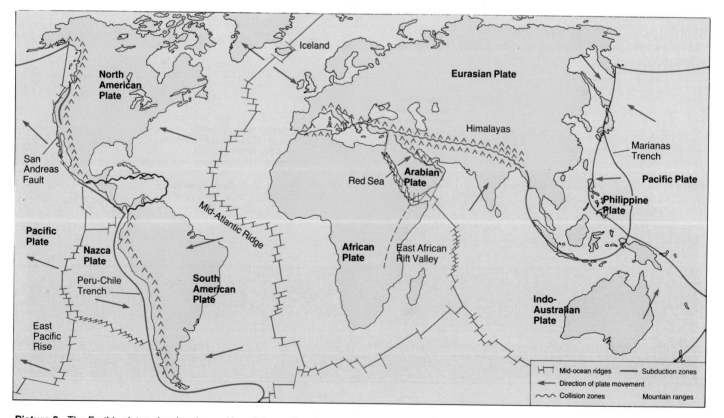

Picture 8 The Earth's plates showing the position of the continents

Plate tectonics

We have seen that the Earth's crust is divided into sections. These sections are called **plates**. Some plates have continents riding on top of them. They all fit together into a spherical jig-saw. Sometimes plates are destroyed, sometimes new plates are made. The plates move their position about the globe, driven by the convection currents within the mantle. All these processes happen very slowly, over many millions of years. The movements of these plates, and the things that go on at their boundaries are called **plate tectonics**. Picture 8 shows the position of these plates today.

Activities

A Using porridge to model convection currents in the mantle

Fill a beaker with made-up porridge and put it over a bunsen flame. Put a few drops of food colouring in the centre. After a while you should be able to see the colouring being carried round in convection currents. Eventually a skin will form on top — this represents the crust. This crust will be 'subducted' at the edges on the beaker where the current moves downwards.

B How did the Earth become layered?

Put the following in a test tube: about 10 ml cooking oil, 10 ml golden syrup, a couple of spatula-fulls of iron filings. Put a cork in the top and shake vigorously. Leave it to settle. Use the results you observe to make a hypothesis about how the Earth gained its layered structure.

You could then compare what you think with the theories of the Earth's formation outlined in topic I3. How accurate is the model of the layered Earth in this experiment? What are its shortcomings?

Questions

1 Copy picture 9 and fill in the labels (you may need to use some of the labels more than once):

subduction zone
mid-ocean ridge
ocean crust
continental crust
mantle
melting
molten rock rising
volcano
convection current
ocean trench

2 Write a couple of paragraphs to explain plate tectonics to your English teacher. Assume he/she has no scientific knowledge. Use your labelled diagram from question 1 to help.

3 Give the words that fit in the spaces labelled A to R.

a The Earth has a __(a)__ structure. The centre of the core is hotter; the outer part of the core is __(b)__. The core is composed mainly of the elements __(c)__ and __(d)__. Between the core and the crust is the __(e)__ which is made of semi-solid rock. There are two kinds of crust, __(f)__ crust 'floats' on top of __(g)__ crust.

b The crust and the top rigid part of the mantle are called the __(h)__. The lithosphere is broken up into irregular segments called __(l)__. Thermal energy travels through the mantle in __(j)__ currents. These currents reach the crust in the middle of the __(k)__ at the __L__ — ocean __(m)__. Here, hot molten rock rises to make new oceanic crust. The ocean crust and top rigid part of the mantle (the lithosphere) float on top of the __(n)__ and travel along the tops of the __(o)__ currents. At the edges of some oceans the ocean crust follows the __(p)__ current downwards, back into the __(q)__. This is called a __(r)__ zone.

4 No-one has ever been deeper than 14 km into the Earth. How, then, do we seem to know so much about its inside?

5 Waves are either **transverse** or **longitudual**.

a Describe these two types of wave (If in doubt, use the index at the back of the book!)

b Explain why one of these wave types cannot travel through liquids.

c Use your answers to (a) and (b) to explain briefly how earthquake waves help scientists to find out what the centre of the Earth is like.

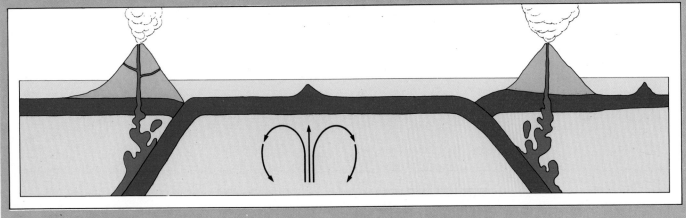

Picture 9

G2 Earthquakes and volcanoes

*Scientists still can't predict exactly **when** natural disasters such as earthquakes and volcanic eruptions will take place. But they can predict **where** they are likely to happen.*

Picture 1   The suffering caused by disasters like the Armenian earthquate is difficult to comprehend

In the middle of the morning of 7 December 1988 it took just 41 seconds for the two Armenian cities of Leninakan and Yerevan to be reduced to rubble. At least 25 000 people lost their lives.

What causes earthquakes?

Look at picture 2. Compare the occurrence of earthquakes with the plate boundaries shown in picture 8, topic G1. You can see clearly that earthquakes occur at the boundaries of the tectonic plates. The Armenian earthquake was caused by movement along an extension of the same plate boundary which formed the Himalayas.

Earthquakes are caused by the movement of rocks inside the crust or in the upper part of the mantle. As the plates of the Earth move about, rocks are put under immense pressure. Eventually the stress becomes so great that the rocks snap along a fault line. As the rocks break, energy is released as shock waves. These shock waves radiate outwards. They travel through and around the Earth and create the shaking effect that causes the damage.

The point at which the earthquake happens (where the rocks snap) inside the Earth is called the **focus**. The point immediately above the focus, on the surface, is called the **epicentre**. It is at the epicentre that the damage is usually greatest.

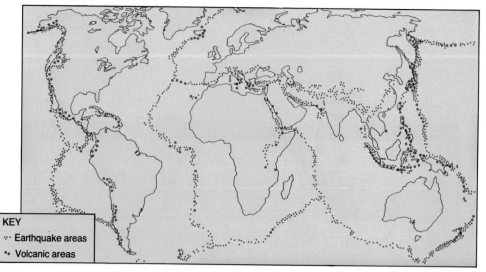

KEY

⊶ Earthquake areas

•⊶ Volcanic areas

Picture 2   This map shows where earthquakes and volcanoes occur worldwide

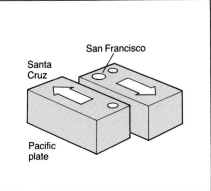

Picture 4 (a) The San Francisco earthquake, October 1989 (left)
(b) California is crossed by many faults like the great San Andreas fault (centre)
(c) the Pacific Plate sliding northwards and tearing against North America (right)

Measuring earthquakes

Earthquakes are recorded on **seismographs**. The **Richter scale** is a measure of the magnitude of an earthquake, that is, the amount of energy released at the focus.

Each step on the Richter scale represents a 30-fold increase in the energy released. So an earthquake of magnitude 8 releases not twice the amount of energy as one of magnitude 4, but 810 000 times as much! But the Richter scale cannot really tell you how severe an earthquake will be terms of damage to people and buildings. The San Francisco 'quake of 1989 and the Armenian 'quake of 1988 (picture 1) both had a magnitude of 7 on the Richter scale. Yet the intensities were very different. There was total devastation in Armenia, where over 25 000 people died. The only major damage in San Francisco was an elevated freeway which collapsed (picture 4), and only 70 people died.

There are three main reasons for this difference:

- The deeper the focus of the earthquake the less the effects will be felt at the surface. The focus of the Armenian earthquake was 5 km down, but in San Francisco it was 19 km down.
- The nature of the rocks at the surface affects the intensity — if they are soft mud or clay, the effects will be worse than firm rock.
- The way people live and build — in a rich country like the USA people can afford to design buildings which can withstand earthquakes. Less-rich countries often have poorer design — for example blocks of flats made cheaply which collapse like a house of cards. Overcrowded conditions can cause secondary effects, such as the spread of fire and disease.

British earthquakes

Britain is many hundreds of kilometres away from a plate boundary. Nevertheless, it does occasionally experience mild earthquakes. The last fairly large one that was felt was on the Welsh borders in 1990. No-one was hurt and slight damage was caused to one or two houses. Britain is criss-crossed with faults, some large, some small. All these faults are the results of ancient earthquakes many millions of years ago. It was one of these faults, very deep in the crust which slipped again slightly to cause the 1990 earthquake.

Picture 6 overleaf shows three different types of fault:

a These are two **normal** faults. They are formed by stretching forces in the crust, the pulling action cracks the rocks, then one side slips down. When two faults act together, they cause a whole block of land to slip down, forming a **rift valley**.

b Reverse faults are caused by compression in the crust. The pushing

Picture 3 The focus and epicentre of an earthquake

Epicentre – the point on the earth's crust directly above the focus

Focus – the point at which the earthquake actually happens

Picture 5 The map shows four faults in Scotland — the Moine thrust, the Great Glen fault, the Highland Boundary fault and the Southern Uplands fault. The occurrence of these faults suggests to many scientists that Britain has been much closer to plate boundaries in the distant past than it is today. Many people believe that the Scottish Highlands and the hills of the Southern Uplands were formed when two ancient continents collided about 400 million years ago.

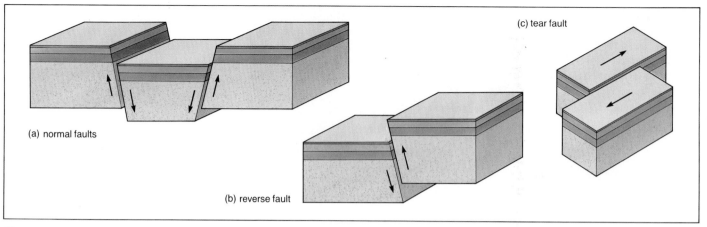

Picture 6   Three types of fault

force causes the rocks to crack. Then one side pushes upwards along the fault and climbs over the other side.

c A **tear** fault is caused by horizontal movement in the crust. The San Andreas Fault in California and The Great Glen Fault in Scotland are both examples of tear faults.

Picture 7   The eruption of an Icelandic volcano

Volcanoes

Volcanoes are fascinating and breathtaking, and yet violent and dangerous. They may lie quiet for years and then suddenly erupt. They pour out molten rock (**lava**), ash or poisonous gases such as sulphur dioxide. All these products are formed because of melting in the mantle or in the crust. There are three locations where volcanoes are likely to occur on the Earth:

1   At mid-ocean ridges

The rising convection current in the mantle causes the upper mantle rocks to melt. Molten rock is less dense than solid rock, so it rises. Iceland (picture 7) is on the mid-Atlantic Ridge and is made entirely from solidified lava. The lava is basalt (see topic G3) which is very runny, so it flows a long way and solidifies into wide, gently-sloping sheets.

2   At the 'ring of fire' — subduction zones

Volcanoes are common all around the edges of the Pacific Ocean. It has become known as the *Ring of Fire*. As the oceanic plate dives down into the mantle it starts to melt. Rocks in the lower part of the crust may start to melt too, molten material rises towards the surface. These volcanoes tend to erupt much stickier, less fluid, lava and clouds of ash and gases. They tend to be more violent and to cause more loss of life and damage to property.

When Mount St Helens in Washington State USA erupted in 1980 half the mountainside was blown away in an explosion. This eruption was one of the few actually predicted by scientists. They monitored the gases at the top of the volcano, the earth tremors and swelling of the mountainside in the months, weeks and days leading up to the eruption. Warnings were given to the public to evacuate the area and though only 100 people died, the loss of life could have been much greater.

Picture 8   The eruption of Mount St Helens

3   Hot spots

Some volcanoes, such as the island of Hawaii, occur in the middle of plates. They occur because they are above 'hot spots' in the mantle. Mushroom-shaped plumes of hot rock rise deep in the mantle and melt the rocks above — either in the upper part of the mantle or in the crust. No-one is certain how these plumes form, but one theory is that the mantle isn't uniform in its composition. There may be pockets in the mantle which are richer in radioactive elements, which make the rocks hotter. (Picture 9).

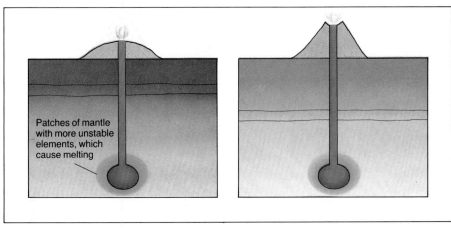

Picture 9 As the ocean plates move over the hot spots, the position of the volcanoes changes. A 'track' in the form of extinct volcanic islands is left behind

Ancient volcanoes

You may never have seen a 'live' volcano, but Britain has lots of 'dead' ones. Many landscape features are formed from ancient volcanoes. Picture 11 shows an extinct volcano in Fife, Scotland.

Picture 10 Most of the ash core has been worn away leaving only the solidified lava in the central pipe.

Activity

Modelling faults

1 Draw coloured bands to represent layers of rock on a block of wood or expanded polystyrene as shown in picture 11. Cut several 'faults' along the length of the block. Move the blocks to form i) normal faults, and ii) reverse faults. Measure the length of the blocks before and after movement in each case.

a Which type of fault would you expect to find in an area of the crust which is stretching?

b Which type of fault would you expect to find in an area of the crust which is being squashed, or compressed?

2 You could try floating the blocks in a tank of water. (The tank should be only slightly longer than your piece of wood.) The wood could represent the crust, and the water the mantle. Are the faults you get normal or reverse? Sketch the shape of the 'landscape' you get at the surface — is it straight or are you getting hills and valleys?

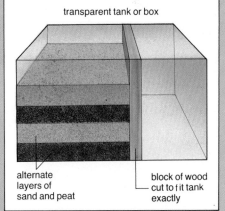

Picture 12

3 Set up the apparatus as shown in picture 12. Push the block of wood **horizontally** towards the layers of sand and peat. What type of fault develops? Draw what you see through the side of the tank. Can you see any other structure as well as the fault?

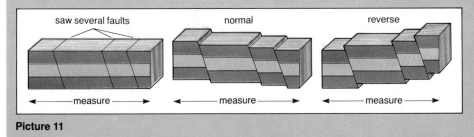

Picture 11

Questions

1 Copy or trace picture 2. Then, using picture 8 in topic G1, which shows the different types of plate, label the volcanoes according to their type.

2 Explain each of the following terms: epicentre, focus, fault, Richter scale.

3 'All volcanoes are equally dangerous.' Do you think that is true? Explain your answer.

4 Why do earthquakes and volcanoes often occur in the same places?

5a Suggest why there are no actual volcanoes, and very few earthquakes, in Great Britain?

b Suggest a possible reason why there were large volcanoes in Britain many millions of years ago.

G3 Rocks and the rock cycle

Wherever you are, you are unlikely to find yourself very far from rock — even if you don't know it.

Picture 1 This front of this bank is made from coarse-grained granite

The bank in picture 1 is faced with **granite**. Because it's been polished you can see quite easily that it's made up of crystals. The overall speckled effect of grey and white shows that it's made of more than one **mineral**.

The statue in picture 2 is made from **limestone**. You can see the remains of living creatures in it — fragments of sea shell that have become **fossilised**, that is, preserved in rock (see *The Living World*, topic F8). It is just possible to see individual grains with the naked eye; they all appear to be made of the same mineral. Parts of the stone look worn or **weathered**.

Many houses are roofed with **slate** (picture 3). Slate has been popular for roofing since the middle of the nineteenth century. It can be split, or **cleaved** very easily into thin sheets. It is usually dark in colour — from purplish grey to green — and it is hard and smooth. You can't see individual grains or crystals in it.

Why are rocks so different?

The three rocks in pictures 1 to 3 were formed in three different ways:

Granite is a type of **igneous** rock. Igneous rocks have solidified from molten rock either in volcanoes, or inside the crust.

Limestone is a **sedimentary** rock. These rocks are usually laid down under water as a **sediment**. Sedimentary rocks are made from fragments worn away from older rocks, or from the remains of living organisms.

Slate is a type of **metamorphic** rock. The word **metamorphic** comes from the Greek — *meta* meaning change, and *morphe* meaning shape. These rocks have been changed by heat or pressure, or by a combination of both. They could originally have been either igneous or sedimentary.

Picture 2 A statue made from Portland stone

Picture 3 Slate roofs

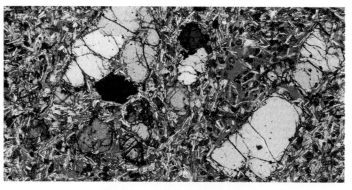

Picture 4 (a) Basalt (b) Basalt seen through a microscope showing different minerals

Igneous rocks

Igneous rocks are formed when molten rock, called **magma**, cools and solidifies. Nearly all igneous rocks are formed of **crystals**. You can see the crystals in the granite in picture 1. Notice that the crystals are **interlocking**.

Molten rock is less dense than solid rock, and so it tends to rise very slowly towards the surface of the Earth. If the molten magma reaches the Earth's surface and is erupted from a volcano, it is called **lava**. When the lava cools and solidifies it becomes rock. **Basalt**, a smooth black rock, is the most common type of lava (picture 4(a)). There are many less common types of lava, and sometimes volcanoes throw out ash (see topic G2). Rocks formed by eruption from volcanoes are called **extrusive** igneous rocks.

With the naked eye, basalt looks smooth and black. But under the microscope, crystals of several different minerals can be seen. These are different kinds of **silicate** minerals. Picture 4(b) was taken through a special microscope which uses polarised light. This gives some of the minerals bright colours which makes them easier to see.

Sometimes magma contains dissolved gases. When the magma reaches the surface, where the pressure is lower, the gases come out of solution to form gas bubbles. You can see the same effect when you open a bottle of fizzy drink. When the lava cools, the bubbles are preserved in the rock. Pumice is a type of lava with more gas bubbles than rock. If you've used pumice stone in the bath, you'll know it can float.

Very often magma gets trapped inside the crust and never reaches the surface. These rocks are called **intrusive** igneous rocks. Because the magma is insulated by the surrounding rocks, it takes a very long time to cool — sometimes millions of years.

Granite is probably the most common intrusive igneous rock. It is different from basalt in that it cools very slowly from molten magma deep in the Earth's crust. Because it takes a long time to cool, the crystals are larger. They are easy to see with the naked eye and are much lighter in colour than basalt. (Picture 5.)

Picture 5 Granite

Picture 6 Layers of sedimentary rock, Burton Bradstock, Dorset

Picture 7 (a) Conglomerate

(b) Sandstone

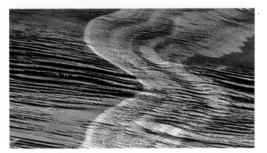

(c) Mudstone

(d) Shale

Picture 8 Travertine like this is used on many shop fronts

Many of the sea cliffs and high moors in Cornwall are made from granite. It has resisted the wearing effect of weather and water for many millions of years.

Sedimentary rocks

One way to think of sedimentary rocks is as 'second-hand' rocks. They are made up either of fragments of older rocks or from the remains of living organisms.

Sedimentary rocks often occur in layers, or **beds**. Picture 6 shows the cliffs at Burton Bradstock in Dorset. These cliffs are made of beds of sandstone which were laid down in the sea about 190 million years ago. Most sedimentary rocks are laid down in the sea, but they may also form in rivers, lakes, and even in deserts.

Sedimentary rocks from particles or fragments

Even the hardest rock will start to crumble and weather as thousands of years go by. Fragments of rock fall down hillsides due to gravity, or get washed down by rain. Rivers will carry the fragments down into deltas, lakes, or the sea. Waves beat against cliffs and knock fragments away. (More about **weathering** and **erosion** is in topic G4.)

Look again at picture 6. You can see where the cliff has fallen to make a pile of boulders on the beach. The action of the waves will eventually break them up into pebbles and sand. These pebbles and sand could become the sedimentary rocks of the future.

Sedimentary rocks are named according to the size of the fragments that they're made of. A rock made of pebbles is called a **conglomerate** (sometimes called a 'pudding stone' because it looks a bit like an old-fashioned steamed plum pudding). A rock made of sand is called simply a **sandstone**. A rock made of fine mud is usually called a **mudstone**, but if it's flaky and breaks easily into layers, it's called a **shale**. Some of these rocks are shown in picture 7.

Conglomerates and sandstones are made mainly of fragments of the hard, resistant mineral **quartz**. In many cases this will have been weathered from granite. **Quartz** is one of the commonest minerals found in rocks (Quartz comes in many colours, from transparent bright purple (amethyst) to opaque grey or brown flint. This can make it tricky to identify.)

Mudstones and shales are made from clay minerals. These vary in colour from creamy white, to reddish brown, grey, and black. The individual grains are too small to be seen even with a hand lens.

How do sand pebbles and mud turn into rock?

There are two ways that loose sediment can be turned into rock. First, as more sediment accumulates on top, everything gets squashed and compressed into rock. Second, other minerals sometimes seep between the fragments in the rock, and 'glue' them together. You can see this very clearly in the conglomerate in picture 7. The spaces between the pebbles are filled with brown quartz.

Organic sedimentary rocks

Most limestones are made of the shells and skeletons of organisms that lived in the water. Sometimes the shells are so small, or broken up into such small fragments, that you can't see them with the naked eye, or even with a hand lens.

Shells and skeletons are almost always made of calcium carbonate, so limestone is made of calcium carbonate too.

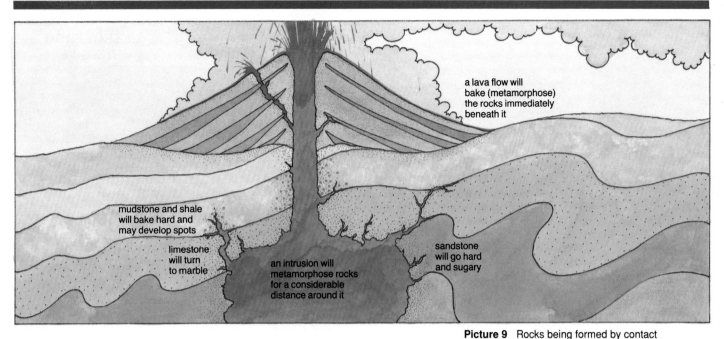

Picture 9 Rocks being formed by contact metamorphism

Some limestones are formed chemically when calcium carbonate precipitates out of a solution (see *The Material World*, topic G4). The solution of calcium carbonate could be in sea water, in lake water, or in a hot spring. Picture 8 shows a chemical limestone, called **travertine**. It's used as an ornamental stone on the front of most 'Macdonald's' hamburger restaurants.

The limestone in picture 2 is a mixture of a chemical limestone and shells. It was formed, around 100 million years ago, in very warm tropical seas.

Metamorphic rocks

Rocks may be changed or **metamorphosed** in two ways: by heat alone or, more usually by a combination of heat and pressure. New minerals form in the rocks. The texture of the rocks is changed too, and any fossils in sedimentary rocks are obliterated.

Heat only

When lava spills out of a volcano onto older rocks, the older rocks get hot, and literally 'cook'. When intrusive igneous rocks get trapped inside the crust they heat the rocks around them. This time the 'cooking' effect is longer and slower, but more pronounced. (See picture 9.)

The heat of the magma tends to make the surrounding rocks harder. The grains in sandstones stick together tightly and form a hard, sugary-looking rock. Limestone changes to a sugary-looking rock too — **marble**. Mudstones and shales turn into hard splintery rocks at high temperatures, but at lower temperatures they develop spots — a sort of geological heat-rash!

Heat and pressure

The most common types of metamorphic rock are formed by heat and pressure deep in the crust at subduction zones or collision zones. (See topic G1 — pictures 6 and 7 — look at them again before you read on.)

Slate is one of the commonest rocks formed in this way. It is a metamorphosed mudstone or shale. Slate can be split very easily along **cleavage** planes. The cleavage is caused by the effect of pressure.

Table 1 Clues to rock types

Igneous rocks
<ul><li>have no fossils,</li><li>have an interlocking crystalline structure,</li><li>are likely to be hard.</li></ul>
Sedimentary rocks
<ul><li>may have layers (bedding) visible,</li><li>will have separate grains (not interlocked),</li><li>may be quite soft (you may be able to rub grains off),</li><li>may have fossils,</li><li>if calcium carbonate is present, will fizz with dilute HCl.</li></ul>
Metamorphic rocks
<ul><li>may be able to be split along a cleavage,</li><li>may be banded or streaked,</li><li>may be hard,</li><li>may have sparkling mica flakes, aligned in streaks or layers,</li><li>may have a 'sugary' texture, or be noticeably crystalline,</li><li>if marble, will fizz with dilute HCl,</li><li>have no fossils.</li></ul>

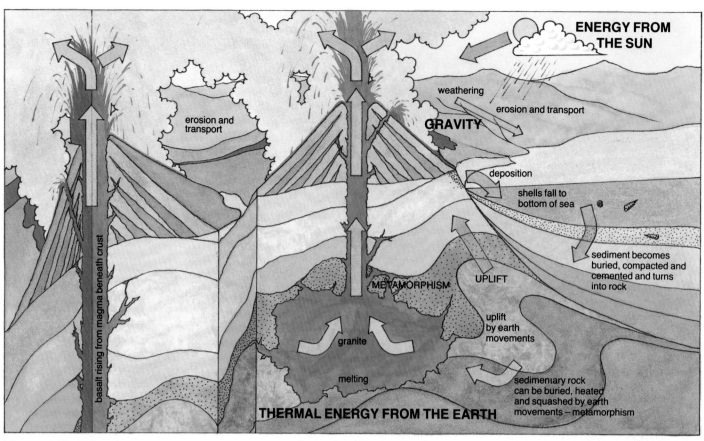

Picture 10 The rock cycle

Limestone will also turn into marble under the effects of heat and pressure. Most limestone is impure. It is these impurities that form the coloured bands and swirling patterns that make marble so attractive.

Rock detective work

Table 1 gives you some clues to help you decide whether a rock is igneous, sedimentary or metamorphic:

The rock cycle

We've already seen that older rocks can be worn away to provide fragments to make newer sedimentary rocks. We've seen how rocks can be changed by heat and pressure into metamorphic rocks. In fact if rocks get heated enough (about 600°C), they can melt to form an igneous magma. So, one kind of rock can be re-cycled to form another type.

Look at picture 10. If you follow the arrows from the top of the volcano in the foregound, you can see igneous rock being formed from magma, then eroded. The fragments are deposited in the sea as sediment. The sediments harden into rock, and can then be buried and metamorphosed. The metamorphic rock may get so hot as to melt and form magma which will cool to form igneous rock. Then the cycle can start all over again. There are many other ways around the **rock cycle**. A summary of the rock cycle is shown in picture 11.

In the next topic, G4, we'll learn more about how rocks weather and about erosion and transport. In G5 we'll consider the vast time span over which the rock cycle has operated since the Earth's crust was first made about 4.3 billion years ago.

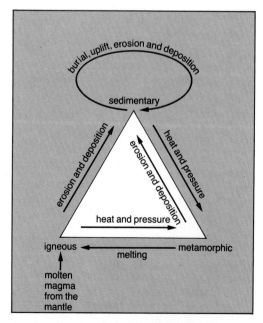

Picture 11 A simplified view of the rock cycle

Activities

A Investigating the rate of cooling and crystal size in igneous rocks.

CARE! Salicylic acid gives off choking fumes, take care not to inhale them.

Take two glass slides. Put one in the freezer for a while, and warm another, either on a radiator or over a very gentle bunsen flame. Melt some salicylic acid in a boiling tube in a water bath. With a **warmed** glass rod, drop a couple of drops of the melt on the cold slide and a couple on the warm slide. The salicylic acid on the cold slide should crystallise almost immediately.

When both have crystallised, look at the crystals through a microscope at low power. Sketch what you see. What do you think your results tell you about the rate of cooling on the crystals in igneous rocks?

B The properties of rock-forming minerals

Investigate the properties of the three minerals in granite: quart, feldspar and mica. (Use either individual specimens of these minerals, or a piece of coarse-grained granite.) Which is the softest mineral? Which is the hardest?

C Igneous, sedimentary of metamorphic?

Get some rock samples from your teacher. Use the clues in the box on page 209 to identify rock specimens. Don't worry if you can't give an actual name to a rock. But do try to write down evidence for the origin of the rock — whether it might be igneous, sedimentary or metamorphic.

You could then do this activity with building stones, facing stone and roofing materials in your local high street. Again, look for evidence to decide whether each rock is igneous, sedimentary or metamorphic. But do not use acid, hammer off samples or damage building stones — this will not make you or your school very popular!

D Looking at gravestones

Work in groups. Make a study of your local graveyard. Try to identify the rocks the gravestones are made of, or at least if they are igneous, sedimentary or metamorphic. Don't try to use acid or damage the stones in any way — treat the graveyard with respect. Note the date on each stone. Look for signs of weathering — give each stone a score of one to ten for its state of preservation. Try to decide which rocks are least resistant to weathering.

E The matchstick/cleavage analogy

Empty a box of matches onto a table — with the matches pointing in all different directions. Then push the matchsticks. The matchsticks will tend to line up at right angles to the direction in which they're being pushed. The same thing happens to the minerals in the slate — tiny flakes of mica (too small to be seen with the naked eye) align themselves at right angles to the pressure in the Earth's crust. These lines of mica flakes form lines of weakness in the rock — the cleavage. (See picture 12.)

F Rock densities

Use one of the methods for measuring density given in topic A to find the structures of as many different rocks and minerals as you can. You should try to use: *rocks* — sandstone, limestone, slate, marble, granite, basalt; *minerals* — haematite, galena, cassiterite, zinc blende.

Is there a pattern in your results? Try to explain it, in terms of the structure of the Earth, and how the rocks and minerals were formed. (Remember that granitic continental crust overlies ocean-type basaltic crust.)

Picture 12

Questions

1 Classify the following as igneous, sedimentary or metamorphic rocks:

limestone, basalt, slate, conglomerate, marble, sandstone, granite, mudstone, rhyolite, gabbro, shale, serpentine.

2 Picture 13 shows a rock which is used as a facing stone in most high streets. It's called **blue pearl larvikite** and comes from Norway. Is it igneous, sedimentary or metamorphic? Say why you have made your choice.

3 Picture 14 shows a rock from the North Pennines. It fizzes when put in hydrochloric acid. Can you suggest what it might be? Write down as much evidence as you can for your answer.

4 Copy the diagram of the rock cycle picture 10 into your note book. Look at picture 7 — which process in the rock cycle do you see taking place? What evidence can you see that some of the beds of rock in the cliff might be harder than others?

Picture 13

Picture 14

5 Look at pictures 8, 9 and 10 in topic G1. Which sort of plate boundaries are likely to produce: (a) basaltic rocks, (b) granitic rocks?

G4
Shaping the landscape

As old as the hills — we think of the landscape as permanent and unchanging. But the landscape is slowly and continually being re-shaped. This topic is about how this happens

Picture 1 A faulted landscape

Pushing and pulling — how the basic shapes are formed

In topics G1 and G2 we learned about how the crust of the Earth is divided into plates. These plates move the crust about, pushing together in some places and tearing apart in others. Volcanoes erupt lava and ash, and can build up into high mountains. Sometimes the rocks in the crust snap, causing earthquakes, and leaving behind cracks or **faults** (picture 1).

Sometimes the rocks don't fault — they bend. Sudden forces tend to cause cracks (faults). Forces that act slowly can cause gradual bending. (See topics A1 and A2). If the rocks in the crust are pushed slowly they bend and form **folds**. This can happen when two plates push together and collide (look back to picture 8, topic G1. About 50 million years ago, Africa started to move towards Europe. The rocks were compressed and folded into a mountain chain — the Alps. The ripples spread many hundreds of miles away, as far away as the south of England. (See picture 2.)

The upward, humped shaped fold is called an **anticline** and the downward, basin-shaped fold is called a **syncline**. The top of the anticline cracked, and the rocks wore away along the line of weakness. This has left lines of chalk hills (called **escarpments**) which form the North and South Downs. The line of chalk hills formed by the northern limb of the syncline forms the Chiltern Hills. (See picture 3.)

Picture 2 A folded landscape

Picture 3 The hills of the Chilterns and the North and South Downs were pushed into shape at the time the Alps were formed.

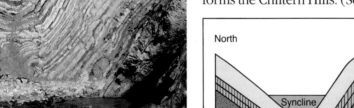

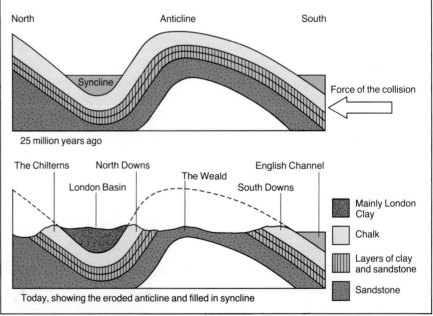

North   Anticline   South

Syncline

Force of the collision

25 million years ago

The Chilterns   North Downs   English Channel

London Basin   The Weald   South Downs

Today, showing the eroded anticline and filled in syncline

- Mainly London Clay
- Chalk
- Layers of clay and sandstone
- Sandstone

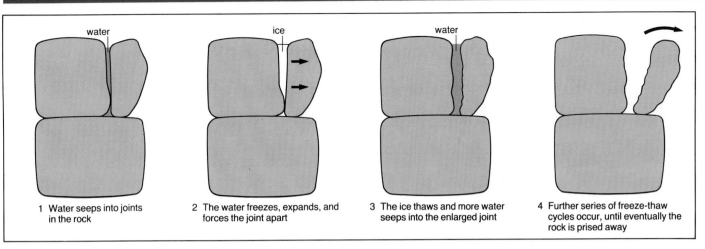

1 Water seeps into joints in the rock

2 The water freezes, expands, and forces the joint apart

3 The ice thaws and more water seeps into the enlarged joint

4 Further series of freeze-thaw cycles occur, until eventually the rock is prised away

Picture 4 Frost shattering

The energy that produces folding and faulting comes from within the Earth. It comes from the internal energy of the Earth which drives the convection currents in the mantle, which in turn drive the movements of the crustal plates.

Weathering and erosion

When rocks are exposed to the weather — to wind rain and frost — they will start to disintegrate and crumble. This is called **weathering**. There are three main types of weathering. They often act together, rather than by themselves.

Physical weathering

This is the loosening of fragments of rock by physical means. There are two important ways that this can happen:

a Some rocks absorb water (see topic G3). If water absorbed in a rock freezes, it expands and push the grains of rock apart. When the rock warms up again, the ice will melt. Picture 4 shows what will happen if this freezing and thawing is repeated time and time again.

b The minerals that make up a rock will expand on heating and contract again on cooling. If this happens time and time again — for example heating up during the day and cooling at night, for months and years on end — the resulting stresses will cause the rock to crack.

Chemical weathering

This can happen in two ways. First, the minerals which compose the rock can break down and change. Often they change into clay which is then easily washed away by rain. Picture 5 shows a china clay pit in Cornwall. In the hot wet climate of the past, the feldspar in the granite reacted with warm water to form clay. A jet of water washes the clay into a processing plant, leaving behind the quartz and mica.

Second, some minerals will react with **impurities** dissolved in rainwater. Rainwater is often slightly acidic and so will react with limestone. (See *The Material World*, topic D3.)

Biological weathering

Plants and animals can cause rocks to break up. Roots can be very forceful in exploiting any weaknesses in the rock as the photograph in picture 7 shows.

Picture 5 China clay being mined in Cornwall

Picture 6 The elements can play havoc with old buildings like this

Picture 7 Tree roots splitting rocks

Picture 8 How potholes are formed in the bed of a stream

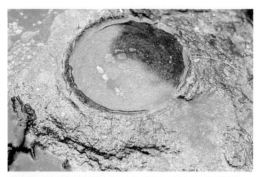

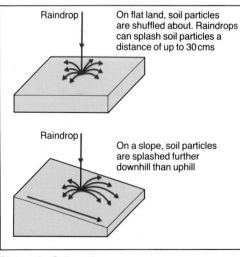

Raindrop — On flat land, soil particles are shuffled about. Raindrops can splash soil particles a distance of up to 30 cms

Raindrop — On a slope, soil particles are splashed further downhill than uphill

Picture 9 Soil erosion

Biological weathering is important in the formation of soil. Weathered rock fragments form the basis of all soils. (To find out more about soils see *The Living World*, topics B12 and B13.)

Transport and erosion

Sometimes the material weathered from rocks is carried (transported) away. This can happen in several ways. **Gravity** will cause material to fall down slopes — for example down the side of a valley.

Water in the form of rain and rivers can wash material away. Wind can blow loose material away. In cold climates and mountainous areas, material can be carried by ice.

This removal of weathered material is called **erosion**. As the material is transported, it can itself cause more erosion. For example, pebbles bumping and shuffling their way along the bottom of a stream will slowly wear away the rocks on the stream bed.

Soil erosion happens where the soil isn't bound together by roots. It is most common where hills have been stripped of their natural vegetation and ploughed for farming. This can have devastating results as shown in picture 9.

Deposition

Eventually the eroded material is deposited as pebbles, sand or mud. This happens when whatever is carrying it — water, wind or ice — slows down. For example, when a gust of wind dies down, any sand or dust it was carrying is dropped. You are quite likely to be able to see this happening on a windy day in your school grounds. Again gravity is the important force here. When the force of the moving wind or water supporting the load stops, gravity takes over and pulls the load downwards towards the Earth.

Rivers

Rivers run from their sources in hills or mountains down towards the sea. Rivers rarely have just one source. Several streams join one by one to form a **river system** flowing in a **drainage basin**.

In the upper part of the river system, the streams carve out deep V-sided valleys for themselves. The gradient of these valleys is often steep. They carry the material they have eroded away in three ways:

- in **solution** as dissolved salts,
- in **suspension** — a cloudy mixture of sand and mud,
- as pebbles and sand pulled along the bed of the river by the water — the **bed load**.

The amount of sediment a river can carry is dependent on two things — the speed of the river and the volume of water in the river.

Upland rivers flow quickly down steep valleys. Lowland rivers flow more slowly along gently sloping valleys. (See picture 10.)

The fast, turbulent water in upland areas causes most erosion. This happens particularly in winter and spring when they are 'in spate' due to heavy rain and melting snow.

The rocks, pebbles and grit carried by a river in flood find weaknesses in the rocks. Bends develop. Water flows faster on the outside of bends than on the inside. Faster water causes more erosion, so the bends are made bigger. This effect causes the **interlocking spurs** seen in picture 11(a).

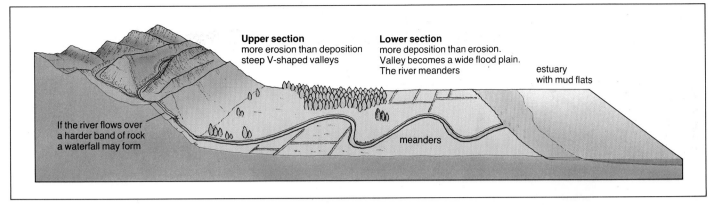

Picture 10 The main features of a river from source to sea

Picture 11 (a) An upland river with interlocking spurs
(b) A lowland river with meanders

In flatter land the river travels more slowly. It deposits its load of sediment carried from the uplands. This make the river shallow and wider. It is liable to spread over the surrounding countryside in times of flood.

At bends, the same effects occur as in upland rivers. Slight bends will grow larger as the 'fast' side is eroded and sediment is deposited. Very large bends called **meanders** are formed, as shown in picture 11(b) and sediment is deposited on the 'slow' side.

When the river floods, more erosion takes place on the 'slow' side, and the river is able to carry more load. If the river bursts its banks, coarse material is deposited on the banks forming natural raised edges to the river called **levees**. Fine mud is deposited across the flood plain — this adds nutrients to the soil, and so is a help to farmers. (See picture 12.)

When the river finally reaches the sea, the rest of its load is deposited. Some of the material will be carried out to sea, but some may be deposited in the river estuary as mudflats, or in a delta sticking out into the sea.

Coasts

Whenever you walk along a seashore, you can see erosion and deposition happening. The waves gnaw away at cliffs; pebbles crash against each other and gradually break into sand, and the sand is deposited on the beach.

The hardness of rocks determines the shape of the coastline. Softer rocks erode faster and bays are formed, while harder rocks resist the battering of the waves better and remain standing as headlands (picture 13 overleaf). Eroding coastlines often have jagged outlines for this reason.

The eroded pebbles and sand may be transported some distance before being deposited. They are often deposited as long, smooth, straight beaches or **spits**.

Glaciation

Although ice is a solid, it can flow very slowly. In a deep layer of very cold ice, the pressure of the upper layers of ice will squash the lower layers and cause the ice crystals to align themselves as shown in picture 14 overleaf. Once aligned, the crystals can slide over each other. Also if the ice is near its melting points, 0°C, a thin film of melted water will develop underneath it. This causes the ice to slide along, lubricated by the water.

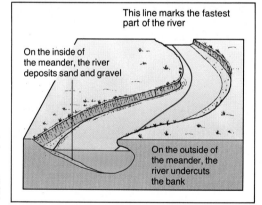

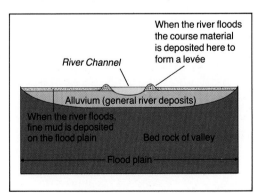

Picture 12 Water covering the flood plain of the River Severn in February 1990

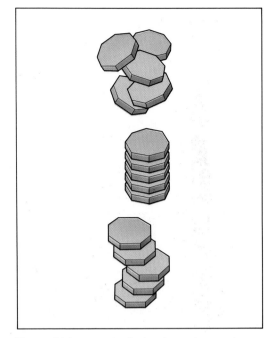

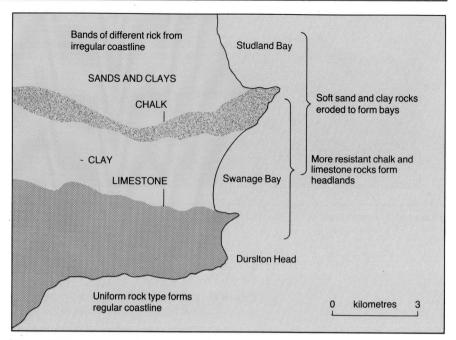

Picture 13 Erosion on the Purbeck coast, Dorset

Bands of different rick from irregular coastline

Studland Bay

SANDS AND CLAYS

CHALK

Soft sand and clay rocks eroded to form bays

- CLAY

LIMESTONE

Swanage Bay

More resistant chalk and limestone rocks form headlands

Durslton Head

Uniform rock type forms regular coastline

0   kilometres   3

Picture 14 Ice crystals aligning themselves so that the ice in the glacier can flow

Picture 15 A moving glacier

Picture 16 Eroded landforms in the Sahara desert

As ice moves, it scours the rocks it is travelling over and picks up fragments. The ice will transport these fragments until it melts. Then the load will be deposited. This is happening today in high mountainous regions such as the Alps (see picture 15). During the ice age, (about 2 million years ago) Wales, Scotland and the Lake District looked at times much like the Alps do today. The glaciers have left behind deep valleys, and steep angular mountainsides.

Britain was covered in ice as far south as Bristol in the West, and London in the East. In the lowland areas, the huge wide ice sheets scoured away soil and rock fragments. All this material was transported by the ice and then deposited when the ice melted. Deposits dropped by melting ice are found over most of Britain. They are an extremely important source of sand and aggregate for the building industry.

Deserts and wind

We tend to think of deserts as being full of sand dunes. Some areas of desert are. But where could all that sand have come from? It must have been eroded and transported from somewhere before being deposited as dunes.

Picture 16 shows a tract of stony desert in the Sahara. Grains of sand blown by the wind have been showered against the rocks for thousands of years, and have 'sandpapered' them away. The softer rocks erode faster than the harder ones. The sandgrains themselves down into perfect spheres.

But deserts are not the only parts of the world where the wind plays a part in the landscape. Even in Britain, the wind can contribute to soil erosion. In East Anglia and parts of Lincolnshire some soils are light and peaty. When they are ploughed, they are unprotected by a layer of vegetation. This means that they can easily be blown away in strong winds. Soil carried in the wind will batter young plants in adjoining fields. This effect is worse where fields have been made larger by the removal of hedges.

This last point illustrates another important force in shaping the landscape — **people**. We alter the landscape by farming, building roads, towns and tunnels, and by trying to alter the course of rivers. People try to make the landscape more convenient for their own needs. But we have to understand that the landscape is always shaping itself, and we have to understand how it works to make the best use of it and to preserve its usefulness for the future.

Activities

A Which rocks are resistant to weathering?

1 *Frost shattering*

You will need two pieces of each of the following type of rock — granite, sandstone, limestone and shale (or mudstone). Soak one set for at least half an hour in water. Then put both sets of rock in the freezer. Take both sets out every day, and allow the wet set to thaw completely. Re-soak the wet set if necessary. Put both sets back in the freezer. Repeat this for several days, or weeks if possible. At the end of the experiment compare both sets of rock. What can you conclude about: (a) the role of water in this type of weathering? (b) which rocks are more resistant and why. (Perhaps you could devise a further experiment to confirm why you think some rocks may be more resistant.)

2 *Temperature changes*

You will need a bunsen burner, tongs and safety glasses for this experiment. Use specimens of the same rock-types you used in **1**. Hold small specimens of each rock in the tongs in the bunsen flame. Then plunge them in cold water. Repeat as often as you can. What conclusions can you draw from your results?

3 *Dripping dilute HCl on rock specimens*

CARE! Use very dilute hydrochloric acid. Design an experiment to find out which rock is most easily eroded chemically. Check your plan with your teacher before carrying it out. Compare your results with those in activity E in topic G3.

Wear eye protection.

B Investigating rivers

1 Wherever you live you won't be far from a river. Find out its name, where it has flowed from and where it is flowing to. If you can visit part of the river, look for evidence of erosion and deposition. Assess what effect the river has on the landscape. Can you see any evidence that people have interfered with the course the river takes, or with its erosion and deposition?

2 If there is a suitable river or stream near you, carry out some practical investigations.

CARE! Be sensible near rivers. As well as the obvious danger of drowning, there may also be diseases carried in the water. Wear rubber gloves to touch the water, and wash your hands before eating anything!

a Use small, light sticks to investigate the speed of the water. Drop them in the middle, and then near each bank and see if there is any difference. Watch to see what path the stick takes each time. Time the sticks over a measured distance and calculate the average speed of the water.

b Observe the river's load. Can you see sand and gravel moving along the bottom?

c Take samples of the river water from different places — the middle, the sides, the bottom and the top. Use bottles and string. Do this from a quiet bridge. You must be supervised by your teacher. Evaporate your samples back in the lab and work out the mass of solid material in your sample.

C Investigating meanders

1 Take a large, old, baking tray. Put it in the sink, and let a trickle of water run along its length. Does the water run smooth and straight, or does it flow in a winding, turbulent manner?

2 Fill the tray with fine wet sand. Draw a straight channel along the sand from top to bottom. Put the try in the sink, and allow a trickle of water to flow down your channel. Does your channel stay straight? What do you think the results of this experiment would tell an engineer working on a project to straighten meanders in a river?

D Modelling rivers in the lab

Take a long piece of guttering — as long as will fit into your lab (about 3–4 metres). Put the lower end in a tank or bath, and the other end under the tap. This will be your basic river. Tip sediment — a mixture of sand, gravel and mud — into the top of your 'river'. Watch what happens. Vary the speed of the water by increasing the inclination of your guttering. Vary the volume of water by turning the tap on more. Investigate how these two variables affect the load of the river.

You could try bending your guttering in the middle to form a lake and seeing what happens to your river's load. Or you could make a dam to fit just halfway across, and observe the effects.

When you have finished this investigation, discuss what you have found out. Could anything you have found be useful to a civil engineer building a dam, or wanting to straighten the course of a river?

E Your local landscape

Find out what has shaped the landscape of your region of Britain. Outline what you feel has been the most important factor.

Questions

1 Imagine you have a party of primary school children visiting your school. Using what you have learnt in both G3 and this topic, write a short story for them entitled 'the adventures of a grain of quartz'. Perhaps it could start off in some granite, and eventually end up in a sandstone, or in a desert. Make it as eventful as you can.

2 Explain the following words: weathering, erosion, transport, deposition.

3 Explain how the following sources of energy have shaped the landscape on the crust of the Earth:

a thermal energy from inside the Earth,

b the Sun,

c gravity.

4 Look at picture 6, topic G2 (page 204) and picture 1 (page 212). What type of fault is shown in picture 1? Give reasons for your answer.

5a What evidence can you find or think of to support the theory that the landscape around where you live hasn't changed very much i) in the last 100 years ii) the last 1000 years.

b Do you think it has changed much in the last million years? Give reasons for your answer.

What shape is the Earth? How big and how old? This topic explores ways of finding out the answers.

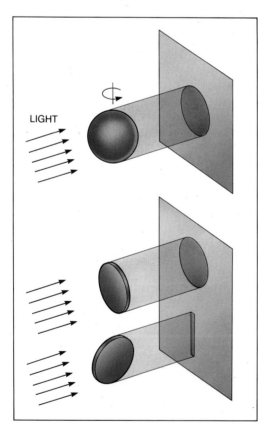

Picture 2 (a) The shadow of a spinning sphere is always a circle
(b) The shadow of a spinning disk is sometimes a circle — sometimes just a thin line!

Picture 1 The Earth seen from space

The shape of the Earth

The fact that the Earth is a sphere has been known for thousands of years. Even so, there are still people who don't believe it! Christopher Columbus knew that the Earth was round and sailed west to get around to India, in the East. But his sailors weren't so sure, and were very worried that their ship would fall of the edge of the world.

The simplest evidence for the Earth being a sphere is that we now have photographs of it, taken from space (picture 1). The Ancient Greeks were the first people to realise that the Earth was round. They proved it to themselves by noticing such things as:

■ when ships sail away from land you first see the hull disappear, then the masts.
■ during eclipses of the moon, the **shadow** of the Earth is round.

The shadow on the Earth is cast onto the moon during eclipses. It is always a circle, which it wouldn't be if the Earth were shaped like a flat disc. (See picture 2.)

How big is the Earth?

The first reasonably accurate measurement of the radius of the Earth was done by a Greek-Egyptian astronomer, called Eratosthenes. Both Greeks and Egyptians were good at geometry, and in about 200 BC Eratosthenes used it to calculate the size of the Earth.

How did he do it?

It so happened that there was a famous well in southern Egypt at a town called Syene (now called Aswan). The well was famous because on just one day of the year, when you looked down into the well, you could see the reflection of the Sun in the water. This was because the sun was directly overhead. The day it happened was midsummer day, when the Sun is at its most northerly position in the sky.

Eratosthenes realised that if he measured the angle the sun made with the ground on the same day, further north in Alexandria, he could calculate the circumference of the Earth. The only other information he needed was provided for him by the army, whose soldiers had been trained to march at a steady exact pace. Picture 3 explains how he did it.

Astronomers have used much the same method until quite recently, but with more complicated astronomy and more accurate instruments. Nowadays much more accurate measurements can be made using instruments carried by Earth satellites.

How old is the Earth?

A bishop once used evidence in the Bible to work out that the Earth was made in the year 4004 BC. But, over the last two centuries, scientists have collected evidence from rocks that suggests that the Earth may be considerably older.

Counting the layers

In some newer rocks it is possible to distinguish lighter and darker layers, laid down in different seasons of the year. So the layers can be counted just like the rings on a tree. Picture 4 shows a rock formed in a Tyneside coal mine during the 1800s. A white mineral, barium sulphate, settled out in a water trough. During working days it was blackened by coal. So, you can see how many days were worked during the week, and count the number of weeks. The rock is artificial, of course. Natural banded rocks like this do occur. But they are very rare and very young compared to the age of the Earth. We couldn't date the whole Earth with them.

How we can tell if one rock is older than another.

We can use common sense. The positions of different rocks in the field gives plenty of evidence about the order in which things happened. Look at picture 5. In a pile of sedimentary rocks, the ones at the bottom must be the oldest — sediment is always added to the top.

The fragments in any sedimentary rock must be older than the rock itself, so pebbles in a conglomerate are obviously older than the rock that contains them.

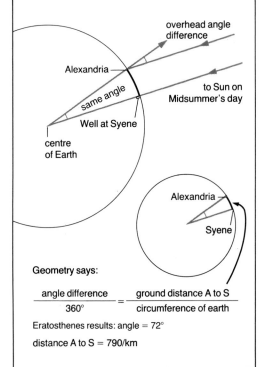

Picture 3 The actual circumference of the Earth is 40 000 km. How near did Eratosthenes get to this?

Geometry says:

$$\frac{\text{angle difference}}{360°} = \frac{\text{ground distance A to S}}{\text{circumference of earth}}$$

Eratosthenes results: angle = 72°

distance A to S = 790/km

Picture 4 The Sunday stone was formed in a Tyneside coal mine in the 1800s

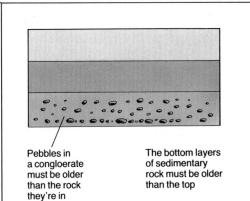

Pebbles in a congloerate must be older than the rock they're in

The bottom layers of sedimentary rock must be older than the top

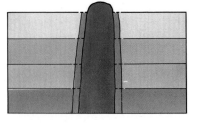

The sedimentary rocks must be older than the igneous rocks

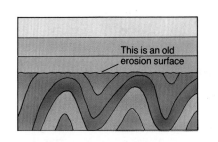

This is an old erosion surface

The lower beds were once flat layers. Then they were squashed and folded. The upper layers were later deposited on top.

Picture 5 How we can tell the age of rocks by looking at them

An igneous rock must be younger than the rocks into which it has intruded.

If unfolded beds of rock lie on top of folded ones, then the unfolded ones must be a lot younger than the folded one. Think of what must have happened. Sometime after being deposited and turned into solid rock, the older rocks were squashed and folded. Then they were eroded, and then, **after that**, the younger set of rocks was laid down.

Using fossils

The remains of living organisms found in the rocks also give information about which rocks are older and which are younger. (More information about fossils is given in *The Living World*, topic F8.) If we know the time when certain fossils lived, using evidence from one kind of rock, then we can use those fossils to date other rocks that contain the same fossils.

The methods described so far enabled the geologists of the nineteenth century to unravel the story of the Earth. But they could only work out the order in which things happened. They gave names to the various stages and events in the Earth's history (see the time chart page 221). But no-one was able to give reliable dates to these events. It was not until the early part of this century that scientists discovered that **radioactive elements** could be used to date rocks.

Radioactive dating

Radioactive elements decay into lighter elements. They decay at a known rate. So, using a special machine called a mass spectrometer, scientists can measure how much of the original radioactive element is left and how much of the new, lighter elements are present. The ratio of the two gives a fairly accurate estimate of the age of the rock. (See topic D5.)

The oldest rock so far found in the world, was found in a remote region of northern Canada and is 3.96 billion years old. It is of sedimentary origin. (Why does this indicate that it couldn't have been the first rock in the Earth's crust?) (Picture 6.)

Radioactive dating has been used to measure the age of meteorites and moonrocks. These measurements suggest that the whole Solar System, Earth, meteorites and the Moon were formed at the same time — 4.6 billion years ago. For the first 4 billion of those years, there was no life as we know it, and people have only been on the Earth for the last 0.003 billion years. We shall learn more about how the Earth was made in topic I3.

Picture 6 Samuel Bowring of Washington University, St Louis, USA, with the oldest rock so far discovered. It is estimated to be 3.96 billion years old

Questions

1 Why must the pebbles in a rock be older than the rock itself?

2a Explain what you understand by *radioactive half life*.

b Uranium 238 has a half life of 4.5 billion years (4.5×10 years). By a remarkable coincidence, the Earth is believed to be about 4.5 billion years old. How much of the original uranium 238 is still on Earth, unchanged?

c Uranium decays, eventually, to a stable form of lead (lead 206). A rock sample was analysed, which showed that the proportion of uranium 238 atoms to lead 206 atoms was 3 to 1.

i) What fraction of the uranium had decayed?

ii) Explain why a reasonable estimate of the age of this rock is about 2.25 billion years.

iii) Would you expect to find a rock in which the ratio of uranium to lead was, say, 1 to 3? Explain your answer.

3 Why is radioactive dating quite accurate for igneous rocks, but is no use for sedimentary rocks?

4 A rock specimen is guessed to be about 100 million years old. Which of the following radioactive elements found in it would be most useful in measuring its age by radioactive dating? Give a reason for your answer.

Element	Half life
uranium 238	4.5 billion years
radium 226	1620 years
lead 205	50 million years
thorium 230	80 000 years

The history of the Earth: a geological time-chart

Now

CENOZOIC				
The Earth's climate became colder, resulting in Ice Ages.	QUATERNARY	Holocene	0.01	— Early Egyptian cities, etc
		Pleistocene	2	Many ice Ages Humans migrate over Asia
This is the age of the mammals, but also of insects and flowering plants	TERTIARY	Pliocene	7	— First human beings (Africa)
		Miocene	26	India meets Asia, Himalayas form First deer, monkeys, dogs, cats Africa, Europe and Asia collide, Alps form
		Oligocene	38	
		Eocene	54	First rodents, elephants, horses
Opening of North Sea		Paleocene	65	Extinction of dinosaurs

MESOZOIC			
This was the age of the dinosaurs	CRETACEOUS	136	— Final break-up of Pangaea Sea flooded to cover more land, Formation of chalk over much of what is now Europe First flowering plants and modern types of insect
The great southern continent of Pangaea broke up, forming most of our modern continents	JURASSIC	190	Beginning of the Atlantic Ocean First birds and mammals
The Earth's climate was warm and pleasant almost everywhere	TRIASSIC	225	Many new species Sea levels fall all over the world

PALEOZOIC			
At the beginning, most life was in the sea. Plants colonised the land in the Silurian era, followed after a few million years by amphibians. Towards the end of this period the first reptiles appeared, as land animals.	PERMIAN	280	Rise of the reptiles Formation of Pangaea
	CARBONIFEROUS	355	Britain at the Equator; coal laid down in many parts of the world. Age of amphibians. First reptiles.
	DEVONIAN	395	Age of the Amphibians First fish, first flying insects Animal life moves on to the land
	SILURIAN	440	Plants move on to the land
	ORDOVICAN	500	sea covers many continents most life in the sea, as shellfish and floating plants
	CAMBRIAN	570	

PRECAMBRIAN			
This covered a huge period of time – over 4000 million years. It began with the formation of the Earth and ended with the first many-celled organisms The first green plants appeared in the sea – and began to put oxygen into the atmosphere	PROTEROZOIC	2500	— First multicellular organisms
	ARCHAEAN	4600	— ? First free oxygen in atmosphere — ? First living organisms — single celled plants and animals — 3960 — Age of oldest known rocks Earth cooling down; formation of tectonic plates

Activities

A The geological history of your area

See if you can collect enough information to write a short geological history of your area.

1 Find a geological map of Britain in an atlas in your library. Use it to find out the rough ages of the rocks in your area.

2 Try to find out if there is a collection of fossils available nearby, perhaps in a local museum, or even in your school. Find out if any of these have been useful in dating the rocks nearby.

3 If there are igneous or metamorphic rocks nearby, try to work out if they are older, younger, or the same age than any sedimentary rocks.

See if you can collect enough information to write a short geological history of your area. (You may find the time chart, page 221 useful for this.)

B Be a geological detective

(You may prefer to work in groups for this.) Picture 7 shows the rocks exposed in two cliffs, 200 km apart.

1 Try to work out the order of the ages of the labelled rocks in each cliff.

2 Looking at the fossils, decide which beds in the two cliffs are the same age.

3 For both cliffs together, try to work out the order of the rocks labelled (a) to (k).

4 How do you think the skills of being able to work out ages of rocks in different areas helps the geologist looking for coal or oil?

The granite was aged by radioactive dating at 300 million years old, and the basalt at 50 million years old. Does that agree with your findings? Look at *The Living World*, topic F8, picture 4, and see if the life forms shown as fossils tie in with these ages.

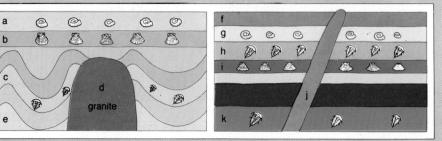

Picture 7

The disaster at Armero

Read the following newspaper article (adapted from the Sunday Times 17 November 1985), and answer the questions which follow:

'A town of 25,000 is gone . . . there is no Armero'

For much of the way on the road to Armero in Columbia, South America there is no hint of the horror that awaits you. Clouds hide the volcano that has erupted. But, the road ends, cut off by mud, about five kilometres from Armero and you have to take to the fields. As you make your way into the hills, you pass a line of refugees coming down. For the most part they are silent, although some are crying. The first man to talk to us has no shoes 'All I have is my trousers' he says, in a daze. All his family lie dead.

As you get nearer to Armero, you see the force of the mud slides that have swept across the fields like some prehistoric sea. A town of 25 000 has gone. In its place is a small pile of junk scattered over a sea of slime.

The driving force of the Nevado del Ruiz eruption was the grinding movement of geological plates that form the Earth's crust. Previously, the site of the tragedy had been a tranquil tree-lined valley. But tremors from the volcano, culminating in a huge explosion, disturbed ash laid down in prehistoric times on both sides of the valley's steep slopes. Combined with snow, melted by the heat of the volcano, and torrential rain, the ash was turned into a highly unstable mud, which began sliding with unstoppable force down the valley towards the town of Armero. It was like the tragedy of Aberfan in Wales, where coal slurry from an old tip buried a village in 1966, except that the Columbian mud avalanche was a natural disaster, not human-made.

The volcano began showing signs of life almost a year ago and has been making noises for months. In September it rumbled enough to leave a coating of ash on towns nearby. There had been a small eruption the day before. But in the early evening villagers had listened to government officials on their local radio station reassure them that there was nothing to worry about.

A family of ten took the advice. Most of them stayed at home to watch a soap opera on the TV, and when they heard the volcano they kept on watching. When the mud drove through Armero, their two rooms collapsed into one. Six escaped onto the roof and were rescued after two days. But four, including three children, died because the family thought they were safe from the volcano.

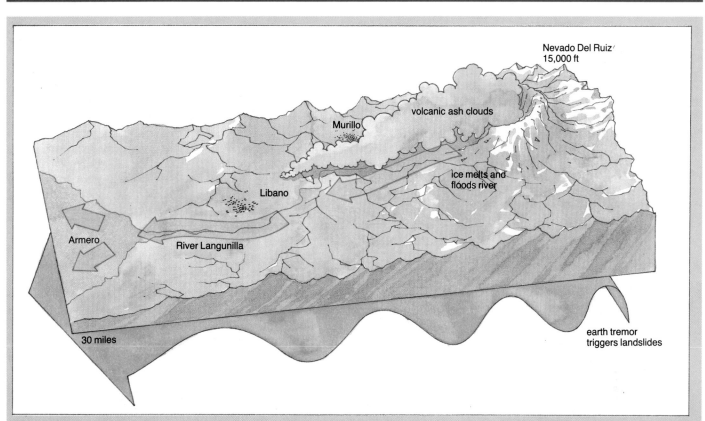

Picture 1 The eruption of the Nevado del Ruix volcano at Armero, Columbia 1985

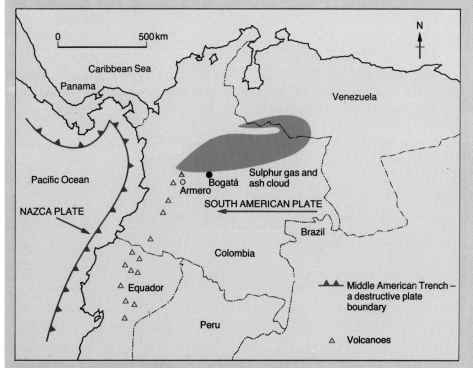

Picture 2 Map of the area showing the factors which led to the disaster.

Now answer these questions.

1 What kind of plate boundary is shown in picture 2? Draw a diagram to show this plate boundary and the formation of the volcano.

2 From the following list of volcanic hazards, select the one(s) you think caused the loss of life at Armero:

Poisonous gases
sticky lava
melted snow
ash and water mixed to make mud
volcani bombs (lumps of solidified lava hurled from the volcano)
runny lava
clouds of ash
fire

3 There was very little loss of life in the 1980 Mt St Helen's eruption. Do you think loss of life could have been avoided at Armero? Explain your answer.

4 Work in groups. Imagine you have just been appointed as a scientific advisory team to the Colombian Government. Write a short report with the following headings: 'Can a disaster like this be prevented?'; 'Can a disaster like this be predicted?'; 'How should Colombians adapt to living with the threat of volcanic eruption?'

5 Find out what you can about two of the following volcanoes, using your school library. How does the damage they have caused to people and to property compare with Armero?

Krakatoa; Mt Etna; Mt Pelee, Martinique; Paracutin, Mexico; Kilimanjaro, Tanzania; Santorini, Greece; The Canary Isles; Mt St Helens, USA; Vesuvius, Italy; Mt Hekla, Iceland.

G6
But they didn't believe him

The story of Alfred Wegener.

Picture 1 Alfred Wegener

A brand new theory

The trouble with scientists is that the more they know, the harder it is for them to change their minds. Of course, other people are like this as well. But now and again someone comes along and turns the scientists' world on its head. Quite often, the person who does this is an outsider. The world of the geologist — the 'earth-scientist' — was overturned by a weather man.

The biggest change in our understanding of the geology of Earth was made by a man called Alfred Wegener (picture 1). He was born in Berlin, Germany, in 1880 and studied meteorology and astronomy. He worked in universities in Germany and Austria, but he took part in expeditions to Greenland. It was on one of these expeditions that he died, in 1930, twenty years before his ideas became accepted.

It's just a coincidence

If you look at a map of the world (picture 2) you can see that the outline of South America looks as if it would fit quite neatly on to the west coast of Africa. Most geographers and geologists thought that this was just a coincidence. No one could imagine that such huge objects, fixed on the solid Earth, could move sideways for thousands of kilometres. But Wegener had the idea that the continents *did* move and had been joined together at one time.

There were no known forces big enough to move them. The most that could happen, they thought was that parts of continents might move, very slowly, up and down to make mountains. The continents were fixed in solid basalt rock, and there seemed no way that they could move.

The arguments

Take a closer look, said Wegener, *there's a chain of mountains in South Africa that exactly match a chain in Brazil. They were made at the same time, and have the same rocks.*

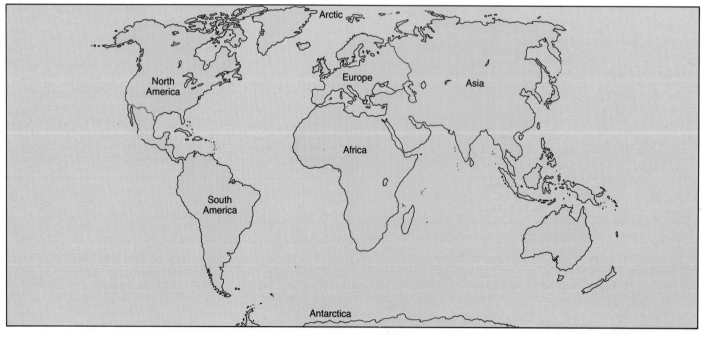

Picture 2 Do the continents fit together?

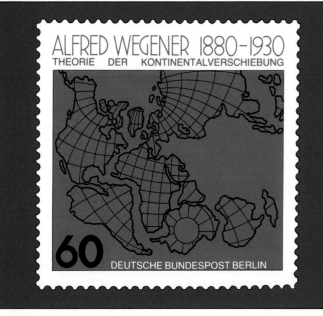

Picture 3 Wegener argued that South America must have been joined to Africa like this

Picture 4 A shrinking Earth could make mountains, like on the surface of a wrinkled apple

But it's not an exact match, the experts said.

Look at mountains — like the Himalayas, Wegener pleaded. *The rocks are folded so much that they must have been pushed sideways.*

Now that is interesting, said the geologists. But the reason, of course is that the Earth is shrinking. It is losing heat, cooling down. Mountains rise, rocks are squeezed because the Earth's surface is getting smaller. No problem. It's just the sort of thing that happens with apples. (Picture 4.)

Wegener pressed on. *Why is there coal in Siberia?* he asked. *Coal is formed from trees, so Siberia must have been closer to the equator at some time. It has drifted north since then.*

Well — climates can change, they said. And maybe trees in cold climates can sometimes grow well enough to form coal.

But look at the animals — and their fossilised ancestors. South American animals are different from African ones. But their fossilised ancestors are almost exactly the same. Their ancestors must have lived on the same continent. And it must have split apart 60 million years ago.

Not at all! There was a thin land bridge connecting Africa with America so that animals could move across. In fact, it must have broken up 60 million years ago. (See picture 5.)

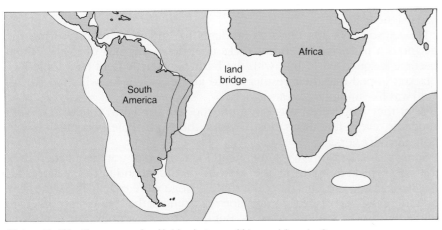

Picture 5 Was there once a land bridge between Africa and America?

Picture 6 The cover of the first edition of Alfred Wegener's book, *The Origin of Continents and Oceans*

Wegener didn't accept the geologists' arguments. He thought that he was right. He collected together all his reasons and evidence and published them in his book *The Origin of Continents and Oceans* (1924) — see picture 6. But the evidence wasn't powerful enough to prove his case. The world of geologists was against him.

How do new theories get accepted?

This is a typical example of what happens in a scientific revolution. At the time, there was no real proof either way. The evidence, as they say, was **circumstantial**. And the man wasn't even a **geologist**. The 'professional' view won the argument.

The moving force

Detectives need to find a motive, a moving force, for a criminal act. What was the moving force that made continents drift?

Wegener didn't know, but just before he died an English geologist put forward a theory for this. Arthur Holmes suggested that they could be moved by convection currents in the hot mantle (see topic G2). Direct evidence for this came in the late 1940s. Advanced echo-sounders developed for submarine warfare were now used to map the ocean floors. They charted the underwater landscape.

The oceans, like dry land, had mountains and ridges — and volcanoes. But it doesn't rain under the sea! So the mountains don't get worn down. The Atlantic, in particular, had some strange features. In the middle of the Atlantic is a long line of volcanoes. Some are big enough to reach the surface — in Iceland and the Azores. Either side of this line is a pattern of ridges and hollows. *And the pattern is the same each side*. See picture 6 in topic G1.

The rocks in each opposite part of the pattern are the same. They are volcanic lava with an overlay of thin sediments from the sea above. It looks exactly as if the bottom of the Atlantic ocean was being made continuously, on a production line. Like two rolls of cloth rolling up from the central volcanic slit and spreading out sideways. Could the middle of the Atlantic be the very top of one of Arthur Holmes' convection currents? (Picture 7 in topic G1.)

The biggest tape-recording in the world

The final proof that Wegener was right came from a theory in astronomy — which turned out to be wrong. In 1950 Professor Patrick Blackett was investigating a theory that would explain why some stars — and planets — had magnetic fields.

He set some of his research students to find ways of measuring the magnetism in rocks. When hot rocks cool down any bits of iron in them become magnetised in line with the Earth's magnetic field (see topic E2). Blackett hoped that their results would give evidence for what the Earth's field was like millions of years ago.

What they discovered was quite unexpected: the Earth's field *flips*. Very suddenly, every few thousands of years, the North Pole becomes a south pole, and vice versa.

Some years later, in 1963, someone thought of looking at the magnetism of the rocks picked up from the bottom of the Atlantic. The patterns of magnetic changes each side of the middle ridge were exactly the same. The bottom of the sea was like a huge tape recording of the changing magnetism of the Earth over millions of years. But the interesting point was that it was the same record each side of the control ridge.

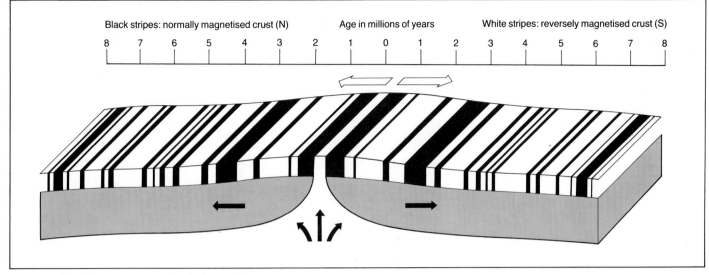

Black stripes: normally magnetised crust (N)   **Age in millions of years**   **White stripes: reversely magnetised crust (S)**

Picture 7   The biggest tape recording in the world

Imagine that you had a large enough cassette player to play these 'tapes'. Each side would play the same 'tune'. (See picture 7.)

This was the final proof that the rocks, now thousands of kilometres apart, had been formed at the same time. Their magnetism had been fossilized into them as they flowed out of the central volcanic vent into the cold sea. But in the course of many millions of years the new rocks had been pushed further and further away from each other. As they moved they carried the continents of Africa and South America with them.

Plate tectonics

Wegener's theory of 'continental drift' wasn't quite right. The continents do not move on their own. It is the huge 'plates' they are fixed to that move. But Wegener was more right than wrong, and his theories have become the science of *plate tectonics*. This science, and its importance, is dealt with in topic G2.

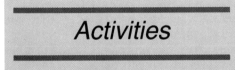

Activities

A Putting a case

Divide into groups of about five or six people. Find out a little more about the theories discussed in this topic. Arrange a debate, with Alfred Wegener and Arthur Holmes on one side, and the geologists who opposed them on the other.

Do continents drift? Let the best arguers win!

B Sharp-eyed research

Look at a good map of the world — or a globe. Can you find any other parts of then world that might once have fitted together like Africa and South America?

Questions

1 What do the following scientists study:

a geologists, b astronomers,

c meteorologists?

2 What piece of evidence in favour of Wegener's theory do you find most convincing?

3 Make two lists, headed

A Points for Wegener's theory,

B Points against Wegener's theory

Find three facts or arguments to put under each heading.

4 Explain how the research into rock magnetism gave good evidence for the theory of continental movements.

5 Explain what the following are:

(a) mountain chain, (b) convection currents, (c) a magnetic field, (d) a land bridge, (e) a fossil.

H1
Why worry about the weather?

It is not easy to predict the weather — but it is very important to try.

Picture 2 Butterflies don't have much effect on the weather — but large changes in the way weather patterns develop can be triggered off by tiny, unpredictable changes in pressure and temperature

The weather is caused by movements of huge blocks of air. Some of the blocks are warm, some cold, some are dry, some damp. To know in advance what the weather will be like tomorrow means somehow calculating where these blocks will be, and whether or not they have changed.

The air is made up of molecules. There are millions and millions of them, moving with an average speed of 500 metres per second. We know the laws of physics that decide what each molecule will do if you give it more energy, or if it collides with other molecules. But there are just too many of them for even the largest computers in the world to apply the laws and work out what happens next — even if they could all work together.

Butterflies and hurricanes

Now suppose the weather forecasters (called **meteorologists**) *could* work it all out, and *did* predict the weather a week ahead. But then, just as everything was happening as predicted, this butterfly comes along. It flaps its wings, and some molecules go somewhere else. They waft over a leaf and carry off some extra water molecules. This cools the leaf a little, and affects the air next to it, causing a draft. As result, the very light breeze that is blowing might change direction. It affects a stronger wind on the other side of the hill. In turn, this changes the wind pattern of the district.

At the end of a long and complicated chain of events a hurricane that has been growing in the South China Sea moves in a slightly different path. It hits Hong Kong instead of a piece of empty coastline, and affects many hundreds of people.

It's not fair to blame the butterfly for this — but the fact is that large changes in the weather can be triggered off by very tiny changes of temperature or pressure. And these small changes are unpredictable.

Topic H2 deals with how we can make rough predictions about the weather, using the idea of *weather patterns*.

Why is weather forecasting important?

It costs £10 000 to spread a mixture of salt and grit on all the roads of the county of Suffolk in England. From the end of November to the beginning of April the county Highways Department has teams of drivers ready and waiting, on call at home, to collect the gritting lorries and start work. They

Picture 1 This is what a tornado looks like. It can crush houses and uproot trees that fall in its path

Picture 3 Snow seriously affects transport

Picture 4 Every year fog or mist causes serious road accidents

work. They will be called out if the Meteorological Office forecasts heavy frost and road icing.

If the forecast is wrong, then either the County wastes £10 000 of their taxpayers' money, or there will be many accidents next morning as cars crash and collide on the slippery roads.

Farmers also need to know what the next day or night's weather will be like, to help them plan their work. Ships need to be given storm warnings, aircraft and airports need to know if snow or fog is likely to make landing difficult or impossible. Unexpected fog can cause accidents on motorways. Some of these weather hazards are shown in pictures 3 and 4.

On the evening of October 17th 1987 the weather forecasters warned of strong winds, even at gale force, for the South of England. They got it wrong. The winds almost reached hurricane strength, felled thousands of trees and caused millions of pounds worth of damage (picture 5). Most of the damage was unavoidable — but had the forecast been accurate the emergency services would have been better prepared, and some precautions could have been taken. Perhaps not so many cars would have been parked under trees.

The winter gales of 1989/90 were almost as strong, but the forecasters had learned some lessons and gave clearer warnings.

Picture 5 Britain has suffered seriously from high winds recently. This picture shows storm damage caused in the hurricane of October 1987

Activities

A   Checking the weather forecasts

Keep a record of the weather forecasts for a week. Collect them from a daily newspaper, or video the TV forecasts. How accurate were they, as the weather developed day by day? Was there a special reason for any poor forecasts?

B   The Great Storm of 1987

Read what you can about the storm of October 1987. Explain how the forecasters got it wrong.

C   Remembering the weather

Carry out a survey of what you parents, grandparents, aunts, uncles and friendly neighbours remember about any unusual weather they have lived through. Do they remember storms? Floods? Snowfalls? Hot sunny weather? Use their stories to make a time chart of The Most Interesting Weather of the Last 30 (or more) Years. Do they think that weather has got better or worse recently? How reliable is their evidence? You could use a library to check.

Questions

1   Give two reasons why it is hard to make accurate weather forecasts.

2   Name tour groups of people who find it very useful to know what the weather will be like during the next 24 hours. For each group write a sentence or two explaining why they would find the information useful.

3   The laws of physics are accurate enough to be used to predict exactly where the Voyager spacecraft would be after 14 years of space travel. Why can't the same laws be used to predict the weather?

4   Describe any example of unusual weather you have experienced. What effect did it have on your life, or the region where you live?

H2 Weather patterns

Weather seems to come in patterns in large, global systems of air flow. This topic explains how these affect our daily weather.

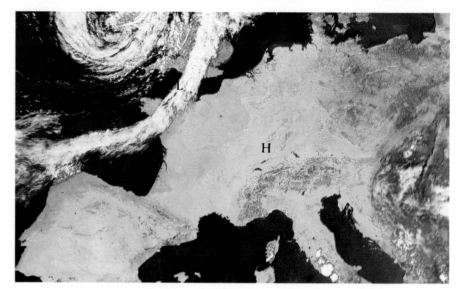

Picture 1 A satellite picture of the low shown on the map in **Picture 5**. Can you see the different cloud types named in pictures 2 and 3 and described in topic H3?

Lows and highs

Picture 1 was taken by a weather satellite and shows clouds over Britain and the North Atlantic Ocean. You can see pictures like this every day on TV. The clouds (in the part marked L) look like cream that has just been stirred in a cup of coffee. 'H' is a region free of clouds, and most of Europe is enjoying warm sunny weather.

Lows

In fact, just like a stirred cup of coffee, the air and the clouds in region L are indeed swirling around. This pattern is called a **low**, because it is a region of low air pressure. The air in it is warm. This means that it has expanded so is less dense than colder air. The whole block of swirling air weighs less than the same sized block of normal air would. The result is that it exerts less force on the ground under it. Pressure is *force per unit area*, so the air pressure in the low is less.

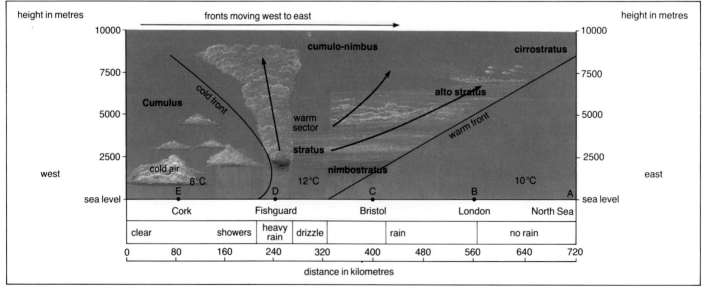

Picture 2 Cross section of a typical low

Clouds

Warm air evaporates water from the sea. This **water vapour** turns back to drops of water (or **condenses**) when it gets colder. It is these tiny drops of water that we see as **clouds**.

This condensation can happen if the air mass rises and moves into a cooler region of the atmosphere — for example, when the moving air reaches mountains and rises to flow over them. It is this effect that makes mountain regions like Wales and the Lake District cloudier and wetter than, say, the flat, eastern parts of Britain.

It can also happen at the edge of the warm air, where it touches and mixes with the colder air surrounding it. This effect causes the clouds that can be seen in picture 1, and the effect is illustrated in picture 2.

Picture 3 Thin, very high clouds, called cirrus or 'mares' tails. The curled 'tails' are caused by strong high level winds, probably the jet stream

Rain

Many of the clouds in the satellite picture are likely to be **rain clouds**.

Not all clouds are rain clouds. When water vapour condenses it form very tiny drops, which are too small to fall as rain. They are held up by the air movements. But if there are enough of them they tend to grow bigger and bigger by joining together. It actually takes about a million 'cloud drops' to form a drop big enough to fall as rain. This means that we don't get rain from every cloud in the sky.

Warm and cold fronts

The boundary between a warm block of air and the colder air surrounding it is called a **front**. The low is moving to the east, and a person on the ground would see the clouds coming along, first high in the sky (picture 3), then lower and thicker and giving rain (picture 4).

Once the boundary is past, the air is warmer. But you would probably only notice this in winter. This boundary is therefore called a **warm front.**

It might not be raining in the middle of the low, but the warm air is damp and it feels muggy or close. Then the low moves away and its boundary passes over us again, with more rain, often heavy.

We are now in colder air, and have passed through the **cold front**. The skies clear, although there may be showers. Then the weather becomes more stable until the next warm front comes along.

Picture 5 shows the weather map for the day shown in picture 1. You should be able to pick out the different parts mentioned above.

Picture 4 Typical rain clouds — nimbostratus

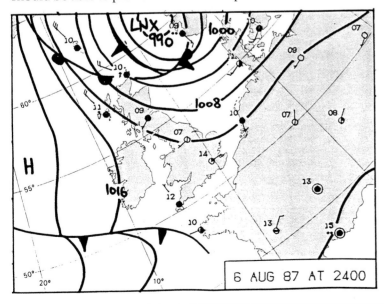

Picture 5 A weather map showing the low in picture 1

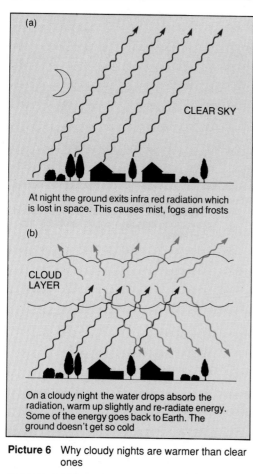

(a)

CLEAR SKY

At night the ground exits infra red radiation which is lost in space. This causes mist, fogs and frosts

(b)

CLOUD LAYER

On a cloudy night the water drops absorb the radiation, warm up slightly and re-radiate energy. Some of the energy goes back to Earth. The ground doesn't get so cold

Picture 6 Why cloudy nights are warmer than clear ones

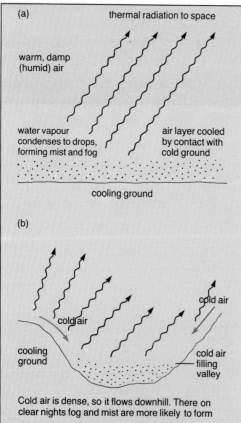

(a)

thermal radiation to space

warm, damp (humid) air

water vapour condenses to drops, forming mist and fog

air layer cooled by contact with cold ground

cooling ground

(b)

cold air

cold air

cooling ground

cold air filling valley

Cold air is dense, so it flows downhill. There on clear nights fog and mist are more likely to form

Picture 7 How fog and mist are formed

The atmospheric pressure is measured by a **barometer**, in units called **millibars** (mb). Normal atmospheric pressure is about 1000 mb. In a normal low the pressure might drop to about 975 mb. The lines on the map are called **isobars**, and are drawn through places at the same atmospheric pressure.

In hurricanes and tornadoes the pressure at the centre is much lower. This not only means stronger winds, but the difference in pressure between the normal pressure inside a building and sudden low pressure outside it can make the building explode!

Highs

Highs are caused by blocks of denser air in the upper atmosphere. Because the air is dense it weighs more, and tends to sink through less dense air. It causes a **high pressure** region. There is very little water vapour in it, and so there is hardly any cloud. It lets sunlight through and if it is summer we enjoy hot sunny weather.

In autumn and winter highs give clear crisp days, with frost in the mornings, before the sun rises to melt it. The frost comes from what water vapour there is in the air. This has probably evaporated from the land during the day.

Because there is no cloud the land gets much colder at night, and frost is more likely to form. Clouds act like a 'blanket', absorbing the thermal radiation from the ground and re-radiating it back. The ground doesn't lose as much energy as it would if the sky was clear. See picture 6.

Fog and mist

In autumn the ground is wet and can be quite warm. We may get light frosts, but the real danger is from fog. There is always more water vapour near the ground than high in the sky, and as the ground cools under a clear sky the water vapour condenses to form millions of tiny drops floating just above ground level. This is **fog** or **mist**. Picture 7 shows how fog and mist are formed.

When the cold water drops touch the even colder ground, or cars and trees, they might freeze to form a **freezing fog**. The combination of fog and icy roads is very dangerous for road traffic.

Winds

Winds are movements of air which travels from regions of high pressure to regions of low pressure. This is like water flowing downhill.

But just as water doesn't move perfectly smoothly, neither does air. When you pull the plug out of your bath the water doesn't flow straight down the plughole. It swirls around to make a small whirlpool. The air flows in a similar pattern as it moves from high pressure to low pressure.

In the northern half of the Earth the winds swirl anticlockwise into a low. They swirl clockwise out of a high — but usually quite gently. The result is that when you stand with your back to the wind, the low pressure region is always to your left, provided you are north of the Equator.

The changing wind directions follow a pattern as the low comes along. As the low reaches us in Britain the winds swing round to come from the south west. In the middle of the low the winds are usually very light. Then as the low moves away and the cold front passes over us the winds go back to the north west. These changing wind directions are shown in picture 5.

The opposite effects are seen for lows below the Equator, in places like Australia and New Zealand. This suggests that the swirling is something to do with the spinning of the Earth. See topic H4 for more about this.

Breezes and gales

The speed of the wind is decided by how big the pressure differences are. Very 'deep' lows have much lower pressure than the surrounding air, and so the winds can be very strong, reaching 'gale force'.

You can tell a deep low on a weather map because the isobars are closer together. This is because the pressure changes so much across the low. This is just like the contour lines that show height on an ordinary map.

Activities

A Weather maps

Find a good weather map in a book or a newspaper. It will use various symbols to show different weather features. Find and draw the symbols for the following features:

1 cloud,   4 an isobar
2 rain   5 a cold front
3 snow   6 a warm front.

B How fast does the weather change?

Much of the weather we get in Britain is caused by the movement of lows and highs across the country. Keep a record of the weather maps for a week or so.

1 Choose a particular low — which might have a letter attached to it on the map — and measure how far it travels in a given time. It might move away in a day or two or take a week. You can use an atlas with a mile or kilometre scale on it to find the distance it travels. Calculate its speed in miles or kilometres per hour.

2 Do the same calculation for a high. Do highs and lows travel at about the same speed?

C Why don't blocks of warm and cold air mix?

You might expect warm and cold air to mix together, or at least that the energy from one flows into the other. But the warm lows seem to last for quite a long time before they mix with a high. This activity is about investigating this effect with water.

Picture 8 shows you what to do. Part A is a model of a cold front, Part B models a warm front. Do both experiments and describe what happens.

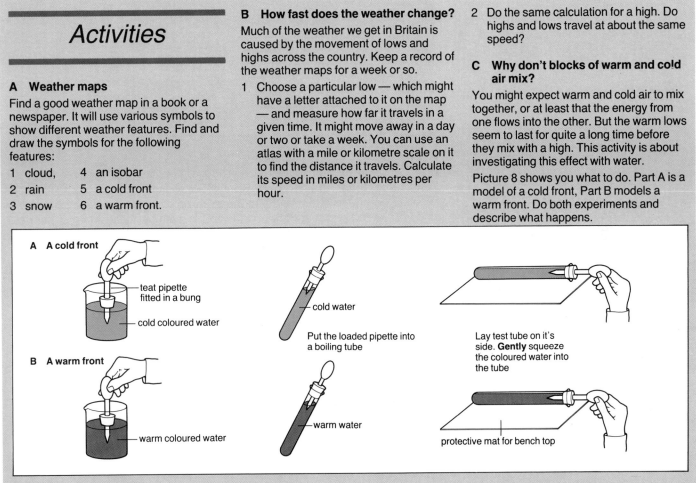

A A cold front
— teat pipette fitted in a bung
— cold coloured water

B A warm front
— warm coloured water

— cold water
Put the loaded pipette into a boiling tube

— warm water

Lay test tube on it's side. **Gently** squeeze the coloured water into the tube

protective mat for bench top

Picture 8

Questions

1 Explain what the following words or phrases mean:

a a low,   b a high,
c an isobar,   d a warm front.

2 Explain why warm air is likely to have more water vapour in it than cold air.

3 The table below shows the pressure readings taken at a weather station over a period of 24 hours. Picture 2 may help you answer some of the questions.

a Plot a graph of pressure against time.

b What do you think the weather was like (dry, rainy, cold, warm) at the following times:
i) 9 am, ii) 1 pm, iii) 4 pm, iv) 8 pm.

c At what time of day would you expect to see: i) thin, high clouds, ii) rain clouds, iii) clear sky?

4 What are clouds made of? Why doesn't it always rain when there are clouds overhead?

Time	9.00	10.00	11.00	12.00	13.00	14.00
Pressure (mb)	1000	995	990	985	980	975
Time	15.00	16.00	17.00	18.00	19.00	20.00
Pressure (mb)	970	970	980	995	1005	1005

H3 Predicting weather from patterns

Red sky at night, shepherds' delight Red sky in the morning, shepherds' warning.

Picture 1   A red sky in the evening. Will it be fine next day?

Shepherds, farmers and sailors learned to forecast the weather from the patterns of clouds and winds. In Britain our weather systems come from the west, across the North Atlantic. In the evening when the sun is setting low in the west its red rays will light up any clouds over our heads — see picture 2. But only if there aren't any clouds out there in the Atlantic! So the chances are that there aren't any rain clouds on the way.

In the morning the rising sun will show up clouds far to the west — which might be bringing rain. The shepherds are warned!

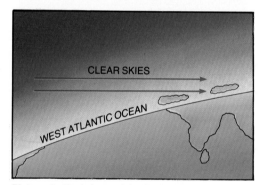

Picture 2   Red sky at night. Red light from the setting sun lights up overhead clouds — but only if the western sky is clear of clouds.

Weather measurements

Clouds are a useful sign of what the weather will be like, but they don't tell us everything. Weather forecasters use instruments like:

- **barometers**, which measure air pressure,
- **thermometers**, to measure the air temperature,
- **hygrometers** to measure how much water vapour the air contains,
- **anemometers** to measure wind speed.

Weather data

Weather forecasters need data. They need to get measurements from as large an area as possible, and as often as possible. In the British Isles the weather can change very rapidly.

Aircraft and ships at sea take measurements far out in the Atlantic. On land there are official weather stations that take measurements, but there are also lots of ordinary people who take measurements as a hobby. These measurements are reported every few hours to the Meteorological Office (see picture 3).

Now and again special weather balloons are sent off (picture 4). These take measurements at definite, preset heights, and the readings are sent back to ground by radio. Our weather is strongly influenced by winds high above the Earth that most of us don't even know about. (This will be explained in the next topic.) Information from these balloons helps to make 'long range' weather forecasts, for up to seven days ahead.

All these measurements are fed into a very large computer and weather maps are made several times a day. From these, the experts produce weather forecasts every hour, for the next six hours. These forecasts are broadcast to aircraft and ships at sea.

Picture 3   Local weather stations can provide useful data for weather forecasters

Weather satellites

Weather satellites take infra red photographs of the Earth beneath them, and these show up clouds and regions of warmer or cooler air. They add extra information, particularly about places far out to sea where there aren't any ships or aircraft.

Picture 4 These balloons can reach great heights. They carry instruments to measure temperature, pressure and wind speed. The readings are sent back to ground by radio.

Looking at clouds

Clouds give very good clues about the type of weather we can expect in the next few hours. Page 230 shows a section through a typical low with different cloud types marked on it. Cloud types are given Latin names: **cirrus** — a curl; **stratus** meaning flat; **nimbus** — a rain cloud; **cumulus** — a pile; **alto** — meaning high.

What clouds do you see as the low passes? Cirrus clouds come first. They are thin, feathery clouds which are usually high in the sky (see page 231). You see them at the leading edge of a warm front and so they are a good sign of rain approaching in a few hours.

Cirrus clouds may thicken to form a thin high layer of cloud called **cirrostratus** — see picture 5. The blue sky becomes a paler blue, as if it is covered by a white veil. The Sun becomes weaker and you might see a halo around it. These clouds are often found in highs but may also form on the leading edge of a low. They are made up of ice crystals and drops of supercooled water.

Cirrocumulus are high thin lumpy clouds. They usually come in a large group with spaces in between, producing what is called a *mackerel sky* (picture 6). They show that the low is now nearer, and they may then join together to form a thicker cloud layer — **altostratus**.

The warm front of the low is now very near and the wind direction changes ('backs') to blow from the southwest as the low level rain clouds race across the sky. These are the **nimbostratus** clouds. We are now in warm tropical air, and the clouds are low and thick (**stratus** clouds — see page 231). It may stop raining or just drizzle. As the centre of the low passes away the wind veers around to the north-west and the clouds break up into separate piles — the **cumulus** clouds (shown in picture 7).

If the temperature difference of the air either side of the cold front is large enough the clouds may pile even higher, towering **cumulo-nimbus** clouds (picture 8). The tops of these clouds are high enough to be sliced off by the jet stream to give them their typical appearance of a blacksmith's anvil. These give showers, sometimes with hail and even thunder. **Fair weather cumulus** clouds form on a fine day when the ground is warmed by the sun (see picture 9). Water vapour rises until it is cold enough to condense back into water drops. Sometimes these clouds grow into cumulo-nimbus and we have thunderstorms.

Picture 5 Cirrus cloud joining to form cirrostratus

Picture 6 Cirrocumulus giving a 'mackerel sky' — it looks like the scales of a fish

Picture 7 Fair weather cumulus clouds

Picture 8 Cumulus and cumulo nimbus

Picture 9 Cumulus clouds with precipitation

Activities

A Measuring the weather

This is a group activity. It can last as long as you like. Use as many as you can of the instruments described in this topic. Take measurements daily and display them on a chart or a graph.

Compare your readings with the official forecasts or results published in your local paper. You may find that your home or school is in a place with its own 'microclimate', with weather slightly different from its surroundings. Write a report every month on the weather changes and patterns that you notice.

B Do-it-yourself weather forecasts

People who are out in the wilds do not get newspapers or TV, and if they don't have a radio they could get into trouble if bad weather comes along unexpectedly. How good would you be at forecasting the weather?

With practice, common sense and the knowledge you have gained from reading this topic and the previous one you should be able to forecast the weather for up to 6, 12 or even 24 hours ahead. Don't cheat by watching or reading any weather forecasts for a day or two. Then look at the sky, note the clouds and which way the wind is blowing.

Write down your weather forecast in a chart like this:

```
Date :
Time of forecast
FORECAST                    ACTUAL WEATHER
(a) For 6 hours ahead       (a)
(b) For 24 hours ahead      (b)
```

Make your forecast at the same time each day — e.g. 9 am. Keep this up for a week or two. How good are you? Did you get any better?

C Local weather

Find out how much rain your area normally receives each year. Is it above or below the national average? Do you get your fair share of sunshine? What effect do these figures have on local industry (e.g. tourism) or agriculture?

Questions

1 What do the following instruments measure?

a a hygrometer,

b a barometer,

c an anemometer?

2 A barometer is useful for weather forecasting, but it can also be used to measure height above sea level. Suggest how it can do this.

3 The normal air pressure is 1000 'millibars'. This is a long winded way of writing '1 bar' or 'one standard atmospheric pressure'. The scientific way of measuring pressure is in newtons per square metre. In these units the normal air pressure of 1 bar is $100\,000$ N/m^2.

a What is the force in newtons, due to atmospheric pressure, acting on a table of area 2 m^2?

b A mass of 1 kg weighs 10 newtons. What is the mass of the air above the table in (a)?

c The roof of your house may have an area of 50 m^2. What is the mass of air above your house?

d The total area of the Earth's surface is about 5×10^{14} m^2

$(500\,000\,000\,000\,000$ m$^2)$. Calculate the mass of the Earth's atmosphere.

4 'Rain is a part of the *water cycle*'. Explain, briefly, what the 'water cycle' is.

5 *Weather forecasting is not scientific.* What do you think? Try to give arguments for and against this statement.

6 Compared with cold air, why is warm air:

a less dense,

b likely to have more water vapour in it?

Thunderstorms

The weather has been nice and sunny for several days. The air is warm. Then one afternoon it gets hazy, and it is hard to see the sun. It seems to be getting dark early. You hear the loud rumble of thunder, and you may see a flicker of lightning in the sky.

Then it starts to rain, maybe very hard indeed. The raindrops are very large and as it gets darker the rain pours down even harder. Each clap of thunder gets louder and you can see lightning flashing from the clouds to the ground.

People start getting worried. They stop work and look out of the window, if they are lucky enough to be indoors. Suddenly hard bullets of frozen water — hailstones — come down. Hail can rip leaves off trees and break windows in greenhouses. It can make dents in car roofs. A hailstone can weigh half a kilogram and be 10 centimetres across. One this size can knock out an adult.

Thunderstorms happen quite often in summer, but in fact they can happen at any time of the year. Picture 2 shows what is going on inside a typical summer thunderstorm. Compare it with the photograph of a thundercloud in picture 1

The main things a thunderstorm needs are damp air near the ground and a layer of much colder air above it. The warmer air rises, taking water vapour with it. The water vapour condenses into water droplets, making a cloud. The cloud grows and soon makes a huge tower that rises up to 10 km high. Even in summer the air at a height of 3 km or so is so cold that water freezes. The top of the cloud is made of ice crystals.

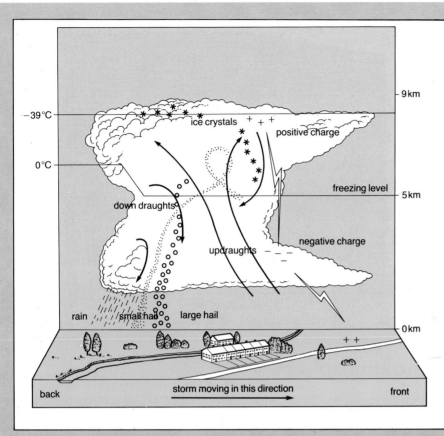

Picture 2 Inside a typical thunderstorm

Picture 1 A thundercloud

Picture 3 A lightning conductor on a church

Rain and hail

Cold air from the upper part of the cloud falls down the outside of the stream of uprushing air (picture 2). This cold air takes frozen 'icedrops' with it, and they collect water around them as they reach the bottom of the cloud. They may melt and fall as rain. But instead, they might be picked up by the rising air and carried back up. They freeze into even bigger 'drops' of ice — becoming small hailstones. These hailstones may be carried up and down the growing storm cloud many times, getting bigger and bigger each time (see picture 2). Sooner or later, they fall.

But why the lightning?

The streams of cold and warm air flowing past each other in different parts of the cloud produce electrically charged molecules (*ions*). This is similar to what happens when you rub a comb against a piece of cloth — both cloth and comb are electrified by friction. The top of the cloud becomes positively charged, the bottom becomes negative. There might be up to a billion volts between the top and bottom of the cloud. And what is more important, there is the same huge voltage between the cloud and the ground.

A lightning flash starts with a fairly small current that jumps between different parts of the cloud or between the base of the cloud and the ground. It zig-zags in steps of 10 to 200 metres in length, following the easiest path where there are lots of ions in the air. As it does so it makes more ions. This allows a much bigger current to flow, *which usually goes from the ground up to the cloud.* This 'return stroke' may carry a huge current, possibly 10 thousand amperes. It lasts less than a thousandth of a second.

As this huge current passes through the air it makes it very hot indeed. The hot air expands explosively, making a clap of thunder.

Lightning conductors

Lightning conductors do **not** conduct lightning. They are far too thin to carry a current of this size (picture 3). If they are struck by lightning the copper strip is instantly vaporised. What they do is more subtle. By being sharply pointed and fixed high in the air at the top of a building they encourage the formation of ions in the air. These ions can move to neutralise the charge on the base of the cloud. Thus the cloud is discharged gently and steadily, so stopping a build up of voltage to dangerous levels.

1 Why does cold air fall from the top of a thundercloud?

2 Describe the path of a large hailstone compared with a small one.

3 Many people think, wrongly, that thunderclouds have just one kind of electric charge. From a knowledge of the basic physics of 'where charge comes from', explain why the cloud must have *two* areas of *opposite* charge.

4 The lightning conductor was invented by Benjamin Franklyn (see page 149). He investigated thunderclouds by flying a wire into them, using a kite. Why is this a dangerous thing to do?

5 Use the formula **charge = current × time** to calculate how many coulombs of electricity are discharged in a typical return stroke.

6 When you are near a thunderstorm you may see the cloud moving slowly towards you. But the wind is blowing *towards the cloud.* Suggest a reason for this. (Hint: look at the diagram and think of convection currents)

7 The top of a large thundercloud spreads sideways in one direction, giving what is called an 'anvil head' (picture 3). What does this tell you about wind conditions 10 000 metres above the ground?

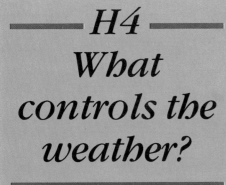

H4
What controls the weather?

The Earth's atmosphere is never still. It is like a huge engine powered by the Sun.

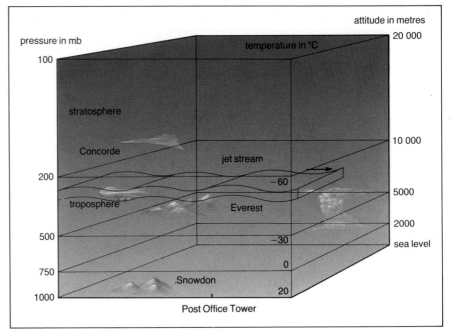

Picture 1 A section through the atmosphere. Weather changes occur in the lower region — the troposphere

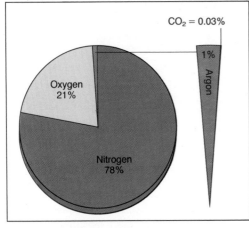

Picture 2 What air is made of (it also contains a small and variable amount of water vapour

Picture 3 (a) The poles of Mars are covered by a thin layer of ice, which melts as each pole turns to face the Sun.

The atmosphere

Picture 1 shows the Earth's atmosphere. Nearly 80% of it is nitrogen, the rest being mostly oxygen. There are also small quantities of other gases, like water vapour, carbon dioxide and the 'noble gases' (helium, neon, argon) — see picture 2. There is more about the composition of the atmosphere in *The Material World*, topic B1.

As far as weather is concerned, water vapour is the most important gas in the atmosphere. But carbon dioxide plays a very important role as a 'greenhouse gas' that helps keep the Earth's surface warm enough to support life. See topic I4 in *The Material World* for more about the greenhouse effect.

Water is a liquid between 0°C and 100°C. The temperature of most of the Earth's surface is just right for water to exist as a liquid. This fact is very important for the development and continuation of life on Earth.

Our nearest planet, Mars, is not quite warm enough for water to exist as a liquid. The main reason for this is that Mars doesn't have much of an atmosphere. Its atmospheric pressure is just 7.6 millibars compared with Earth's 1000 mb.

Also, Mars is further away from the Sun than Earth, and so gets less solar radiation energy. But there would be enough energy to sustain life if it weren't for the fact that Mars has such a thin atmosphere. Without an atmospheric 'blanket' to trap some of the energy, the surface of the planet re-radiates too much back into space. Thus Mars has a permanent frost, although the thin ice-caps you can see in picture 3 do melt from time to time when the poles face the Sun.

The planet Venus is our other neighbouring planet. It is almost the same size as Earth, and it is closer to the Sun. When it was formed (see topic I3) it was probably just too hot for water vapour to condense into rain. At that time, Mars, Earth and Venus probably all had atmospheres rich in carbon dioxide. But because it didn't rain on Venus the carbon dioxide in its atmosphere did not get washed out.

Carbon dioxide is a very effective 'greenhouse gas', so its blanketing effect on Venus is immense. Solar energy is trapped very well. The temperature has risen to over 475°C so that the surface actually glows at night. Lead and tin would melt at this temperature, and mercury would boil!

Carbon dioxide dissolves easily in water, and reacts with other elements, like calcium and magnesium, to form a carbonates. Carbonates, like calcium carbonate, are a major part of the rocks on Earth (see picture 4). They form limestone and chalk (calcium carbonates) and dolomites (magnesium carbonates).

In its early days the Earth was cool enough for liquid water to exist. As it fell as rain it washed carbon dioxide out of the atmosphere, which lost most of its carbon dioxide as result. It has been converted into the billions of tonnes of carbonate rocks that we find on Earth. So the Earth didn't have such a powerful greenhouse effect and become an overheated planet like Venus.

Picture 3 (b) Venus, surrounded by thick sulphurous cloud

Pressure in the atmosphere

At ground level air is crushed by the weight of the air lying above it. This compression makes the air at sea level quite dense — it has a mass of about 1 kilogram per cubic metre. At a height of 10 km the pressure is a quarter of the pressure at sea level. The density of the air has also decreased to a quarter of its sea level value.

The top of Mount Everest, the highest point on Earth, is 8.8km above sea level. The pressure is less than half of the sea level pressure. There is just enough oxygen to support human life — if you don't work too hard. Most Everest climbers have taken oxygen with them, but some have climbed it unaided, but very slowly.

The weather is decided by what happens in the bottom 10 km or so of the atmosphere.

Powering the atmosphere

The energy in a typical storm or hurricane is immense. Topic H1 has discussed how much damage they can do. This energy has come from the Sun, carried by the electromagnetic radiation we call **infra red** radiation (or 'heat rays'). As it happens, most of this radiation passes straight though the atmosphere. Its energy is absorbed by the Earth's surface, which in turn warms up the air in contact with it.

In the tropics the Sun is almost overhead for most of the year. At the poles the Sun is low in the sky even in summer. In winter it disappears below the horizon for months at a time. The energy received by a given area of ground in the tropics is thus much greater than the energy reaching the same sized area in the polar regions. Picture 5 shows that even in an Arctic summer the energy in a given beam of sunlight is spread over a much larger area than it is at the poles.

This means that there is a great temperature difference between tropical air and polar air. It is this difference that causes the great air movements that result in the weather patterns described in topic H2.

Picture 4 The White Cliffs of Dover — the carbon dioxide that used to be in the Earth's atmosphere is now locked up in billions of tons of carbonate rocks, like these

Airstreams

When air near ground level is heated it expands and becomes less dense. It then floats upwards through the cooler, denser air above it. Of course, more air has to flow in to take the place of the air that has floated upwards. This effect is called **convection**. As the warm air rises higher into the atmosphere it cools, gets denser again and begins to fall towards the surface. On Earth huge blocks of air move in this way, up from the tropics and down again nearer the poles. You can model this effect by using water in a beaker, or air in a special square tube (see activity A).

Picture 5 The energy in each beam is the same. But at the poles it has to cover a much greater area of ground.

Picture 7 The jet stream follows the boundary between two sets of convection currents in the atmosphere — the cold polar air (polar cell) and the warm tropical air (tropical cell). Britain lies north of latitude 50° and is usually just under the jet stream.

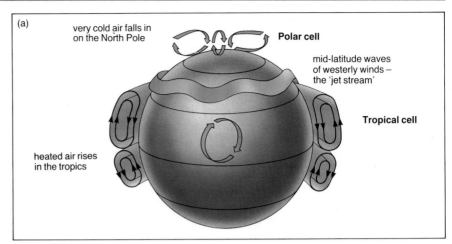

(a)

very cold air falls in on the North Pole

Polar cell

mid-latitude waves of westerly winds — the 'jet stream'

Tropical cell

heated air rises in the tropics

The effect of the Earth's spin

But the Earth is more complicated than a beaker of water. In the first place, the Earth is *spinning*. As the warmed air rises it spins more slowly than the Earth. To us on the ground, spinning along with the Earth, it seems as if the air is moving sideways a little. The blocks of warm, rising air seem to move in a curved path, like a ball on a moving table. Energy is released as water vapour condenses. This energy increases the speed of the winds. When these spinning blocks of warm air reach ground level again they form the **lows** that you have studied in topic H2. See picture 6.

This is a very simplified description of how the air moves. To some extent the air is dragged along with the Earth surface, held by the gravity field. Also, the difference in the rate of the spin speed at the equator compared with the poles means that the air spins faster as it moves north.

The mathematics of air movement is quite complicated!

Latent heat

Much of the energy from the tropical sun is used to evaporate water from the sea and the land. The energy needed to change the state of a liquid or solid is called **latent heat**. It is the energy needed to separate the closely packed molecules of water, for example, as they spread apart to form water vapour. There is more about this in topic C1 of *The Material World*.

Thus warm damp air contains a lot more energy than you might think, just from measuring its temperature. There is a vast amount of energy 'hidden' (or **latent**) in the water vapour it contains. Activity C is about the energy changes involved when substances change state.

When the water vapour condenses to form drops of water, this hidden energy is released. It raises the temperature of the air mass and also contributes to the wind energy of lows. Warm air flowing over a warm sea picks up lots of water vapour. Tropical storms — typhoons and hurricanes — get a lot of their energy from condensing water vapour. The British wind-storms of autumn 1987 and winter 1990 were linked with a warm Atlantic.

If the Earth didn't spin the high tropical air flow would go straight to the North

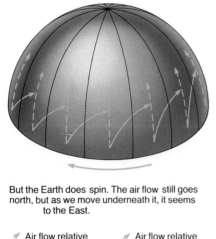

But the Earth does spin. The air flow still goes north, but as we move underneath it, it seems to the East.

↗ Air flow relative to the atmosphere
↗ Air flow relative to the moving Earth

Picture 6 This diagram shows the main air movements in the Northern half of the Earth. There is a similar pattern in the Southern hemisphere. This pattern is powered by solar energy and strongly affected by rotation of the Earth

Jet streams

As you have seen in topic H1 cold air and warm air do not mix very easily. The cold air flowing towards the equator from the poles keeps to itself. So does the warm air rising towards the polar regions. This effect, together with the fact that the Earth is spinning helps to produce a set of very fast high level

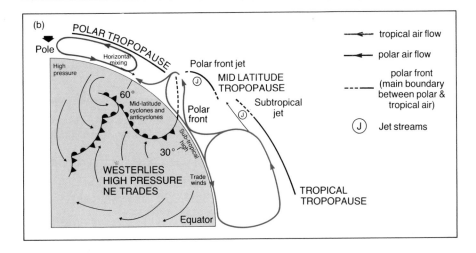

winds that circle the Earth at a height of between 8 and 10 kilometres. These are called the **jet streams** — see picture 7.

The weather we actually get in Britain is carried towards us by one of these streams. High above Great Britain the jet stream flows around the world from west to east. The middle of this huge river of air is travelling at over 200 km per hour. It is positioned just at the boundary between the warm tropical air and the cold polar air.

The stream is like a long snaking tube, and it wobbles up and down and from side to side. When it moves north the air underneath becomes less dense, and allows the warm tropical air to move a bit further north. This produces our typical **low**. When it moves the other way it brings colder, denser air to form a **high**. This effect is shown in picture 8. The mixture of lows and highs underneath the jet stream give us our typical rainy weather.

How to have a nice hot summer

The British Isles are usually just underneath one of the most important jet streams — there are others in other parts of the world. But our jet stream moves about. When it is well to the north we are firmly in a mass of warm ex-tropical air. This usually gives good weather. This is what happened in the hot summers of 1976 and 1989.

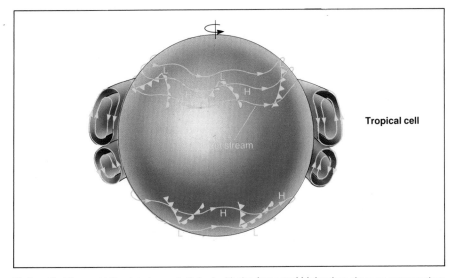

Tropical cell

Picture 8   The wobbling jet stream is linked with the lows and highs that gives us our weather

Picture 9 The position of the Polar Jet Stream strongly affects our weather

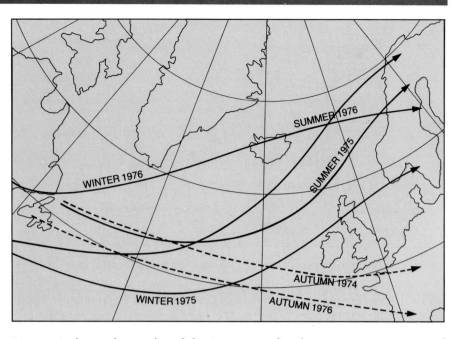

Picture 9 shows the tracks of the jet stream for the summer, autumn and winter of 1976.

In the summer of 1976, all the lows were channelled under the jet well to the north of the British Isles. This allowed blocks of stable tropical air to move up over Britain. These highs (**anticyclones**) stayed there for most of the summer, blocking the lows and keeping them away from us. With warm air and little cloud the summer of 1976 was hot and dry.

One result of this long period of hot weather was that the North Atlantic near Britain was warmed up by an average of 2°C. In the autumn, when the jet stream moved south, the lows were able to reach Britain once more. More water vapour than usual came from the warm sea and the autumn was unusually wet, filling the reservoirs that had become low in the hot summer.

Looking for the jet stream

In Activity B of topic H1 you can plot the movement of a low as it passes from the Atlantic to the east of the British Isles. When you do this you are also plotting the position of the jet stream.

In winter the jet steam moves further south. This means that the lows move to the south of Britain and go across France and Germany. These countries then get plenty of snow because the ground and the air close to it are usually quite cold in the mountain regions.

When the jet stream is in this position Britain is in the grip of cold polar air from the continent of Europe. This mass of cold air might stretch as far away as Moscow and even into Siberia. We get cold easterly winds, clear skies and high pressure. The high pressure air once again acts a **blocking** anticyclone which keeps away the lows coming across from the North Atlantic.

Sooner or later the jet stream wobbles north again. It then brings damp warm air to mix with our very cold polar air and we too get snow. This usually happens in early spring (February or even March).

Long range weather forecasts are based on trying to predict the path of the jet stream. These streams were only discovered during the Second World War, when the new, high flying military aircraft found that they could cross the Atlantic from west to east much more quickly than when they travelled the other way.

The hot air balloon flights across the Atlantic rely on the jet stream to give the balloons their speed (picture 10). They have to travel high above the Earth. At lower levels the balloons would travel much more slowly, and tend to go round the lows in huge circles.

Picture 10 Hot air balloons can travel great distances very quickly in a jet stream

Activities

A Convection currents

1 You can show convection currents in water by dropping some coloured crystals (e.g. potassium permanganate) into a beaker of cold water. Place them to one side of the beaker, and use a bunsen burner with a low flame to gently heat the water just by the crystals. Watch and then describe what you see happening.

2 The apparatus shown in picture 11 is one way of showing convection currents in air. Your teacher may be able to show it to you.

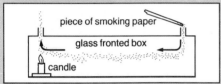

Picture 11 The candle flare heats the air above it. The air expands and rises, drawing cooler air into the box in a convection current.

B Looking for the jet stream

This could take some time — and needs patience. Use newspaper weather maps collected over a period of a few months to plot the paths of lows across the British Isles. You will need a 'blank map' to plot the positions of the centre of each low, day by day. When you have finished you should be able to link the position of the jet stream with the general weather pattern of a season.

C Investigating latent heat

1 Solid to liquid and back again

When a solid melts its molecules lose their crytalline order and separate into a jostling chaos. It takes energy to break the particles away from each other. This is the latent heat of melting. This energy has come from the source of

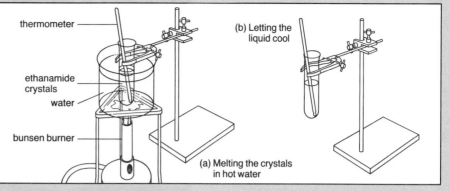

Picture 12

heating. When the liquid cools this energy is returned to the surroundings. Investigate this effect by measuring the temperature changes when a crystalline solid is melted and then cools back down to become solid again.

Picture 12 shows a way of doing this. One third fill a boiling tube with ethanamide crystals and put the tube in a clamp stand. (You could use octadecanol instead.) Melt the crystals by putting the tube in a beaker of water and heating the water to boiling. **CARE:** hot water and molten ethanamide can cause scalds.

Put a thermometer in the melted ethanamide while it is still in the hot water. Have a stop watch ready. Turn off the bunsen burner and remove the tube from the hot water. Stir gently.

Take measurements of the temperature of the ethanamide every minute as it cools and solidifies. Plot the results on a graph of temperature against time. Explain the shape of this graph.

2 Liquids — boiling and evaporating

This is a quick, rough experiment to compare the energy needed to raise the temperature of water with the energy needed to change it from water to steam. Your teacher may be able to let you use apparatus which guives a more accurate result.

Put 200 cm³ of water in a graduated

400 ml beaker. This is, near enough, 200 g of water. Use a medium bunsen flame to heat the water. Take readings of the temperature every minute.

a What do you expect to happen when the water reaches a temperature of 100°C?

When the water is boiling keep taking readings until half the water is boiled away.

b What was the energy from the bunsen burner being used for whilst the water was boiling?

c It takes 4200 J of energy to raise the temperature of 1000 g of water by 1°C. Use your results from the first part of the experiment to calculate how much energy the bunsen delivered per minute. You can do this by seeing how much the temperature rose in a fixed time, say 10 minutes.

d How long did it take 100 cm³ of water to boil away?

e Use your answers to parts (iii) and (iv) to calculate how much energy the bunsen delivered to boil away 100 g of water.

f Now calculate how much energy would have been needed to boil away 1000 g of water. Compare your result with the accepted value of 2.3 MJ per kilogram for the latent heat of vaporisation of water.

g What has the latent heat of vaporisation of water to do with hurricanes?

Questions

1 Why do high-flying aircraft get from New York to London quicker than when they fly from London to New York?

2 Explain the following:

a How warm air gets from the tropics to Britain.

b Why cold 'Siberian' air rarely gives us snow.

c Why lows heading straight for Southern England often turn north and hit Scotland instead.

3 Why is knowledge of the jet stream important in making long range weather forecasts?

4 When air rises it moves into a lower pressure region. It then expands and cools. Use this fact to explain why the

tops of mountains are often covered in clouds, even on a fine day.

5 Warm damp air (air with lots of water vapour in it) contains a lot more energy than the same amount of warm dry air at the same temperature.

a Explain why this is.

b What is the importance of this fact for the weather?

11
Sky patterns

The stars are fixed in their constellations. We learn about the Solar system by studying the movements of the Sun, the Moon and the planets against the background of the fixed stars.

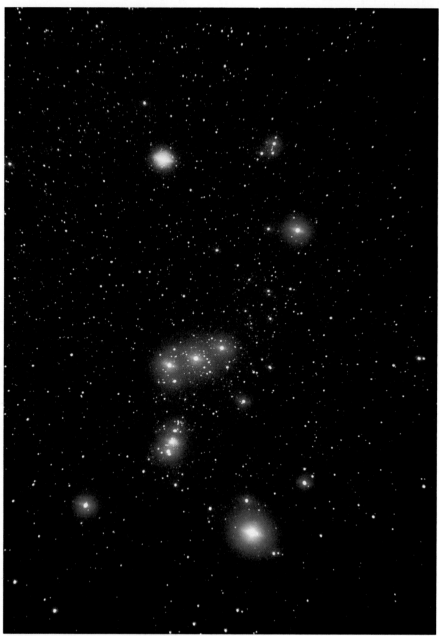

Picture 1 Orion is one of the easiest constellations to find

Picture 2 The Pole star makes a circle around the Celestial Pole. This picture was taken over several hours using a camera pointing at the Pole Star

Star patterns

On a clear starry night you might be able to see five thousand stars, if you have good eyesight. To make sense of them star watchers have arranged them into groups — the **constellations**. Picture 1 shows the constellation called **Orion**. Most of the constellations are named after ancient Greek heroes or heroines. Orion is named after a Greek hero who was a mighty hunter.

You can see Orion from autumn to spring, in the southern sky. If you look at it every hour or so in the winter you will notice that it moves from east to west. You will see other stars rising in the east, like the Sun, and setting in the west.

All the stars follow this same pattern of movement. But if they are high enough in the sky they never rise or set. Instead they circle around a fixed point in the northern sky. (See picture 2). Close to this point is a star, called

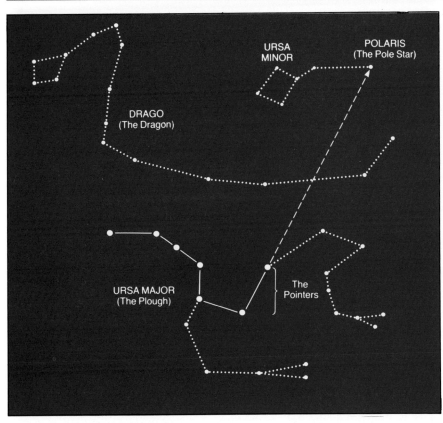

Picture 3 How to find the Pole Star

Picture 4 The Moon seems to change its shape, 'waxing and waning'

the **Pole Star**, or **Polaris**. Picture 3 shows how to find this star by using the two 'pointer' stars in the constellation of the **Plough (Ursa Major)**.

The Pole Star is directly above the North Pole of the Earth, and so it is very useful in navigation. As you may remember from earlier work, this movement of the stars across the sky is caused by the fact that the Earth is spinning on its axis.

The Earth spins on its axis once every 24 hours, and as we move under the stars they seem to be passing overhead. It is this spinning of the Earth that causes the Sun to rise and set, of course.

The Moon

The Moon, too, rises and sets. But it also changes shape to give the **phases** of the Moon (picture 4). The rising and setting is because of the spinning of the Earth. Its changing shape is caused by the fact that the Moon actually does move on its own. It is a **satellite** of the Earth. It moves around the Earth in an orbit, taking 28 days to complete it.

This is shown in picture 5, which also explains how its shape appears to change. What we see is only the part of the Moon that is in sunlight.

Because it moves on its own, the Moon doesn't keep pace with the stars. If you look at the Moon at the same time every night you will see it in a slightly different part of the sky each night. It seems to slip back against the constellations.

The stars are further away from us than the Moon is. We can tell this because as the Moon moves through the star patterns it cuts them off. We never see stars in front of the Moon.

But the Moon does move in front of the Sun. When this happens the sunlight is cut off. We then have an **eclipse** of the Sun, or **solar eclipse**. It can only happen when the Moon is 'new', lying directly between us and the Sun.

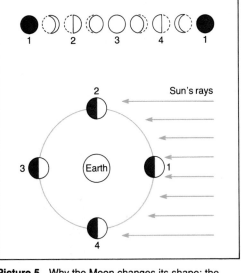

Picture 5 Why the Moon changes its shape: the phases of the Moon

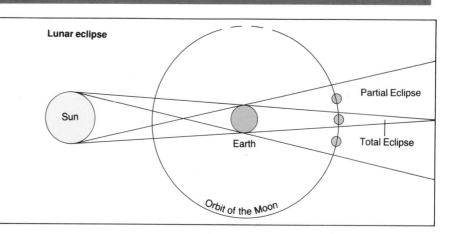

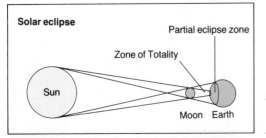

Picture 6 Eclipses of the Sun and the Moon

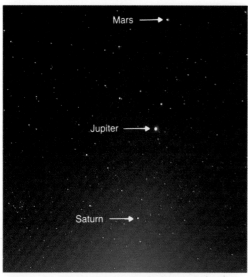

Picture 7 The two planets look close together but Jupiter is much further away from us than Mars.

Sometimes the full Moon gets dark. This happens when the Moon is in the Earth's shadow, so we have an eclipse of the Moon, or **lunar eclipse**. Picture 6 shows how these eclipses are produced.

The planets

The planets look like stars but they are in fact very different. They are much smaller than stars, and do not shine by their own light. We can only see them because they reflect the light of the Sun.

They were first noticed in ancient times. Like the stars, they rise and set. But careful starwatchers picked them out because they also move against the background of the ordinary 'fixed' stars.

The ancient astronomers discovered five of these 'wandering stars'. We still call them by the names given to them by the ancient Greeks, who named them after their gods: Mercury, Venus, Mars, Jupiter and Saturn.

The easiest ones to see are Jupiter, Venus and Mars. Mars and Venus are easy to see because they are the planets closest to Earth. Jupiter is a lot further away but it is the biggest planet of all. We now know that there are three more planets: Uranus, Neptune and Pluto (which was only discovered in 1934).

Uranus and Neptune are visible to the naked eye, but quite hard to find unless you are an experienced observer. Mercury is very close to the Sun and you can see it best just before dawn — if you know exactly when and where to look.

The zodiac

The Sun, Moon and planets all seem to move across the sky through a belt of stars called the **zodiac**. The stars in the zodiac are grouped into 12 constellations — the 'signs of the zodiac'. You will recognise their names because they are often listed as 'birth signs' in newspapers and magazines in their astrology column.

Don't confuse *astrologers* with *astronomers*. Astronomers observe and measure the positions and properties of stars and they are very accurate. Astrologers claim that they can predict people's future and personality from the positions of the stars and planets. But they get it wrong about as often as they get it right.

Your 'birth sign' is decided by where in the zodiac the Sun was when you were born. If you were born on 12 June, for example, you are a 'Gemini'. At this time the Sun is between us and the constellation of the **Twins (Gemini)**. Picture 8 shows the zodiac and the Sun's position at this time.

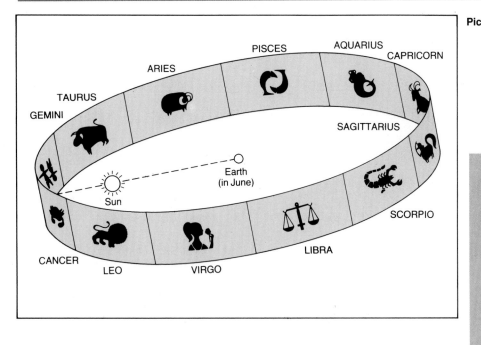

Picture 8 As the Earth goes around the Sun, the Sun appears to move against the background of the stars. The star groups it moves in front of are called the Signs of the Zodiac

The Milky Way

The Milky Way is the part of the sky where stars are most crowded together. It is can be seen all through the year. It is a wide band of stars that runs through the constellation of Cassiopeia, which never sets. In the winter it is close to Gemini and Orion, and in summer it is a splendid sight as it crosses the sky from Cygnus to Scorpius.

Look at any part of the Milky Way through a small telescope or pair of binoculars. What to the naked eye looks like a faint whitish blur becomes a mass of stars. Find out more about the Milky Way in topic I4.

Questions

1 Give two differences between stars and planets.
2 Explain what the following words mean:
   zodiac, eclipse, orbit, astronomy, astrology, planet, star, lunar, satellite.
3 The Moon goes around the Earth once every 28 days or so. Why don't we get eclipses of the Sun and Moon every month?
4 We don't get eclipses of the Sun when the Moon is a 'half' moon. Explain why not.
5 Your young brother refuses to believe that the Earth is round. 'It looks pretty flat to me', he says.

   What would you say to him, or show him, to prove that the Earth isn't 'flat'?

Activities

A Looking for the constellations

Use a simple star map. These are published every month by some newspapers (*The Times, The Guardian,*

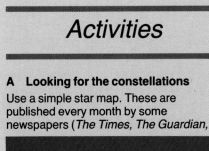

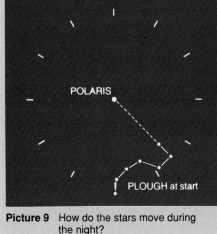

Picture 9 How do the stars move during the night?

The Daily Telegraph, The Independent) and you should be able to get a photocopy from your library. The best time of the year is when it gets dark fairly early (autumn to spring). Choose a moonless night. Try to find the following constellations: Great Bear (Plough), Cassiopeia, Gemini, Leo, Cygnus, Pleiades, Orion, Perseus with Andromeda and the Great Square of Pegasus.

B Do the stars move?

Draw or photograph the positions of stars near the Pole Star.

1 Drawing: find the Great Bear and mark its position on a 'clock diagram' (picture 9) at, say, 7 pm. Mark its position every half hour until 10 pm.
2 Taking a photograph: you need a camera with a 'B' button that lets you keep the shutter open as long as you want. Put the camera on a firm surface pointing at the Pole Star. Open the shutter and leave it open for an hour or so. It is important not to move the camera. Make sure that there aren't any bright lights nearby.

C The moon

1 Use a telescope or a good pair of binoculars to look at the Moon. The best time is when the Moon is between 'half' and 'full'. Can you see the 'mountains'? Draw what you can see as carefully as possible. A good encycliopaedia will have a 'Moon Map' you can use to identify the main features. Try your school library.
2 Draw the shape and position of the Moon (compared with the star patterns) every night for a week — or every other night for a fortnight.

D Looking for planets

Use a newspaper map (see activity A) to find some planets. Draw where they are compared to nearby stars. Do this for a week or two. Comment on what you notice.

E The Milky Way

Use a small telescope or a pair of binoculars to look at the Milky Way. Describe what you see. What does this tell us about what the Milky Way might be?

12
Patterns in space

The Sun, Earth and planets are part of a pattern in space called the Solar System.

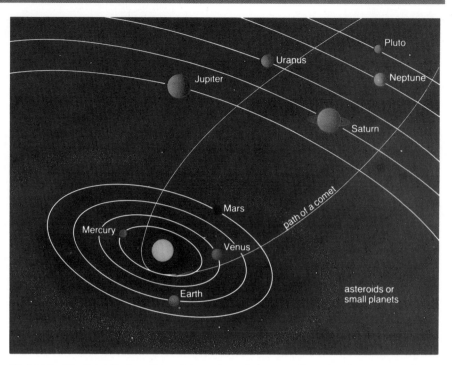

Picture 1 The Solar System. It is not drawn to scale because the outer planets are so far away from the Sun compared with the inner ones

The Solar System

The Sun is the centre of the Solar System. It contains 99.8% of all the mass in the System, and so produces a huge gravity field that holds the planets in their orbits around it. (See topic B8.) Picture 1 shows the Sun and the planets, viewed from outside the System. The actual masses and distances of the planets from the Sun are given in table 1.

The Sun

The Sun is a star. It is quite a small one, as stars go, and it isn't very bright. It is the brightest object in the sky because it is so close to us, compared with other stars. The energy it provides supports all life on Earth. This energy comes from the Sun's own mass, which it uses up at a rate of 4 million tonnes per second.

Table 1 Planetary data

	mass (Earth = 1)	diameter (km)	density (tonnes per m³)	surface gravity field (N/kg)	distance from Sun (Mm)	period 'year'	'day'
Mercury	0.05	4880	5.4	3.6	58	88d	59d
Venus	0.81	12112	5.25	0.87	107.5	224d	243d
Earth	1	12742	5.51	9.8	149.6	365	23h 56m
Mars	0.11	6790	3.95	3.7	228	687d	24h 37m
Jupiter	318	142600	1.34	25.9	778	11.9y	9h 50m
Saturn	95	120200	0.70	11.3	1427	29.5y	10h 14m
Uranus	14.5	49000	1.58	10.4	2870	84.0y	10h 49m
Neptune	17.5	50000	2.30	14.0	4497	165y	15h 48m
Pluto system	0.003	2284	2	—	5900 (variable)	248y	6.4d

The Sun emits energy at the rate of 400 000 000 000 000 000 000 000 000 watts (4×10^{26}W). Only a tiny fraction of this energy reaches Earth. Most is radiated into empty space.

Question 6 is about using the Einstein equation $\boldsymbol{E = mc^2}$ to check these amazing figures.

The energy is produced by a process called **nuclear fusion** (See topic D7). The Sun is mostly hydrogen. Hydrogen nuclei in the centre of the Sun are under a huge pressure and at a very high temperature. Some of the hydrogen nuclei collide to form helium nuclei. In doing this they lose a tiny fraction of their mass which is converted to the kinetic energy of the particles that are left.

Picture 2 shows the surface of the Sun, photographed during an eclipse. The surface is a gas at a temperature of about 5500°C, and the picture shows a typical 'Sun storm' in which hot gas is hurled far out into space. These storms reach a peak every 11 years. At times of peak activity the Sun sends out far more ionised particles than usual, which affect our atmosphere and so the weather on Earth.

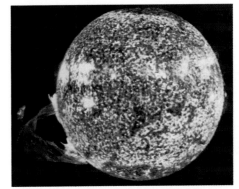

Picture 2   A sun-storm. On the scale of the picture the Earth would be about the size of one of the dark blobs or about half a millimetre across

The planets

The planets move in orbit around the Sun. They all move in the same direction, which is anticlockwise when viewed from above. The nearer the planet is to the Sun the faster it moves. Thus the orbital speed of Mercury is 55 metres per second while Neptune travels at a tenth of that speed, about 5.4 m/s.

Neptune has a larger orbit than Mercury and travels much more slowly. This means that Neptune takes much longer than Mercury to complete its path around the Sun. This orbit time is called the planet's 'year'. A 'year' for Mercury is just over 12 Earth weeks, while Neptune takes 247.8 Earth years to make one orbit around the Sun.

Mercury is so close to the Sun that its surface is heated up to over 400°C, which is hot enough to melt tin. It spins on its axis rather slowly, managing to get three spins ('days') for every two orbits ('years') around the sun. It is too small to have an atmosphere.

The next planet out from the Sun is **Venus** (picture 3). It is almost exactly the same size as the Earth and is only a little less massive. You might think it would be a good place to go for a sunshine holiday, but you would be wrong. Its atmosphere is mostly carbon dioxide, so the greenhouse effect is so great that it is even hotter than Mercury, with a surface temperature of 460°C.

The atmosphere of Venus is very corrosive, containing hydrogen chloride and hydrogen fluoride. No one has seen the surface of Venus, as it is hidden by clouds of concentrated sulphuric acid and particles of pure sulphur. Also, the atmospheric pressure is 95 times what it is on Earth, so that a very

Picture 3   Venus

	surface temperature (°C)	number of moons	atmosphere
Mercury	350	0	none
Venus	460	0	Thick; carbon dioxide, sulphuric acid
Earth	20	1	nitrogen, oxygen
Mars	−23	2	Thin: carbon dioxide
Jupiter	−120	16, 1 ring	Hydrogen, helium, ammonia methane
Saturn	−180	17, plus rings	Hydrogen, helium, ammonia, methane
Uranus	−210	15, plus rings	Hydrogen, helium, ammonia, methane
Neptune	−220	8	Hydrogen, helium, methane
Pluto system	−230	1	None — frozen

Picture 4   A Martian volcano

well-designed space suit would be needed. When the Russian Venus Probe landed by parachute on the surface of Venus in 1972 it managed to survive for only about 30 minutes in these very nasty conditions.

Bypassing planet **Earth** (described fully in section G), we next reach the planet **Mars**. At one time it was thought that Mars could support life. Indeed, a nineteenth century astronomer was convinced that the markings he could see on its surface were canals. He was wrong; they were simply lines of craters like the ones we see on the Moon.

But spacecraft sent to Mars have taken photographs which seem to show old water channels. It may be that at one time there was enough surface water on Mars to support life. It also looks as if Mars once had volcanoes (picture 4). This suggests an internal source of energy which might have been able to support some kind of life. Mars even has an atmosphere, but it is much thinner than the Earth's.

In 1976 the American Viking Spacecraft managed to 'soft-land' probes which were able to take pictures of the Martian surface.

These probes also looked for signs of life. They scooped up some Martian soil and tested it to see if carbon dioxide or any other chemical signs of life were present. The first experiment seemed to give a positive result. But when the experiment was repeated many times no more of the chemicals were detected. Scientists now agree that there is no life on Mars.

Asteroids

Between Mars and the giant planet Jupiter astronomers have discovered hundreds of stony objects. These range in size from the largest (Ceres) which is just over 700 km in diameter to rocks which are less than a few kilometres across. There are probably thousands of others too small to see. These objects should be called *planetoids* (little planets) rather than **asteroids** (little stars).

The asteroids are mostly clustered in their orbits between Mars and Jupiter. But some of them wander away from this region. They climb high above the plane in which the planets move, or have strange orbits which bring them closer to the Sun even than Earth. It is calculated that one of them, Hermes, might one day pass between Earth and the Moon.

Picture 5   Jupiter's Great Red Spot

The Giant Planets — Jupiter, Saturn, Uranus and Neptune

Beyond the asteroids lies the largest planet, **Jupiter** (picture 8). It is big enough to hold 1300 Earths. If it was just a little more massive it would turn into a star. As it is, the energy generated as it slowly collapses on itself creates huge storms in its atmosphere. One of these storms, a huge hurricane 48 000 kilometres long by 11 000 wide, has probably existed for thousands of years. This is the famous Great Red Spot, which you can see in picture 5.

At the visible 'surface', Jupiter's gravity field is 2.6 times stronger than Earth's. But what we see is not the planet's real surface, but the top of its atmosphere. This is made of swirling clouds of hydrogen methane and ammonia. The bands on its surface are huge **jet streams** (see topic H4). Below the atmosphere is a very deep 'sea' of liquid, metallic hydrogen. This covers a comparatively small solid core. The planet's core is very hot. It might be rocky, or even white hot, solid hydrogen.

Like Jupiter, the next three planets are large and have a low density (see table 1). They are also likely to be made mostly of hydrogen and helium.

Saturn is famous for its 'rings', shown in picture 6. These are made of small rocks, pebbles and grains which orbit the planet, all together in the same plane. The Voyager spacecraft discovered that both **Uranus** and

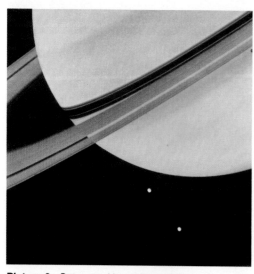

Picture 6   Saturn and its orbiting rings

Neptune (picture 3, page 31) had rings as well. They are not so large or as clearly visible as Saturn's.

All these outer planets have several moons. Some orbit so close to their planet that they are in danger of being pulled apart by its gravity forces. One of the moons of Jupiter, **Io**, is being shaken up so much that it is hot enough inside for volcanoes to exist.

The odd one out

Pluto is the furthest known planet of the Solar System. But it isn't always the furthest — it moves in an orbit that cuts inside the orbit of Neptune.

Pluto is small and very hard to see. It was not discovered until 1930, 14 years after its existence had been predicted. (Just like Neptune had been, because of its effect on the movement of the other outer planets — see topic B3).

It seemed to be a very small planet, probably smaller than Earth. We now know that it is a very strange object indeed. It is a 'double planet', with a moon almost as big as itself.

Activities

A Holidays in space!

Imagine that it is the year 2020 and you are an advertising agent for holidays on the planets. Make up an advertising slogan and —

■ draw a poster, or
■ produce a TV commercial, or
■ write and record a radio commercial for "A Holiday on Mars" (or Venus, or Jupiter, or . . . .)

B Exploring the Solar System

Use a library to find out what you can about the probes that have been sent out to investigate the planets. Examples are: Venera (to Venus); Mariner (to Mercury and Venus); Pioneer (to Venus); Viking (to Mars); Voyager (to the outer planets).

C Looking for planets

At the beginning of each month some newspapers print star maps which show where the planets are in the sky for that month. You should be able to get one of these maps from school or from a public library. Use it to find some planets. Watch them over a few weeks to track their movements against the background of the fixed stars.

Questions

1 Use table 1 on page 248–9 to answer the following questions.

a Which is the most massive planet?

b Which planet has the greatest gravity field at its surface?

c Which planet has its 'year' shorter than its 'day'?

d Which planet has the shortest day?

e Which planet has the longest day?

2 Why is it hard to get information about:

a Venus,.

b Pluto?

3 Four planets have much higher densities (around 5 tonnes per cubic metre) than the others (see table 1).

a Which planets have this higher density?

b What is the reason for this high density?

c Get a sheet of paper at least 50 cm long and 5 cm wide. Draw a simple scale diagram of the Solar System to a scale of 100 Mm (megametres) to 1 mm. What does the attempt to do this tell you about the high-density and low-density groups of planets?

4 Use the data in table 1 to plot one or more of the following graphs. Put the name of the planet against its point on your graphs. Comment on each graph you draw.

a Plot the density of the planets against their distance from the Sun.

b Plot the mass of the planets against their distance from the Sun

c Plot the mass of the planets against their density

d Plot the orbital period ('year length') of the planets against their distance from the Sun

e Plot the surface temperature of the planets against their distance from the Sun.

5 Use the data in table 1 to make bar charts of the following:

a the sizes (diameters) of the planets,

b the number of moons for each planet,

c the day lengths of the planets.

6 The Sun radiates 3.9×10^{26} joules of energy every second. This energy comes from the conversion of its mass into energy, given by the Einstein equation $E = mc^2$. Use this equation to check that to produce this energy output the Sun has to lose 400 tonnes (4×10^9 kg) of matter every second.

7 The asteroids are the 'bits' of a planet that was never made. They orbit between Mars and Jupiter. Use the data in table 1, and any other ideas that you have to answer the following questions.

a What would such a planet be like?

b Suggest its 'planetary data' — day length, size, distance from Sun, temperature, etc.

c Suggest a name for this planet.

13
How the Earth was made

No one really knows how the Earth, the Sun and the planets were made. Scientists think that it was probably like this . . . .

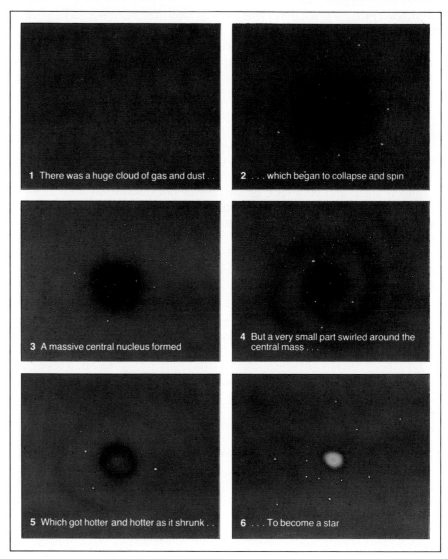

1 There was a huge cloud of gas and dust . .

2 . . . which began to collapse and spin

3 A massive central nucleus formed

4 But a very small part swirled around the central mass . . .

5 Which got hotter and hotter as it shrunk . .

6 . . . To become a star

Picture 1 The formation of the Sun

Making a sun

It would have been about fifty thousand million years ago. On the outer edge of the Milky Way a thin, invisible cloud of gas and dust was collapsing inward, pulled by the force of its own gravity. The gas was mostly hydrogen, but about a fifth of it was helium. These are the two lightest elements, and make up most of the Universe.

As the cloud fell together it began to spin, and the smaller the cloud got the faster it spun (picture 1). When the cloud reached the size of the present solar system it began to collapse very quickly. It took just twenty years to reach the size of the Earth's orbit.

At this size the molecules of gas were crowded close enough to smash into each other. As the particles fell they lost gravitational potential energy, and the gas got hotter.

The first glow

As the centre of the cloud became hotter and hotter it gave out radiation. At first the radiation was invisible infra red radiation. As the temperature rose the cloud began to glow red. It was almost — but not yet — a **star**.

The light did not escape easily from the hot centre. The cloud of hydrogen was not pure. Mixed in with it were particles of other elements, mainly iron, nickel and silicon. There were also compounds: water, methane, silicates. These particles clumped together to make fine dust, or even small 'stones.' The frozen water was mixed with silicates to make lumps of what astronomers call 'dirty ice'. All these stopped most of the visible radiation escaping.

The gas continued to collapse, but much more slowly now that it was so hot. The gas and dust particles were moving very quickly at this temperature. Like any mass of hot gas it was trying to expand. There was a delicate balance between the force of gravity and the tendency of the gas particles to escape.

The birth of a star

Time went on. The gravitational energy of the gases falling into the centre of the cloud made it hotter and hotter. It went from being red hot to being white hot. In the hottest part the atoms smashed into each other so hard that their electrons were stripped away, leaving the nuclei of the hydrogen and helium atoms quite bare.

The mixture of electrons and nuclei were squashed together even more under the enormous pressure of the collapsing cloud. They became hotter and hotter, moving and colliding with greater and greater energies.

A hydrogen bomb

Fifty million years after the cloud started to collapse the centre of the cloud reached a temperature of 10 million degrees Celsius. The hydrogen nuclei were now moving so fast that they began to stick to each other, producing helium nuclei. This is called **nuclear fusion.**

Now one of Einstein's ideas had come into play, the equation $E = mc^2$. The helium nucleus has less mass than the separate hydrogen nuclei, and the lost mass (**m**) became energy (picture 2).

Nuclear fusion like this is exactly what happens in a hydrogen bomb. The result was a sudden huge release of energy at the centre of the cloud. But the hot centre of the gas cloud did not explode like a bomb — it was far too heavy to be blown apart. It simply kept on working as a giant nuclear reactor, and has been doing so for at least the last ten billion years. It had become the star we call the Sun.

The force of light

One of the strangest properties of electromagnetic radiation — including ordinary light — is that it can exert a force when it shines on an object. The force that comes from a torch is very, very small — and you are not in danger of being knocked over by just the headlights of a car. But the Sun was now producing so much radiation that it began to have an important effect.

First, infra red radiation was absorbed by the frozen water and gases, so that they became hot enough to melt and then evaporate. The fog of 'dirt' and ice became clearer. The radiation reached out to exert its pressure on more distant particles. The lighter material was slowly pushed away, leaving behind the heavier rocky and metallic pieces that were to form the inner planets (picture 3).

Radiation pressure pushed away a lot of the dust and gas. The Sun now contained only a small fraction of the original cloud, but it was still an enormous mass. It was enough for its gravity field to hold all the heavy particles and a lot of the gas molecules in orbit. The steady pressure of its radiation carried on the work of pushing the lightest particles into more distant orbits.

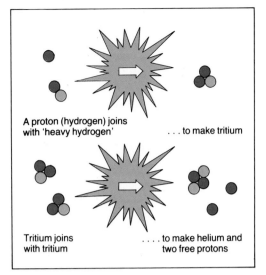

A proton (hydrogen) joins with 'heavy hydrogen'
. . . to make tritium

Tritium joins with tritium
. . . . to make helium and two free protons

Picture 2 Radiation evaporated water and frozen gases, and drove them away from the Sun

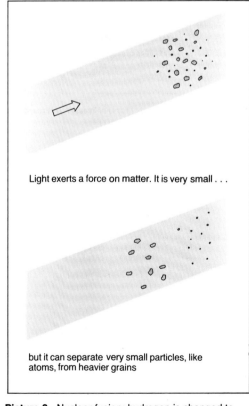

Light exerts a force on matter. It is very small . . .

but it can separate very small particles, like atoms, from heavier grains

Picture 3 Nuclear fusion: hydrogen is changed to helium

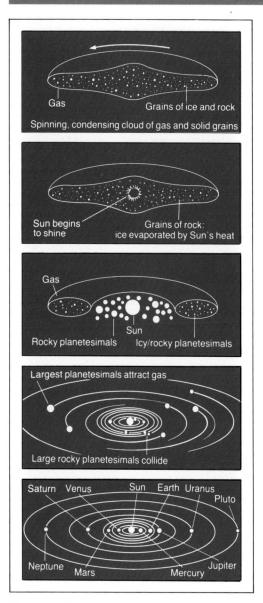

Gravity pulled inward, radiation pushed outward. The balance between these forces made the lighter material form a doughnut-shaped cloud, which was mostly hydrogen and helium, with ice and some solid grains of rock left from the original gas cloud.

This doughnut was about a thousand million kilometres away from the Sun (picture 4). Between the doughnut and the Sun was a chaotic crowd of grains and pebbles, each in its separate orbit.

Making the planets

Even the smallest particles produce a gravity force, and sooner or later this caused the countless trillions of particles, grains and pebbles to move closer together. When they collided they stuck together.

The larger lumps had a larger gravity field. The bigger they grew the more they were able to attract the smaller particles. Millions of rocky 'planetoids' grew, but in the end the larger ones swept up the smaller ones to make the **planets**.

Most of the particles — nearly three-quarters of them — eventually formed themselves into one very large lump. This was almost big enough to become a second Sun. But its central temperature never became quite as high as the 10 million degrees needed for a hydrogen-bomb explosion. Instead, it settled down to become the largest planet in the Solar System — Jupiter.

Most of what was left became the giant planets beyond Jupiter: Saturn, Uranus and Neptune (see topic I2). These large planets are mostly made of light materials, hydrogen, helium and methane. The heavier materials closer to the Sun formed the **inner planets:** Mercury, Venus, Earth and Mars. Between Mars and Jupiter is a belt of unsuccessful mini-planets which never quite made it — the **asteroids**. The orbits of the asteroids are being disturbed by the gravity field of their huge neighbour, Jupiter. This seems to stop them collecting together into a planet. It also means that some of them may be made to pass very close to Earth.

Comets

Far out in space, beyond the orbit of Pluto, there still exists the shell of the gas cloud that formed the Solar System. It contains perhaps 100 billion lumps of rocky ice, up to 8 km across. Now and again something disturbs their slow, distant orbits and one of them is propelled along a very long path towards the Sun. As it enters the Solar System the Sun melts the ice. The pressure of the Sun's radiation and the stream of particles it sends out affect the comet. A long streamer of gas, ice and dust is formed. The sunlight reflects off this tail and we see a **comet**.

Picture 4 How the planets might have been formed

Picture 5 Halley's comet

The most famous comet is **Halley's Comet** (see picture 5). It is a comet that has settled into a fixed orbit. It sweeps in a huge ellipse around the Sun, taking 76 years to complete each orbit. New comets appear quite often, for example, Comet Austin, which appeared in April and May 1990.

The Earth

The Earth is the largest of the inner, dense planets. Even so, its mass is only 0.3% the mass of Jupiter. It finished being formed about 4.5 billion years ago, according to measurements of the age of the oldest rocks.

The Earth can be called an iron planet. Its core is mostly iron, with some nickel mixed in. It has a very thin layer of lighter materials on top, forming the Earth's crust (see topic G2). It is just the right distance from the Sun for life as we know it to exist. It is not too cold — so water doesn't freeze all the time. It is not too hot, so water doesn't boil.

The Earth's crust is amazingly thin, compared with the size of the Earth, and so is the atmosphere. When you trace the outline of a coin with a pencil to represent the Earth, the thickness of the line would cover both crust and atmosphere. It is this crust, and the thin layer of gases surrounding it, which provide the materials that all life on Earth needs.

In topic G2 you can learn a great deal about the surface of the Earth's crust. You can find out how it is changed by the action of wind, water, earthquakes and the mysterious upwellings of new rocks from deeper inside the Earth. The energy to move winds and water on the Earth comes from the Sun, as described in topic H2.

The energy that produces volcanoes and earthquakes comes from deep inside the Earth. This is dealt with in topic G2.

How do we know all this?

The layered structure of the Earth was discovered by the study of earthquake waves. Just as radar waves bounce off different kinds of objects, so the shock waves from earthquakes bounce off the different layers of the Earth. See topic G2 for more details.

Radioactive dating

The best evidence for the age of the Earth comes from measurements of the radioactivity of minerals in rocks. This is also explained in topic G2.

Picture 6 The daytime meteorite of 10 August 1972. It weighed 1000 tonnes and could have destroyed a small village

Using this and other methods scientists have worked out that the oldest unmelted rock found on Earth is about 4 billion years old. The oldest rocks brought back by the American astronauts from the Moon were 4.5 billion years old. The Earth and Moon were probably formed at the same time, but the Earth is so active geologically that we are unlikely to find the oldest rocks still unchanged.

Meteorites

Meteorites are what we call 'shooting stars'. They are bits of dust and rock that enter the Earth's atmosphere from time to time. They travel at high speed and most of them are burnt up by friction before they reach the surface. They are the remains of the original cloud from which the Solar System was formed. Measurements on meteorites that have landed on Earth also give an age of about 4.5 billion years.

The surface of the Moon, and of many of the moons of other planets show the effects of meteorites. In the last stages of the formation of moons and planets meteorites weighing many thousands of tonnes collided with them. They made huge craters. With a good telescope you can see these on the surface of the Moon. They are also shown on picture 4 on page 245 of topic I1.

The oldest parts of the Earth also show large craters, probably formed by huge meteorites. Picture 7 shows Meteor Crater in Arizona. This was made by a meteorite of iron 2000 years ago.

Picture 6 on page 255 shows a meteorite burning up in the Earth's atmosphere on August 10 1972. It was estimated to have a mass of 1000 tonnes, and its trail was large enough to be seen in daylight. Luckily it burnt out before reaching the ground. The Earth was formed by the collisions of planetoids and meteorites like this. It is in fact still growing, at a rate of 400 tonnes a day. Most of this is due to tiny meteorites which burn up in the atmosphere.

Picture 7   The Barringer Meteor Crater in Arizona. It is over a kilometre wide and 200 metres deep. The largest we know on Earth is 26 km across

How old is the Sun?

It is harder to work out the age of the Sun. Scientists have to rely on the theories of the **astrophysicists**, the scientists who apply the laws of physics as we know them on Earth to what we can see happening in distant stars.

Give or take a few million years, the Sun is probably about ten thousand million years old. When it runs out of its hydrogen fuel it will, quite suddenly, cool down and become a **red giant** star. It will grow so big that the Earth will be orbiting inside it, a red hot cinder. Fortunately, it contains enough hydrogen to keep going for another ten thousand million years or so.

Activities

A   Looking for stars

The gas clouds which turn into stars are usually invisible, although they can detected by the radio or infra red radiation they emit. But a small telescope or pair of binoculars will show you a gas cloud which is glowing brightly with the light of new stars that have just formed. This is the cloud ('nebula') in the constellation of Orion. Picture 8 shows a photograph of

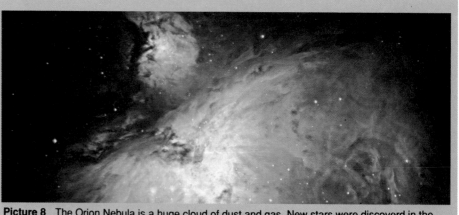

Picture 8   The Orion Nebula is a huge cloud of dust and gas. New stars were discoverd in the Nebula in 1955 that hadn't been there 10 years earlier.

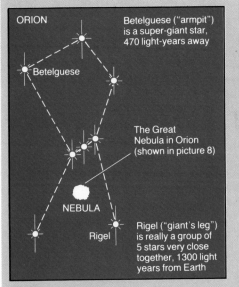

ORION

Betelgeuse ("armpit") is a super-giant star, 470 light-years away

Betelguese

The Great Nebula in Orion (shown in picture 8)

NEBULA

Rigel ("giant's leg") is really a group of 5 stars very close together, 1300 light years from Earth

Rigel

Picture 9

this cloud taken with a large telescope, and picture 9 shows where it is.

Use a star map to find Orion. You can get a map from your library, or from the newspapers which publish star maps near the beginning of each month. Use the map to find the nebula.

B Looking for meteors

Meteors are meteorites that burn up completely in the atmosphere. Most of them are no bigger than grains of sand. They are likely to be the bits that get left behind by comets. Each year the Earth passes through these old comet paths and we get an extra supply of meteors — **meteor showers**.

Try to see some meteors (or 'shooting stars'), during the following periods:

April 12 — 24: linked with Halley's Comet, best on April 22

July 20 — Aug 19: linked with Tuttle's Comet, best on August 11

Oct 11 — 30: another pass through Halley's trail, best on October 19

Oct 24 — Dec 10: linked with Temple's Comet, best on November 13

Dec 5 — 19: best on December 12

C Catching up

The exploration of the Solar System is still taking place. New theories and evidence are appearing every year. This mostly comes from deep space probes like the Voyager and Galileo spacecraft.

There are plans to send probes, and possibly manned spacecraft to Mars.

The Hubble Space Telescope is a satellite-based telescope that should revolutionise our understanding of distant galaxies.

Find out all you can about one of these and prepare to make a presentation to your class about it.

Questions

1 What are: (a) comets, (b) asteroids, (c) satellites?

2 The Solar System began as a gas cloud. When the gas cloud collapsed it was squashed together by gravity. It got hotter. Give an example from your own experience where squashing a gas makes it hot.

3 Explain why the 'inner planets' are made of denser material than the outer ones.

4a A million years is a long time. Were you alive a million *seconds* ago? What were you doing a million minutes ago?

b A billion is a thousand millions. Where were you a billion seconds ago?

c What was the world like a billion days ago?

5 How does the Sun get its energy?

6 Explain, using a diagram, why the energy the Earth gets from the Sun is only a tiny fraction of the energy the Sun emits. (**Hint**: the Earth is 150 million km from the Sun).

Planets, atmospheres and life

This question brings together ideas from a number of topics, as well as from *The Material World* and *The Living World*. The table below shows what the atmospheres of three planets are like. It also shows what the atmosphere of Earth was probably like, 4 billion years ago. The figures for the gases are fractions of the atmosphere, in percentages.

	Planet			
Atmosphere	Venus	Young Earth	Mars	Earth now
(%) carbon dioxide	98	98	95	0.03
nitrogen	1.9	1.9	1.7	78
oxygen	trace	trace	0.13	21
rare gases	0.1	0.1	2	1
temperature on surface (°C)	477	300	−53	13
surface pressure	90	60	0.064	1

1 What are the main differences between the atmosphere of young Earth as it was and the Earth now?

2 One gas, (very important for life!), has been left off the Earth data. What is it?

3 Where do you think the oxygen on the Earth has come from ?

4 What has happened to the carbon dioxide that was in the old Earth's atmosphere?

5 Give two reasons why you would not expect to find life on Venus.

6 Would life be possible on Mars? Give some reasons for your answer.

14
Outer space

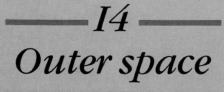

When you look up the night sky you are looking across immense distances. You are also looking back into very ancient times.

Time and distance

On September 5th 1977, a space craft was launched from Cape Canaveral, Florida, by the National Aeronautics and Space Administration (NASA). It left Earth at a speed of over 11 kilometres a second. It needed this high speed to escape the pull of Earth's gravity field.

As it climbed away from Earth it slowed down, just like a ball slows down when you throw it up in the air. But after a few minutes the space craft had left the Earth's atmosphere and thrown away the empty shells of its rockets and fuel tanks. *Voyager 2* had begun a journey to infinity.

Since 1977 Voyager 2 has been climbing through the gravity field of the Sun. The NASA space engineers have been clever. They have used the gravity fields of the planets to help Voyager 2 on its way. As it gets closer to a planet the space craft speeds up because of the increased gravity pull. Its approach is carefully angled so that the space craft changes direction, swinging around one planet to move off exactly in the right direction to get to the next one. This is shown in picture 1.

Some of the pictures of the planets in topic 13 were taken by Voyager 2.

After exactly 12 years, Voyager 2 shot past Neptune, the last of the giant gas planets in the Solar System. As it leaves the Solar System, It will miss Pluto, the outermost planet, because it is in a different part of the sky.

Space travel?

Voyager 2 is now moving into Outer Space. This is a very empty place. By comparison, the Solar System is crowded with planets, comets, meteorites and asteroids. There are also bits of dust and molecules left over from its formation, ten billion years ago.

When will Voyager 2 reach the stars?

When it leaves the Solar System Voyager 2 will be travelling at a speed of about 10 km per second, or 36 thousand kilometres an hour. It will be moving towards the brightest star in the sky, **Sirius**. This is in the constellation of Canis Major which you can find just off the Milky Way, low in the sky below Orion.

Picture 1 How Voyager 2 used the planets to move out to Neptune

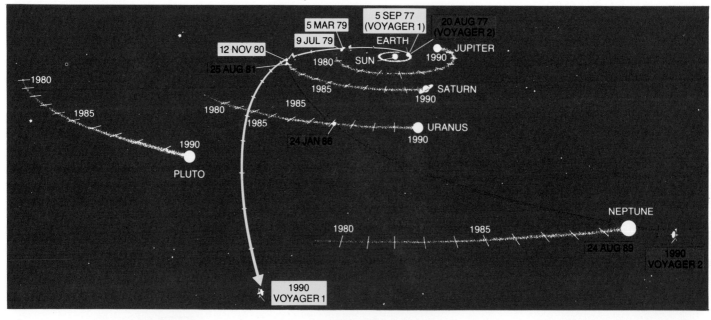

Voyager will take nearly 300 000 years to get there. Barring accidents, it will reach Sirius in the year 296036. Sirius is one of the stars closest to the Solar System.

Space travel using rocket engines and space craft as we know them today is possible — but only to nearby planets. Travellers to the nearest stars will not return. If all went well, their descendants would come back, half a million years later.

The signals sent back from the planet Neptune by Voyager 2 took about 4 hours to reach NASA on Earth. They were of course radio signals, travelling at the speed of light. Signals from our nearest star, **Alpha Centauri**, would take 4.3 *years* to reach Earth. Messages from Sirius will take 9.7 years to arrive.

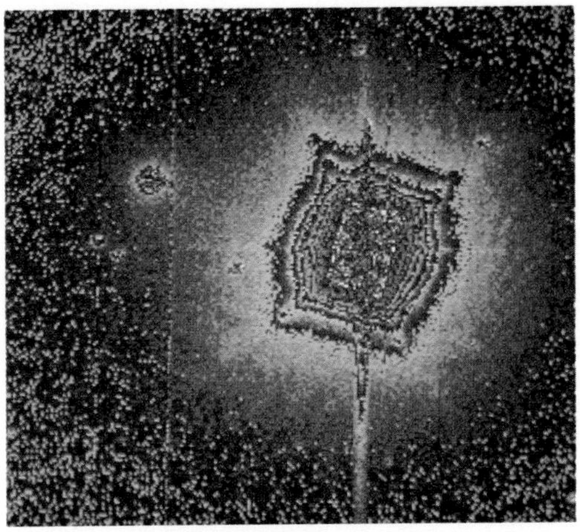

Light years

The stars are so far away that astronomers measure their distances in terms of how long light or radio waves take to travel from them to Earth. A **light year** is the distance covered by light in one year. As light travels at a speed of 300 000 000 metres per second (3×10^8 m/s) in empty space, this is a very large distance. It is 9.5 thousand million million kilometres which we can write as (9500 000 000 000 000 km, or 9.5×10^{15} km).

Table 1 shows the distances of some well-known stars and galaxies from Earth.

Measuring the universe

We can use quite simple physics to finding out what stars are like and what they made of. We can measure how far away they are using much the same methods as surveyors use to make maps on Earth. The problem is that the stars are so far away, and the Earth is so small, that the work has to be done very carefully and with great accuracy.

Signals from space

Information about stars comes to us at the speed of light. It is carried by electromagnetic waves (see topic C8), and astronomers now use almost the whole range of radiation, from the very short X-rays to the long radio waves, to find out about stars.

When we look through a telescope our eyes can detect only visible radiation — light. Special photography can record invisible radiations, like

Picture 2 The 'Red Rectangle' star taken in red light with a CCD camera

Table 1 Distances of stars and galaxies from Earth

Object	Name	Distance from Earth (light-years)	
star	Alpha Centauri	4.3	Nearest star to Earth
star	Sirius	8.7	The brightest star we can see, the 'dog star' in Canis Major
star	Canopus	650	In Carina, used by air navigators
star	Polaris	142	The Pole Star, in Ursa Major
star	Betelgeuse	520	A red supergiant, the brightest star in Orion
galaxy	M31	2 200 000	The Andromeda galaxy, our nearest galaxy
	M81	10 000 000	In Ursa Major (The Plough)
	M87	42 000 000	The Sombrero Galaxy, in Virgo
clusters	Virgo	78 000 000	In the constellation Virgo
of galaxies	Hydra 3	3 960 000 000	In the constellation Hydra
quasar	3C9	32 000 000 000	

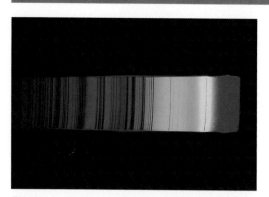

Picture 3 The dark lines in the spectrum of light from the Sun. These lines show what elements are present in the Sun's atmosphere

X-rays, Ultra violet and infra red. But most radiations from space are now detected and recorded electronically, and displayed on computer screens (see picture 2 on previous page).

When it arrives on Earth the radiation from a star is all mixed up together. For it to be useful it has to be split up into a **spectrum**. Picture 3 shows a part of the spectrum of visible light from the Sun. Each dark line running across the spectrum is a clue about what elements the Sun contains.

A spectrum is produced when light passes through a **prism**, or through a diffraction grating. Different wavelengths in the radiation are separated out and so they can be photographed and studied.

Light from stars also shows these spectrum lines, and this is how we know what elements a star contains. Astronomers can also work out how hot the star is, by studying its spectrum. Activity A is about looking at spectra.

Stars also give out radio waves, which are picked up by huge radio **telescopes** (see picture 4). Radio waves tell us a lot about stars and also about the dark material we find in between the stars.

The scale of the universe

Human beings and indeed many animals are quite good at telling how far away things are. If one object seems to pass in front of another when it moves it must be nearer. This effect can also be seen when you move your head, and one object *appears* to move in front of another. (Just try looking out of the window and moving your head.)

How much the object appears to move gives our brains a good clue as to how far away it is. This effect is called **parallax**, and is used by astronomers to help measure distances of stars — see opposite.

Picture 5 shows how triangulation works. Knowing the angles and the length of the baseline, a surveyor can calculate how far away the tree is. Stars are much further away than trees.

So astronomers use the longest baseline they can. This could be the diameter of the Earth, using two telescopes on opposite sides of the Earth. For very distant objects they use the diameter of the Earth's orbit, giving a baseline of 300 million kilometers.

Using this method, astronomers can measure the distances of stars as far away as about 300 light years or so. But this is just a very tiny part of the Universe. There are lots of stars with no measurable parallax which must be even further away.

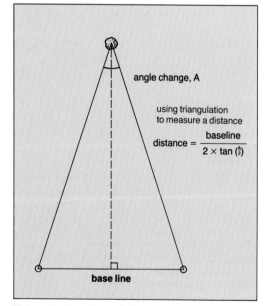

angle change, A

using triangulation
to measure a distance

$$distance = \frac{baseline}{2 \times \tan\left(\frac{A}{2}\right)}$$

base line

Picture 5 How triangulation works

Picture 4 The Jodrell Bank radio telescope. Astronomers now use the whole electromagnetic spectrum to observe the Universe.

Using brightness

Some stars are brighter than others. This might be because they really do give out more light energy — or they might look bright because they are closer to us. The star Sirius, for example, is many times brighter than the Sun. But the Sun looks brighter because it is closer to us.

If we know how bright stars really are we can work out how far away they are. This is where the spectrum of starlight comes in again. Stars of different real brightness have different spectra. By looking at the spectrum of a star we can work out its real brightness, and so its distance from Earth.

This method allowed astronomers to take measurements out to nearly all the stars you can see in the night sky. But if you use a telescope you can see far more stars (picture 6). And some of what you see are not stars at all, they are in fact huge groups of stars — called **galaxies**.

Galaxies

The nearest galaxy to Earth is in the constellation of Andromeda. You can see it with the naked eye, on a fine summer night. Picture 7 (right) shows it as seen through a large telescope. To begin with astronomers thought that objects like this were clouds of gas — and called them **nebulae**, from the Latin word for a cloud.

We can now see that each speck of light is a star. Astronomers have discovered many thousands of these groups of stars, and we believe that the universe contains many millions of them.

Edwin Hubble was an American astronomer who was put in charge of a new telescope at Mount Wilson in California. The first thing Hubble did was look at the Andromeda Nebula. He was the first astronomer to see that the 'cloud' was not a cloud of gas, but was made of millions of stars.

He was amazed to find that the stars in the Andromeda Nebula were further away than any other known object. The Nebula was over two million light years from Earth. He calculated that it contained more than a billion stars.

Galaxies upon galaxies as far as the telescope can see

Since then we have discovered many more galaxies. Some are 'quite close', at about the same distance as the Andromeda nebula. This is our *Local Group* of galaxies: there are 19 of them. Some are quite small, with just a few tens of thousands of stars. Others contain a few thousand million stars, like Andromeda and our own galaxy — the **Milky Way**.

We now know that the Universe is made up of many **groups** or **clusters** of galaxies, separated from each other by huge regions of empty space (picture 8). One of these clusters, in the constellation of Coma Berenices, contains 100 000 galaxies.

Picture 6   Star field in the Milky Way

Picture 8   The Virgo cluster

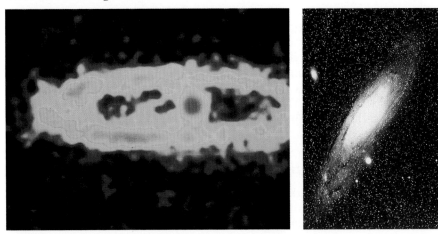

Picture 7   The nearest galaxy to Earth — the Great Nebula in Andromeda

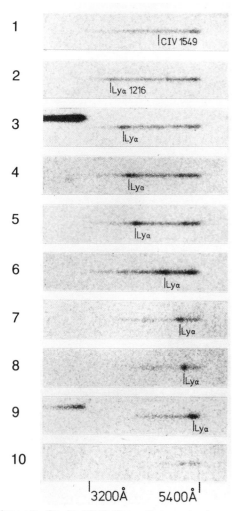

1 |CIV 1549

2 |Lyα 1216

3 |Lyα

4 |Lyα

5 |Lyα

6 |Lyα

7 |Lyα

8 |Lyα

9 |Lyα

10

|3200Å   5400Å|

Picture 9 The Red Shift. These photographs show the spectra of very bright 'stars' called quasars. Going from top to bottom (1 to 9), the same spectrum line (Lyα) shifts across. It moves towards the red end of the spectrum – the 'red shift'. Quasars with the largest red shift are thought to be furthest away. The bigger the red shift, the faster they are moving. This fitted in with Hubble's theory of an expandig universe.

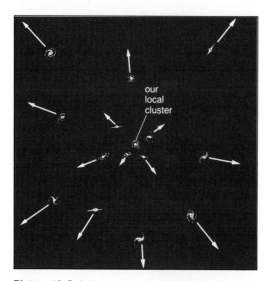

Picture 10 Galaxies are moving away from other galaxies

The expanding universe

After Andromeda, Edwin Hubble then turned his telescope on other nebulae. He measured them to be even further away.

Hubble also discovered something else. The universe is expanding. In fact, it is expanding in all directions. He found that every galaxy in the universe is moving away from every other galaxy. And the further apart they are, the faster they are travelling. It looked like the result of a gigantic explosion that took place billions of years before.

The red shift

The evidence for this expansion came from the spectra of the galaxies. But the spectra had all been somehow distorted. The lines did not appear at the same wavelengths as they did in the Sun. They had all shifted towards the red end of the spectrum — the **red shift** (see picture 9.)

To an experienced astrophysicist like Hubble this was not a great surprise. He knew that waves from objects change their wavelengths and frequencies if the object is moving. It is the well-known **Doppler Effect** that you hear when ambulances or police cars move past you with the siren blowing. The effect is used in radar speed traps to measure the speeds of cars.

What was a surprise was the *size* of the change, and what it told astronomers about the distances of these further galaxies.

The size of the universe

With his new telescope and new techniques Hubble measured the distances of some galaxies that were not too far away. They too showed the red shift. He discovered a simple pattern between red shift and distance. This is illustrated in picture 10. It is known as **Hubble's Law**: *The further the galaxy is from Earth, the bigger is its red shift.*

But what did this mean? The red shift is caused by the galaxy's movement, so what his law really said was:
The further the galaxy is from Earth, the faster it is moving.
But this also means:
The faster it is moving, the further away it must be.

Thus objects with a large red shift must be very far away. This is now how we find the distance away of the most distant objects in the Universe.

Time and the speed of light

The telescopes focused on these distant galaxies are seeing light that left them 9 billion years ago. This was just about the time the Sun was being formed.

We can never find out what they are like now. All the objects we see in the sky may already have changed! When you look at the Andromeda Nebula you looking 2.5 million years into the past. Even the sunlight we see has taken 8 minutes to reach us. The night sky is a historical museum of what the Universe was like at any time from 4 years to 9 billion years ago.

The beginning — and the end

There is now a lot of evidence to back up the idea that the Universe was all together in one very small atom-sized place about 15 billion years ago. That was when time started, and also space itself. There was what astronomers call the **Big Bang**, and everything started to happen.

Before that there was a kind of unimaginable nothing, not even empty space. But out of this nothing came the vast system of space-matter-energy-time that we now see as the Universe.

So what happens next? For a very long time — nothing much. In the very very distant future, long after the Earth has been swallowed up by an exploding Sun, one of three things might happen:

1 The Universe will reach a steady state and stop expanding.
2 The Universe will keeps on expanding for ever and ever.
3 The Universe will stop expanding and start to fall back in on itself.

These options are illustrated in picture 11.

Gravity will decide which of the above actually happens. The Universe is held together by gravitational forces. These are produced by the combined mass of all the stars, dust and gas in all the galaxies and all the spaces in between them.

If there is not enough of this mass, the Universe will keep on expanding. If there *is* enough mass, its gravity force will stop the expansion and make the Universe collapse again, perhaps producing another Big Bang.

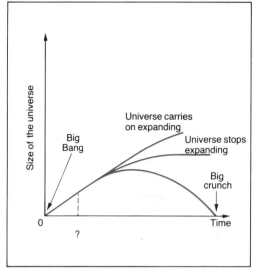

Picture 11 Three possible fates of the Universe

Activities

A Looking at spectra

For this you need a prism — or better still a *diffraction grating*. You can look through it and see the light from different sources being spread out into a spectrum. Use it to do the following.

1 Look at the light from a 12V lamp as the voltage applied to it is changed gradually from 2V to 12V. Explain how astronomers might be able to use a spectrum to tell the temperature of a star.

2 Look at a lamp that has a straight filament. What do you see when you put different coloured filters in front of the lamp? What do you see when you put a solution of chlorophyll in the way? If your teacher can arrange it, look through a tube of iodine vapour. This should help you understand how scientists can detect different elements by means of an *absorption spectrum*.

3 Look at the light given out by some or all of the following:

a a neon tube,
b a helium tube,
c salt being heated in a bunsen flame,
d other chemicals being heated in a bunsen flame — try salts of copper, potassium and barium.

4 Look at a thin line of sunlight reflected on a sheet of white paper. **CARE! Never look directly at the sun!** You may be able to see the dark lines (Fraunhofer lines) that are shown in picture 3.

B In a different universe

There is a respectable scientific theory that says there may well be many different universes, with different properties. Imagine a Universe in which the speed of light was only 1 metre per second.

As a group activity, work out some of the effects that this would have. For example, on TV broadcasts, on team games, traffic etc.

C Looking things up

Use a good book on astronomy, or an encyclopaedia, to find out what the following astronomical objects are:

1 a black hole, **2** a pulsar, **3** a neutron star, **4** a quasar, **5** a globular cluster, **6** a radio galaxy.

Questions

1a The speed of light is 300 000 km/s. How long, in kilometres, are the following distances: (i) a light-second, (ii) a light-minute, (iii) a light-day?

b Why is the 'light-year' a more useful unit for measuring astronomical distances than the metre?

2 Look at table 1, showing the distances of stars from the Earth. Canopus is the second brightest star in the sky, and is just about half as bright as Sirius.

Which of these stars has the greater real brightness? Give a reason for your answer.

3 Put the following in order of size, with the smallest first:

star asteroid galaxy planet meteorite

4 The energy for the expansion of the Universe probably came from the 'Big Bang'. The rate at which the Universe is expanding is getting less. What could be making it slow down?

5 Russia and America are planning to send astronauts to the planet Mars and back. It is the second nearest planet to Earth.

a What are the problems that they might have to solve so that people could make this journey safely?

b This expedition will be very expensive, costing perhaps billions of dollars. Is it worth it?

6 Design a space ship that could take people to Sirius (see topic I1).

7 'Space: the final frontier!' How far will humans go?

8 Outline the problems of setting up a permanent human settlement on the Moon. Suggest some solutions to these problems.

Index

Acknowledgements

The publishers are grateful to the following for permission to reproduce photographs. While every effort has been made to trace copyright holders, the publishers will be pleased to make the necessary arrangements at the first opportunity.
(The numbers cited refer to page and figure numbers respecitvely.)
Aerofilms: 229.4
Allsport: 21.3/Bob Martin, 25.3/Adrian Murrel, 34.1, 114.1/Yann Guichaoua
J Allan Cash: 7.4(a), 22.7, 39.5, 94.2, 106.1, 140.1, 229.3
Alton Towers: 53.4
Heather Angel: 164.1, 213.6
Associated Press: 204.8
Atomic Energy Research Establishment, Harwell: 124.1
BBC: 82.2
BBC Radiovision: 197.4, 204.7, 211.12, 213.5
BICC Photographers: 170.9
British Cement Association: 6.2
British Geological Survey: 212.1, 212.2, 214.7, 215.11
British Library: 226.6
British Museum: 197.4, 207.4, 207.5, 208.7, 208.8, 211.14
British Steel: 8.5 (left), 191.5
CEGB: 168.1
Bruce Coleman Ltd: 12.1/W E Townsend, 14.6, 139.9/Colin Molyneux 164.2/Hans Reinhard,
De Beers: 197.3
Department of Transport: 27.9, 28.10
Edinburgh Butterfly House: 228.2
East Anglian Daily Times: 118.1
Ever Ready: 165.6
Fodens: 8.5(b)
GEC Research/Hirst Research Centre: 167.2
Anna Grayson: 205.10, 207.6, 208.7(b)
Griffin and George: 21.5, 141.3, 142.6, 189.10
GSF Picture Library: 214.8
Robert Harding Picture Library: 228.1/Photri, 229.5, 234.1/Sarah King, 239.4
Hawker Siddley. 9.9
Michael Holford Picture Library. 44.1/Gerry Clyde, 55.3, 196.1
The Hulton-Deutsch Collection: 30.2(top), 59.4, 59.5,
Hutchison Library: 104.6/Andre Singer, 214.9/Philip Wolmuth
The Image Bank: 137.2, 235.5
Imperial War Museum: 88.6
Frank Lane Picture Agency: 50.7/NASA, 231.3/L West, 235.6, 235.7/M Nimmo
London Weather Centre: 231.5
Met Office: 231.4/Miss S M Postle, 235.8, 235.9/M Nimmo
Glyn Millhouse: 92.9, 97.4
NASA: 167.1, 249.3, 250.4, 250.5, 250.6, 256.8
Natural History Museum: 224.1
Pilkington: 6.1
C W Pope, British Rail Research: 47.1, 47.3
Renewable Energy Enquiries Bureau: 130.2, 130.3, 131.1

Richmond upon Thames College Dept of Photography: 90.1, 90.3, 91.6, 97.3
Chris Ridgers: 2.2, 5.10, 7.4(b), 13.5, 14.7(a), 16.1, 16.2, 21.3, 38.1, 69.3, 69.4, 75.7, 74.6, 76.9b, 78.4, 79.8, 80.11, 106.2, 111.3, 112.4, 115.4, 132.1, 132.2, 133.5, 144.1, 148.2, 153.3, 154.6, 156.1, 157.5, 157.6, 158.7, 175.4, 175.6, 176.8, 180.5, 182.1, 186.1, 206.3, 234.3, 237.3,
Ann Ronan Picture Library: 30.2 (bottom)
Rowenta: 159.1
Royal Observatory, Edinburgh: 256.7, 261.1/and Anglo-Australian Telescope Board, 262.9
Salters: 4.6
Science Museum: 151.12
Science Photo Library: 2.1/Professor Harold Egerton, 14.7 (below)/Dr Jeremy Burgess, 15.9/Friedrich Michler, 24.1/ Professor Harold Egerton, 24.2/European Space Agency, 30.1, 31.3/NASA, 44.2/Dr Jeremy Burgess, 44.3, 48.1/ Novosti, 48.2, 49.4/NASA, 50.8/Earth Satellite Corporation, 54.2/55.4/J-L Charmet, 56.5/Bill Sanderson 56.6, 56.7/ Novosti, 57.9/US National Archives/Hank Morgan, University of Massachusetts, Amherst/Patrics Loiez, CERN, 75.8/Dr Tony Brain, 84.9/CNRI, 96.1/David Parker, 98.6/ Richard Folwell, 98.7/Ralph Eagle, 100.2/Department of Physics, Imperial College, 101.1/CNRI, 101.3/Max-Planck Institute for Radio Astronomy, 102.2/Barney McGrath, 119.4, 127.6/Los Alamos National Laboratory, 128.1/John Hesletine, 128.2/Martin Bond, 145.6/Martin Dohrn, 148.1/ JKoivula, 148.4/James Stevenson 150.11/Gordon Garradd, 153.4/Hank Morgan, 160.1, 162.10/Phil Jude, 168.3/Alex Bartel, 173.12/US Department of Energy, 176.10, 177.11/ Chris Priest, 203.4 (centre)/David Parker, 208.7(d)/US Department of Energy, 216.15, 216.16, Ronald Royer, 218.1, 230.1/European Space Agency, 235.4/Phil Dauber, 237.1/Keith Kent, 238.3/Dr Jean Lorre, 239.3/NASA, 244.1/ John Sandford, 244.2/David Parker, 245.4 (top)/NASA, (bottom)/Martin Dohrn, 246.6/George East, 246.7/John Sandford, 249.2/NASA, 255.6/Baker/Milon, 256.7/John Sanford, 259.2/Royal Greenwich Observatory, 260.3/ Department of Physics, Imperial College, 260.4/David Parker, 261.6/Ronald Royer, 261.7/Max Planck Institute for Radio Astronomy/NASA
Scope Optics: 84.5
Sharp: 193.11
Dr Sharp, Loughborough University: 178.1
G Sherlock: 208.7(a)
SIPA Press: 242.10/C Brown
Frank Spooner Pictures: 202.1/V Shone, 203.4(b)/P M Downey
Telegraph Colour Library: 183.2, 246.6(b)
Transport and Road Research Laboratory: 180.7
John Urling Clark: 3.4, 13.3, 68.1, 72.1, 73.3, 73.4, 77.11, 136.1, (bottom), 154.7, 162.7, 162.9, 193.12, 206.1, 206.2, 219.4, 255.4, 226.6,
Washington University Photographic Service: 219.4
Watford Stadium: 13.4
Wild Leltz (Heersbrugg), Germany: 86

The following artworks were based on material supplied by
the following:

Association of Science Education (SATIS T29): 26.6, 27.7
British Telecom: 60.6
British Nuclear Fuels plc: 120.9, 125.4, 129.3
The Department of Transport: Table 1, page 27
The National Grid Company: 169.5
New Scientist: 254.4

Cover photographs supplied by Image Bank.